FOREST GENETICS AND SUSTAINABILITY

FORESTRY SCIENCES

Volume 63

The titles published in this series are listed at the end of this volume.

Forest Genetics and Sustainability

Edited by

CSABA MÁTYÁS

Institute of Environmental Sciences,
University of Sopron,
Sopron, Hungary

KLUWER ACADEMIC PUBLISHERS
DORDRECHT / BOSTON / LONDON

A C.I.P. Catalogue record for this book is available from the Library of Congress.

ISBN 0-7923-6011-7

Published by Kluwer Academic Publishers,
P.O. Box 17, 3300 AA Dordrecht, The Netherlands.

Sold and distributed in North, Central and South America
by Kluwer Academic Publishers,
101 Philip Drive, Norwell, MA 02061, U.S.A.

In all other countries, sold and distributed
by Kluwer Academic Publishers,
P.O. Box 322, 3300 AH Dordrecht, The Netherlands.

Printed on acid-free paper

Printed in the Netherlands.

Contents

Contributors

Co-authors of chapters, if sharing the same employer, are listed under the main author's name.

Boonchoob **BOONTAWEE**	Technical Forest Research Office, Royal Forest Department, Bangkok 10900, Thailand
Timothy J.B. **BOYLE**	United Nations Development Programme (UNDP) DC1-2364, One, UN Plaza, New York NY10017
Gabriele **BUCCI**, Giovanni G. **VENDRAMIN**	Istituto Miglioramento Genetico Piante Forestali, National Research Council, Institute of Forest Tree Breeeding, Via Atto Vannucci 13, 50134 Firenze, Italy
Rowland D. **BURDON**, Michael J. **CARSON**	New Zealand Forest Research Institute Ltd., Private Bag 3020,Rotorua, New Zealand
Jeffery **BURLEY**	Oxford Forestry Institute, Department of Plant Sciences, University of Oxford, South Parks Rd., Oxford OX1 3RB, United Kingdom
Richard E. **DICKSON**, Judson G. **ISEBRANDS**	U.S. Forest Service, North Central Forest Experiment Station, Forestry Sciences Laboratory, 5985 Highway K, Rhinelander, Wisconsin 54501 USA
Yousry A. **EL-KASSABY**	Department of Forest Sciences, Faculty of Forestry, University of British Columbia, Vancouver, Canada V6T 1Z4
Reiner **FINKELDEY**, Gábor **MÁTYÁS**	Swiss Federal Institute for Forest, Snow and Landscape Research, Zürcherstr. 111 CH-8903 Birmensdorf, Switzerland
Jeffrey C. **GLAUBITZ**, Jimena **STRK**, Gavin F. **MORAN**	Molecular Genetics Group, CSIRO Forestry and Forest Products, P.O. Box E4008, Kingston, ACT 2604, Australia
Jusheng **HONG**	Chinese Academy of Forestry, Wan Shou Shan, Beijing 100091, China
David F. **KARNOSKY**	Michigan Technological University, School of Forestry and Wood Products, 101 U.J. Noblet Forestry Building, 1400 Townsend Drive, Houghton, Michigan 49931-1295 USA
Veikko **KOSKI**	Finnish Forest Research Institute, Vantaa Research Centre, P.O. Box. 18, FIN-01301 Vantaa, Finland
Antoine **KREMER**, Valérie **LE CORRE**, Stéphanie **MARIETTE**	INRA Laboratory of Genetics and Breeding of Forest Trees, B.P. 45 33611 Gazinet, Cedex France

Howard B. **KRIEBEL**	Professor Emeritus, Ohio State University, 625 Medford Leas, Medford NJ 08055-2258, USA
Chai T. **LEE**, Norwati **MUHAMMAD**	Forest Research Institute Malaysia, Kepong, 52109 Kuala Lumpur, Malaysia
Yves **LESGOURGUES**	Center of Forest Productivity of Aquitaine, Bordeaux, France
Blanka **MANKOVSKA**	Forest Research Institute, T.G. Masarykova Street 2195, 960 92 Zvolen, Slovakia
Csaba **MÁTYÁS**	Institute of Environmental Sciences, Faculty of Forestry, University of Sopron, P.O. Box. 132, H-9401 Sopron, Hungary
Stephen **MIDGLEY**, Doug **BOLAND**	Tree Improvement and Genetic Resources Program, CSIRO Forestry and Forest Products, P.O. Box E4008, Kingston, ACT 2604, Australia
Gerhard **MÜLLER-STARCK**	Chair of Forest Genetics, Faculty of Forestry, University of München, Am Hochanger 13, D-85354 Freising, Germany
Gene **NAMKOONG**	Department of Forest Sciences, University of British Columbia, Vancouver, Canada V6T 124
Kevin E. **PERCY**	Canadian Forest Service, Atlantic Forestry Centre, P.O. Box 4000, Fredericton, New Brunswick, Canada E3B 5P7
Gopi K. **PODILA**	Michigan Technological University, Biological Sciences Department, 1400 Townsend Drive, Houghton, Michigan 49931-1295 USA
Wickneswari **RATNAM**	Department of Genetics, Faculty of Life Sciences, Universiti Kebangsaan Malaysia, 43600 UKM Bangi, Selangor, Malaysia
Rodolfo **SALAZAR**	Research and Education Center of Tropical Agriculture (CATIE), Turrialba, Costa Rica
Scott E. **SCHLARBAUM**	Department of Forestry, Wildlife and Fisheries, University of Tennessee, Knoxville, Tennessee, USA 37901-1071
Ronald R. **SEDEROFF**	Forest Biotechnology Group, Department of Forestry, North Carolina State University, Raleigh, NC 27695-8008, USA
Pierre **SIGAUD**	Forest Resources Development Service, Forest Resources Division, Forestry Department, FAO, Via delle Terme di Caracalla, 00100 Rome, Italy

Tore **SKRØPPA**, Øystein **JOHNSEN**	Norwegian Forest Research Institute, Høgskoleveien 12, 1432 Ås, Norway
Alfred E. **SZMIDT**, Xiao-Ru **WANG**	Department of Forest Genetics and Plant Physiology, Swedish University of Agricultural Sciences, S-901 83 Umeå, Sweden
Eric **TEISSIER DU CROS**	INRA, Mediterranean Forest Research Institute, Avenue Vivaldi, 84000 Avignon, France
Martin **ZIEHE**, Hans H. **HATTEMER**, Rommy **MÜLLER-STARCK**	Institute of Forest Genetics and Tree Breeding, University of Göttingen, Büsgenweg 2, D-37077 Göttingen, Germany

Acknowledgement

Most of the chapters of the present volume are based on invited lectures presented during the 4th International Consultation on Forest Genetics and Tree Breeding in Beijing, China, August 22-28, 1988, organized by the International Union of Forest Research Institutions in cooperation with FAO.

Considerable support was provided in selecting the best authors by the chairs of the thematic sessions: Prof. Jusheng Hong (Beijiing, China), Prof. Jeffery Burley (Oxford, England), Prof. Gerhard Müller-Starck (Freising, Germany), Prof. David Karnosky (Michigan, USA), Prof. Gösta Ericson (Uppsala, Sweden), Prof. Huoran Wang (Beijiing, China), Prof. Francis Yeh (Edmonton, Canada), Pierre Sigaud (FAO, Rome, Italy), Prof. Howard Kriebel (Medford, N. J., USA) and Prof. Ladislav Paule (Zvolen, Slovakia).

The reviewing of the chapters has been carried out by the fellow authors of the volume, by the above mentioned session chairs and by the following experts: Paul Anderson (Rhinelander, USA), Christel Palmberg (FAO, Rome), Kurt Pregitzer (Houghton, USA), Jozef Turok (IPGRI, Rome) and Wolf Ruetz (Teisendorf, Germany).

Gábor Varga (Sopron, Hungary) assisted in the technical preparation of the texts for the publishing.

Their generous support and selfless help is gratefully acknowledged.

The editor

1. INTRODUCTION

Forest genetics and tree improvement research in IUFRO[1]

Jeffery Burley

On behalf of all my colleagues on the International Council and Executive Board of the International Union of Forestry Research Organizations (IUFRO), I have great pleasure in welcoming you to this important conference of the Division that has been consistently one of the most active in IUFRO. Some of you may not be aware that IUFRO is 106 years old and one of the oldest non-governmental international organizations; it has approximately 700 member institutions in over 100 countries with some 15,000 scientists. These scientists work in 276 Research Groups and Working Parties whose networks cover virtually all the scientific disciplines relevant to forests and forestry; some of these are species-specific, others are topic-specific and some have very narrow geographic themes, but all seek to gain the benefits of collaboration and to bring the results of their research to a wider audience of peer scientists, government and international policy makers, and the public in general.

For the IUFRO Congress in Kuala Lumpur, Malaysia, during August 2000, the Executive Board aims to prepare major state-of-knowledge reports summarizing our information and needs for new research on several issues that have global significance. This present meeting is a major step towards achieving that aim in that it brings together some of the leading scientists addressing the major themes within genetics; it will review past achievements and suggest future contributions of genetics to the sustained management of global forest resources.

As we enter a new decade, century and millennium, the rate of development of new technologies has reached its highest point but, at the same time as these technologies offer new insights into genetic processes and controls, they also offer challenges to ethicists, policy makers and the public at large.

---

1 Opening address of the President of IUFRO at the IUFRO World Consultation in Beijing, August 22, 1998

C. Mátyás (ed.), Forest Genetics and Sustainability, Volume 63, 1–2.

IUFRO does not wish to make land use or forest policies but it does hope that such policies are based on available knowledge or the recognition that relevant knowledge is not available and that thus new research is required. Although policy decisions by definition are made for "political" reasons, the decision maker should be aware of the possible impacts of the decision. This is particularly relevant to the current debates on genetically modified organisms but it is also important for the continued initiation, maintenance and expansion of classical and modern tree improvement programmes. In China 6-7 million hectares of plantations are established; the opportunity to gain from breeding is immense.

Many people have contributed to the organization and success of this conference but in particular I would like to thank and congratulate Professor Jiang Zehui, the chairman of the Conference Organizing Committee, my colleague on the IUFRO Executive Board, Professor Hong Jusheng (vice-chairman of the COC and co-chairman of the CSC), and Professor Csaba Mátyás (chair of the Conference Scientific Committee and co-chair of the COC) who has been the prime mover in Division 2; they have devoted great personal and institutional efforts to this meeting. I am sure that they will find opportunities to thank the many other people who have contributed to the organization but I would like to thank also the authors of papers and the other conference participants who ensure the real success of any meeting. Personally I am sorry not to stay for the entire meeting but I have to proceed to Melbourne, Australia, where the IUFRO Task Force on Sustainable Forest Management has a major meeting examining the needs of all stakeholders in designing research. The output of both these meetings, and of three other major IUFRO activities this year in Chile, Japan and Korea, are diverse in their science but closely allied in their need to portray that science for the benefit of peer scientists, policy makers and the public.

Lessons from hundred years of international research in forest genetics and breeding

Csaba Mátyás

Introduction

There are few areas of science - probably with the exception of nuclear physics - that underwent in this century a similarly rapid expansion as genetics and provoke comparably ambivalent and emotional reactions.

Applied genetics and breeding became increasingly an integral part of forest research and extension; after a "heroic period" following the second World War, progress in this field accelerated in the last two decades when molecular methods became widely accessible for forest genetic research. What did silviculture benefit from the century of genetics? It is a strange development that while modern genetics offers an unprecedented wealth of new insights and of subtle analysis methods applicable in forest tree breeding, the acceptance of any interference in natural genetic processes and the plantation management of improved forest trees is being increasingly criticized both in the public and in professional circles.

It appears that the main attention of present-day forest genetics is focused on the support of close-to-nature management. However, considering the global state of forest resources, it seems at least doubtful that forest genetics could be restricted to basic research and conservation issues. There are at least four very serious challenges to mankind, which are related to forestry in general and – directly or indirectly - also to forest genetics, and demand active involvement:

- catastrophic depletion of forest resources (including genetic resources) and the degradation of forest sites, hand in hand with population increase in the developing world;
- acute shortage of firewood in the arid and semiarid tropics, the energy crisis of under-developed regions;
- environmental threats to forests in industrialized regions; and
- impending effects of predicted climatic changes.

All the four are of global character and forest genetics and breeding cannot ignore these problems. Besides the surveying and analysis of processes in natural populations, the research in possibilities of increasing yields and the improvement of adaptability and resistance through breeding cannot be neglected. The rapid transfer of applicable knowledge to regions where it is needed most should be encouraged therefore by all possible means.

C. Mátyás (ed.), Forest Genetics and Sustainability, Volume 63, 3–8.

1. A century of international cooperation: the beginnings

Although documented initiatives to investigate the heritable variability of traits in forest tree species have a long history and can be traced back into the 18th century, the birth of forest tree genetics as an international research field can be dated to an event exactly a century ago.

During the meeting of the newly founded International Union of Forest Research Institutions (IUFRO) held in September 1900 in Zurich, Switzerland, Prof. Heinrich Mayr from the Bavarian Forest Research Institute (Munich) proposed the initiation of „international comparative trials (later called provenance trials) to elucidate the effect of the seed provenance". The proposal was based on results published by the Austrian scientist A. Cieslar on experiences with conifer plants from various altitudes grown in common garden tests. Prof. Mayr wrote..." these experiments touch a very important aspect of plant biology and agronomy. The question of the heredity of growth potential is tightly linked to other problems, such as... the questions of adjustment to climate extremes, of acclimation to conditions outside the natural distribution; how much of this adaptation, if existing at all, can be transferred to the progeny?"

These are remarkably modern thoughts, considering the fact that this was the year of the rediscovery of Gregor Mendel's laws of heredity, and it took still another decade before Morgan identified the chromosomes as the site of the hereditary material.

A direct consequence of the 1900 resolution to initiate internationally coordinated experiments in forest genetics, was the start of IUFRO provenance tests in 1904 and 1905 with larch, Norway spruce and Scots pine. Many more followed and created an innovative basis for transboundary cooperation and for the present-day research groups linking continents. Although considered by many as outdated and old-fashioned compared to the elegance and sophistication of molecular methods, I believe that these international experimental series are among the most important achievements forestry may offer to biological sciences. Apart from their original aim to identify best-adapted sources for planting, the growth and survival of these unselected, wild populations offers many conclusions and insights into quantitative aspects of genetic variation and of adaptive response. These possibilities are by far not fully utilized yet in forest genetics and are barely appreciated in biology or ecology.

2. Looking ahead: a World Consultation in China

In the 1960s and 1970s, three World Consultations on Forest Genetics and Tree Breeding have been organized in Stockholm (1963), in Washington (1969) and in Canberra (1977). The locations of the conferences indicate that in the booming phase forest genetics and tree breeding was an issue predominantly pursued by developed nations. Although numerous thematic and regional meetings have been held since then, in the last two decades no conference was organized with a comparable global scope.

At the doorstep of a new millennium, and preparing for the first IUFRO World Conference of the 21st century in Kuala Lumpur, Malaysia, it seemed appropriate for IUFRO's thematic divisions to review achievements, mark out the state of progress in their scientific areas and identify future tasks and needs. Leading scientists of IUFRO's Division 2, "Physiology and Genetics" decided to organize an all-division conference on forest genetics and tree improvement in 1998 in China. The author accepted the task of designing the concept and scientific program of the conference, while Prof. Hong Jusheng (Chinese Academy of Forestry, Beijing) initiated the local conference organization.

The aims of the conference were to present and discuss accomplishments, current trends and expected future developments, to redefine the role of forest genetics and breeding in contemporary forestry, to set priorities for future research and development, and to present a comprehensive diagnosis of the situation of forest genetics and tree breeding research, identifying main problems and offering solutions. The meeting was organized in a consultative, interactive manner. Key speakers of world reputation introduced thematic areas, followed by discussions, aiming at the formulation of concrete recommendations. The chapters in this volume have been prepared on the basis of the invited lectures presented there.

There were many reasons to organize the fourth Consultation on Forest Genetics and Tree Breeding in Beijing. Bringing this conference to China emphasizes the fact that the majority of the world's population still has wood and timber in short supply. In view of the ongoing depletion of forest resources, forest geneticists have to offer their expertise to help to solve this problem in a sustained manner. In this field, the developed countries have accumulated a great deal of knowledge that needs to be shared with the rest of the world. Among politicians and decision-makers of many parts of the world it is still not sufficiently understood that sustainability of forests cannot be achieved without taking into account the results in the field of genetics.

China is a positive example for a country where these tasks have been taken seriously. Information on afforestation and rural tree planting indicates that the importance of forests and trees for the demands of a rural society is well understood. The application of tree breeding and genetics in these programs seems natural and inevitable, giving evidence of the profound social and economic dimension of this rather theoretically looking discipline. These developments are documented in detail in the chapter on China in this volume.

3. Looking back for lessons

After a period of surveying the potential in forest genetics up to the 1950s, postwar economic expansion triggered an unforeseen boom in tree breeding projects in the developed world, fueled by a firm belief in unlimited growth. The high expectations were only partially fulfilled with the advent of new ideas on nature (biodiversity) conservation and sustainable development about three decades later. The re-thinking of aims and perspectives in the mid- 1980s coincided with the appearance of new molecular methods in forest genetics. The last 15 years brought an unparalleled expansion in knowledge, and fundamental forest genetics research became probably the fastest-growing discipline serving silviculture. This offered excellent conditions for the support of tree breeding programs. Instead, beginning in the 1980s an increasing shift of research emphasis occurred in the developed world toward molecular genetics and conservation. At the same time, the public support for tree breeding programs has been fading. The reasons and consequences of this shift are extensively dealt with in other papers of this volume.

As a result of this change in research emphasis, much of present-day forest genetics research is serving first of all the understanding of natural processes, the preservation and restoration of natural forest ecosystems, and the development of close-to-nature forestry. The goal of breeding trees to increase timber yield, or producing improved planting stock for man-made forests has earned in the recent years a dubious reputation both in professional circles and in public opinion - at least in the developed world. The yield of forests set aside for nature conservation purposes in the Northern Hemisphere is probably comparable to the potential of tree breeding in that region.

Need for public acceptance and long-term thinking

The long life cycle of trees facilitates the drawing of some conclusions from the past hundred years of forest genetics and tree breeding. One lesson of the ups and downs in forest genetics may be summarized as „trees live longer than research concepts" (D. Lindgren, pers. comm.). The difficulties with frequent changes in priorities and policy affects all disciplines, but forestry suffers more due to the long time period involved. Among specific problems of forest genetic research, the lack of long-term commitment is certainly the most critical; the scientist community is obviously unable to influence decision-makers and society to make this claim understood.

Inadequate support, even among professional foresters, is partly due to the esoteric nature of genetic research. Although it seems unlikely that substantial changes can be achieved in the short term, elements of long-term thinking have to be strengthened both in research funding policy and in public acceptance. *The emphasis should be not only on the sustainability of forest resources: forest genetic research must equally be sustainable!* To achieve this, forest geneticists and breeders have to develop a greater sensitivity towards political and social processes in the modern society. Support may be expected only if it can be unquestionably proven that forest genetics and breeding can be integrated in the concept of sustainable development, without presenting any threat to the diversity and stability of natural forest ecosystems. Forest genetics has the unique tools to prove this, if only the proper ways and means can be found to convey the message to professional decision-makers and the public. Public media should be utilized to a larger extent and interdisciplinary encounters should be sought with other branches of sciences, including the social sciences.

Need for practical links

These messages should be very clear, simple and understandable to be heard in a world overloaded with information. This applies also to the communication with the potential users, namely forest owners and managers. The dependence of research on short-term project funds often isolates it from practical approaches and isolates forest genetic research from the users. Additional funding from the forestry sector – even if generating further administrative problems and distracting energies from programmed work – helps to build a longer-term user commitment and the orientation of research objectives. Just as adaptive changes in tree populations serve the maintenance of species identity, the scientific community has to constantly check actual trends to adjust priorities and aims.

Decisive trends of the present are the maintenance of biological diversity and conservation of genetic resources. There is still not enough awareness in ecology that genetic diversity is the basic element of diversity and a precondition of the stability of forest ecosystems.

International agreements and initiatives may provide support. Funding agencies accept recommendations and proposals easier if formulated by quasi-neutral international bodies or organizations. International networking of research may not only help to find useful collaboration but also sensible sharing of costs as well as additional funding.

Need for a synthesis

The wealth of results and information in contemporary genetics is so large that it is hardly comprehensible to outsiders. It becomes increasingly difficult to understand or decide what is the value of the new information for the practice, how relevant and valid a research finding is. The temptation is great to look ever deeper into details of the hereditary material instead of trying to reach a synthesis. In order to maintain the clarity of purpose of forest genetics research, it is absolutely necessary to be able to identify really important research areas and knowledge gaps. This clarity has increasingly become necessary to justify the financing of further research. The pressing need for synthesis is most evident in the knowledge gap between the results of molecular genetics research and the expression of quantitative traits of adaptive significance.

Need for field experimentation

Doing research with trees means dealing with long-living organisms. In spite of all difficulties, long-term observations in field experiments are indispensable and cannot be replaced by shortcut methods. These experiments will be limited in number by financial and other constraints. It cannot be guaranteed that even the most carefully selected research concept will be valid for the whole life of a long-term field test. But if the design is sound, the documentation as complete as possible and the demarcation in the field well maintained, such tests will prove to be extremely valuable sources of information that cannot be modeled by computer or simulated in growth chambers. Not even the most sophisticated program can fill in the complexity of "real life". The problem of laboratory and modeling work is certainly not the lack of precision; rather it is the magnitude of the investigated effect in a synergistic system of interactions, which can only be assessed under natural growing conditions. This is so even if a large part of such tests is lost or damaged due to a diversity of natural and man-caused impacts.

Need for an ethical approach – the responsibility of the international scientific community

The current practice of research funding and of excellence assessment does not necessarily exclude elements of counterproductivity. With the increasing complexity of science, there are fewer and fewer experts with both an in-depth understanding of specialized areas of research and the connections between them. It is the responsibility of the international scientific community to act as a guiding force to promote creative and synthetic research and to discourage proliferation of repetitive and confirmative work.

It lies in the nature of the investigated subject that securing the continuity of basic concepts has also ethical aspects. Looking back on past achievements helps to understand present trends and to remain humble with expectations, promises and forecasts.

New ethical problems are emerging in connection with the advent of genetic engineering in forestry. It has to be conceded that not even the relatively simple control of traditional forest reproductive material trade is followed to a satisfactory level. A similar statement can be made about the general state of conservation and utilization of genetic resources (see for details in the respective chapter of this volume). One may be justified in doubting that regulations governing the introduction of genetically modified organisms alone can effectively withstand pressure from commercial interests. Again, only the scientific community has the expertise necessary to provide objective information for decision-making. Scientific judgement will only be respected if the public can be convinced about the ethically acceptable and sound guiding principles of forest genetics. To maintain and strengthen this approach is a basic task of the scientist community. Its importance cannot be overestimated.

Acknowledgement

The assistance of the secretary of IUFRO, Mr. H. Schmutzenhofer, in tracing the protocol of the Zurich meeting cited in the text, as well as the valuable comments of Prof. Howard Kriebel (USA) on the manuscript are gratefully acknowledged.

2. SITUATION AND CHALLENGES FOR FOREST GENETICS AND BREEDING

Situation of forest genetics: research and extension

Eric Teissier du Cros

Introduction

This paper is based on an inquiry sent by the author to co-ordinators of Research Groups and Working Parties of Division 2 „Physiology and Genetics" and to other colleagues around the world. The questions of the inquiry are reflected by the main headings of the paper.

At this moment I want to thank all those who have provided me with written information and reprints of published and unpublished papers which illustrate points of the inquiry. Their name is given in the Acknowledgement at the end of this paper. Several regions of the world are not covered in answers but the convergence of certain answers may perhaps allow some interpolation.

1. Definition

To clarify what the title covers let us base our definitions on Ledig's (1997).

Forest genetics is the study of variation and inheritance in forest trees. The science of forest genetics provides the knowledge necessary to breed trees through traditional methods of selection and hybridization and also through the newer biotechnologies.

Tree breeding reflects all efforts made to provide superior varieties for reforestation: via mass selection, provenance and progeny testing, hybridisation, etc.

2. History

„The first known document on tree breeding [in Sweden] is dated 1749 by the Marine command concerning choice of seed source for *Quercus robur*

C. Mátyás (ed.), Forest Genetics and Sustainability, Volume 63, 9–22.

(pedunculate oak) grown for ship construction" (Tore Ericsson 1998, pers. comm.). The first provenance trial (in France) was laid out in 1840 by Philippe-André de Vilmorin when he compared different „races" of Scots pine (*Pinus sylvestris*) in a single place (Bouvarel 1986). Provenance tests have developed extensively during the first quarter of the 20th Century. Later, during the third quarter of the 20th Century, some tests involving exotic species not only included seed sources from the natural range but also land races.

„Modern" breeding started after the Second World War. International or regional co-operation and networking developed rapidly. For instance, the first FAO-IUFRO World Consultation on Forest Genetics was held in 1963, followed by the second and third ones respectively in 1969 and 1977 (Palmberg-Lerche 1993b). During that period important regional conferences were also organised and should be mentioned at least because of their historical reference value: tropical provenance and progeny research (Burley and Nikles 1973), tropical trees (Burley and Styles 1976), forest tree improvement in Southern USA (Thielges 1975) and more recently the Nordic Tree Breeders Meeting (Denmark, 1990). This list is certainly not complete since it does not include meetings of North American Forest Tree Breeders, the European Association of Forest Tree Breeders and others.

Selection traits will be developed further in this paper but one of them should be stressed because of its global value: resistance to diseases and insect pests; this subject was the topic of three conferences in Europe by 1980 (Heybroek *et al.* 1982). Similar ones were held in the USA in 1964 and 1969.

A list of research institutions across the world involved in forest tree breeding was made by Toda in 1974 which included a detailed activity description. At that time Toda's book mentioned the activity in 23 countries (Table 1). However, obviously many countries were not included. An FAO inventory suggests that about half the countries of the world have some kind of a forest tree improvement programme : i.e. about 100 countries (Palmberg-Lerche 1993a)

Table 1. Countries with forest tree improvement activities (as listed by Toda 1974)

Region	Countries	Country names
Europe	14	UK, Ireland, Norway, Germany, The Netherlands, Belgium, France, Poland, Czechoslovakia, Austria, Hungary, Italy, Yugoslavia, Greece
Africa	2	East Africa, Malawi
Asia and Oceania	6	Pakistan, India, Indonesia, Korea, Japan, Australia
N. and S. America	3	Canada, USA, Argentina

During the 1970s IUFRO established a number of Working Parties of scientists involved or interested in forest genetics and tree breeding of various species. One Working Party (2.04.09) has recently published the World Directory of Forest Geneticists and Tree Breeders (Ledig and Neale 1998): the number of scientists involved/interested in forest tree improvement is about 3,000.

The global activity in forest genetics and tree breeding is reflected in the last issue of IUFRO News (IUFRO 1998). It shows the dichotomy between breeding and genetics:

- two Research Groups on Breeding and Genetic Resources, including respectively nine conifer and five broadleaf Working Parties,
- one Research Group on Forest Genetics with five Working Parties.

It also shows the practical (not scientific) separation between breeding and selection for resistance. IUFRO's Division 7 includes a Research Group: 7.01.00 „Physiology and Genetics of Tree/Phytophage Interactions" where Working Party titles all include the word „resistance".

Many Division 2 Working Parties hold meetings more or less regularly. They usually publish proceedings of high scientific, technical and historical value. Two examples can be cited : the IUFRO Conference on breeding theory, progeny testing and seed orchards in 1986 (Hatcher and Weir 1986) and the IUFRO Conference on population genetics and genetic conservation of forest trees (Baradat *et al.* 1995).

3. Strategy

The first prerequisite for practical breeding is genetic variation of economic traits. This prerequisite was felt 2.5 centuries ago but only systematically tested since the beginning of the 20th century and more so since the 1970s. Provenance testing, sometimes combined with open pollinated progeny testing, was developed as the first step of any genetic improvement programme and in many cases as the only one. Common garden provenance testing is now assisted by diversity studies using genetic markers, essentially proteins and DNA.

The second prerequisite for breeding is the inheritance of individual traits and many studies have demonstrated reasonably high heritabilities for a number of features in a wide range of species. Huge efforts and long periods of time are needed to convince users that current breeding is going to provide them with a really positive benefit, but there are now many highly successful examples of programmes, that combine this classical mass selection with more advanced technologies. So most intraspecific breeding schemes now include progeny testing either to estimate genetic parameters and/or to establish seedling seed orchards with subsequent genetic rogueing (at the family level) or to estimate the genetic value of clones in clonal seed orchards.

A third prerequisite for tree breeding is an understanding of breeding systems and genetic compatibility among species, varieties and races and the potential values of, or barriers to, crossing among such taxa.

A fourth prerequisite is a knowledge of the limits of adaptation and genotype environment interactions. The limits on geographic movement of existing or improved genetic materials must be assessed, particularly if multiple population breeding strategies are considered.

The fifth prerequisite, as for any other enterprise, is an understanding of the costs and benefits together with the risks of tree breeding. Breeding has proven to be a costly venture that requires major investments of cash, personnel and land, but it also has resulted in substantial returns on investments. To be a candidate for tree breeding, a forest tree species, variety or race must have high commercial value in order to convince governments or industries to provide substantial long-term resources; perhaps this accounts for FAO's estimate, that only 25 commercial species out of the 50,000 forest tree species in the world have a documented breeding strategy (Pierre Sigaud pers. comm.).

Advanced breeding may involve methods like line breeding or interspecific recurrent breeding (Bisoffi and Gullberg 1996, Kang *et al.* 1998) which need not be developed here but which are used on a limited number of forest tree model species like the south-eastern pines in the USA, radiata pine in the southern hemisphere, eucalypts in tropical regions, willow in Sweden, a couple of conifers and birch in Scandinavia, maritime pine in France, poplar in Italy and France, and a few others.

Simultaneously with these breeding strategies and genetic improvement programmes, research is continued in quantitative genetics, particularly to estimate genetic parameters and genetic gain. Also some attempts are being made to integrate classical and molecular breeding.

4. Selection traits

Selection traits are presented in order of economical and biological importance:

4.1. Adaptation

The term means the ability to withstand certain abiotic factors of the environment such as soil, winter cold, spring frost and summer drought (Teissier du Cros 1998). Its elucidation requires multi-site testing of genotypes to determine:

- in which part of the natural range populations should be sampled for subsequent selection and improvement;
- in which stands and geographic regions seed should be collected if a breeding scheme includes only provenance selection;
- the flexibility (plasticity) of a variety;
- the number, location and attributes of breeding populations to be included in a breeding programme (genotype x site interaction).

4.2. Yield (productivity)

This term generally implies volume productivity, often indirectly estimated by height and diameter increment. Volume yield is of primary importance for most species in most countries when wood production is the target, but there are exceptions. One can think of the now classical case of the Pacific yew (*Taxus brevifolia*) the genetic resources of wich are protected and which has even resulted in some kind of breeding because it contains taxol, an anti-cancer compound (Ledig 1997). Several other compounds and products produced by forest trees have justified or still justify selection : rubber, gum, tannin, resin, leaves for fodder, etc.. The productivity of many of those is generally correlated with volume productivity but there are probably several exceptions.

4.3. Quality

External quality is influenced primarily by qualitative traits. These include:

- bole straightness on maritime pine (Baradat and Pastuszka 1992) or eucalypts (Tibbits *et al* 1997);
- branching habit in beech (Teissier du Cros and Thiébaut 1988) or in Douglas-fir (Bastien and Roman-Amat 1992).

They also include intrinsic qualities of wood including:

- density, particularly in Douglas-fir (Bastien and Roman-Amat 1992) or eucalypts (Tibbits *et al* 1997);
- fiber length;
- absence of spiral grain (Teissier du Cros *et al.* 1980).

Wood quality as a whole is capturing an increasing interest in advanced breeding programmes all round the world.

4.4. Resistance

Resistance to pests and diseases involves many traits for selection in most tree species in their natural range. However, it sometimes becomes essential in intensive plantation forestry with exotic species or when varieties have been developed on a narrow genetic base. Conversely, there are many examples (e.g. eucalypts) where pests are a problem in plantations in natural areas whereas exotic species have no problems. A striking example, when a pest or disease is introduced to a new environment, is poplar in western Europe. A new leaf rust race is observed roughly every four years in plantations established with clones that are all more or less related and that all have a susceptible common ancestor (Lefèvre *et al.* 1994, Villar *et al.* 1996). The current trend to keep in front of a „flexible adversary" (Gerhold *et al.* 1982) is to work on a durable, usually polygenic tolerance based on the knowledge of possible variation of the pest or disease.

5. Results and extension

As discussed above, tree breeding is a long-term, costly investment requiring faith and patience. Most varieties need at least 20-30 years research and testing before being put on the market. With the exception of eucalypt, poplar and a few other fast growing species, variety superiority is based on a genetic gain which has been estimated through research, not in demonstrations or extension tests. Gains predicted from research need to be verified through practical demonstrations lasting one half to three-quarters of a generation. Since intensive breeding started less than 50 years ago, a large amount of confidence in improvement programmes is needed to buy seed from seed orchards or cuttings from selected clones for reforestation.

Important gains in volume productivity and substantial financial returns due to selection, improvement and use of improved varieties have been demonstrated in several countries. As an example, from 30 to 50% gain in volume in many species have been reported (Teissier du Cros 1998). Similarly 22 to over 80% financial returns have been mentioned in China by FAO (Palmberg-Lerche 1993a).

Another evidence of acceptance of results of tree breeding is the amount of improved reproductive material used in reforestation : most to all needs in Greece and Finland, 80% in Italy, a majority in Sweden, more than needed on *Pinus nigra* var. *laricio* in France. The reasons why only part of the needs is covered with improved varieties are diverse:

- higher cost of improved varieties (France);
- limited number of varieties (point of view of certain scientists);
- need for diversification (Scandinavia);
- competition with less expensive seed from selected seed stands (Western Europe).

The last consideration on genetic improvement of forest trees is that it essentially involves conifers, with major exceptions like eucalypt and poplar. Genetic improvement programmes on broadleaves have been initiated all around the world but they lag behind those involving conifers in terms of age, sophistication and output.

6. Reaction to breeding and use of improved varieties

In a few parts of the world, such as Germany, Netherlands, Norway and western USA, the public reacts to exotics and homogeneity of large planted areas. Part of the public even denies the need for improved varieties and for plantation forestry (especially of conifers).

Luckily enough there are also positive examples. Genetic improvement is underway and accepted in about half the countries of the World (Palmberg-Lerche 1993a) and large companies are running their own breeding programme. Better still, some of them have created scientific, technical and financial links with universities and other institutes dealing with research and extension on improvement (co-operatives). Such well-known joint ventures exist in south and west USA dealing respectively with slash and loblolly pines and Douglas-fir. Several others can be mentioned in the southern hemisphere. However, a new model, which joins on the one hand a large amount of small forest owners, including state forests and, on the other hand, research and extension institutes, is described as an example by Y. Lesgourgues in this volume.

7. Future of forest genetics and tree improvement

Diversification is probably the master word for the future.

7.1. Strategies

Sophisticated multi-generation strategies are and should only be applied on a limited amount of species. A given country or region cannot afford much more than one, two or maybe three programs which suppose large planting areas and adapted silviculture favouring the expression of genetic gain.

However, to favour diversification, simple strategies have been or will be applied to a larger number of species. In Queensland and New South Wales, Australia, 25 species are mentioned (Nikles and Lee 1998). In France the similar figure is around 16 (Teissier du Cros 1994). In the Pacific Northwest of USA and Canada it is less than 10 (Tom Adams pers. comm.).

7.2. Species

Larger efforts will be devoted to broadleaves. Not only fast growing ones, such as eucalypts and poplars, but also species of high value. In Europe, for instance, strategies combining management, conservation and breeding are now applied to *Sorbus, Acer, Juglans, Prunus, Fraxinus* and several other scattered or social broadleaves.

7.3. Tools

A rumour says biotechnologies have developed because quantitative genetics had no more future due to declining interest in forest tree improvement. Do not believe that entirely ! However it is partly true. Most young geneticists have no future after their PhD if they have not looked into DNA or located a few genes in the genome. Steve Strauss, in the analysis of an inquiry he made for a recent talk in UK on environmental risks of transgenic forest plantations, says that in a more sensible but perhaps more cruel manner : „[markers] are used mainly for their scientific elegance and power, and short term publication opportunities." (Steve Strauss pers. comm.). I hope this does not offend anybody.

Having said that I must stress that one should be confident in biotechnologies provided they are used in an appropriate way. For example quantitative trait loci (QTLs) cannot be detected without a powerful breeding programme involving segregating F_1s, F_2s or backcrosses. Thus, there are four types of reaction towards the use of biotechnology in forest tree improvement.

- I am a 100% tree breeder. Quantitative genetics still has a long way to go. I do not need biotechnology (Palmberg-Lerche 1993a); they should not compete [in attracting funds] with classical [quantitative] methods.
- I am a forest tree geneticist or maybe a molecular biologist. I am assisting – and dependent upon – advanced multi-generation improvement programmes developed on species with high economic value. I must use biotechnology (micropropagation, somatic embryogenesis, markers) whenever needed and probably develop new tools. This is fundamental research. It is needed, otherwise science will die. The corresponding cost is affordable.
- I am involved in conservation of forest gene resources. I could not develop any sound research without genetic markers.
- I do not think that biotechnology will have any application in tree improvement in the short term.

To conclude in a prudent manner, let us consider biotechnologies as a useful tool for tree breeders, if they are lucky enough to conduct expensive research based on old and advanced breeding schemes.

8. Conservation of forest gene resources

Two points must be emphasized with regard to gene conservation.

- Most, if not all sound breeding strategies include conservation of gene resources (Palmberg-Lerche 1992). Those which do not, have failed or will fail because of the lack of gene resources to exploit when a new question occurs, for instance a new abiotic or biotic adversary.

- The necessity for information, training (Sigaud 1998) and networking (Palmberg-Lerche 1997, Patiño-Valera 1997, Souvannavong *et al.* 1994) and the co-ordination role of international, regional or national organisations is recognised in the community of forest tree breeders.

9. Funding

A general statement is that governmental money devoted to forest tree improvement is recurrent but it may only cover salary costs of national research institutes and some of the background costs like accommodation, communications, etc. This type of money seems to have decreased in availability (Aravanopoulos for Greece, pers. comm.) and research teams have been "downsized" (Bey 1996).

On the other hand research money, when available, tends to be contracted on a short term basis (2-4 years) (Nuño Borralho for Portugal, Tore Ericsson for Sweden, pers. comm.). Instead of originating from one single source, research funds often come from different ministries (Agriculture, Forestry, Research or Environment). The management of funds becomes increasingly complex. In Europe, the European Union has also been an important fund provider for research connected to genetic improvement but activities like laying out field tests are no longer accepted and classical forest tree breeding looks less attractive to scientific experts than biotechnology.

Private funding used to be considered as a short-term resource. The situation has improved considerably with funding from co-operatives and similar joint ventures. Large companies are running long-term breeding programmes (Bey 1996, Garth Nikles pers. comm.) and providing necessary money, land and logistical support.

10. Conclusion and prospects

Most of the opinions described hereunder have been suggested by respondents to my inquiry.

First a very positive point of view expressed in the conclusions of 26th Meeting of the Canadian Tree Improvement Association whose title is similar to that of the Beijing conference : Tree improvement. Its contribution to sustainable development (Beaulieu *et al.* 1998). It says:

„while these specialists all had different backgrounds [tree improvement, ecology, socio-economy, forest industry] and in some cases experience not directly related to tree breeding, they came out with a strong consensus on the importance of tree breeding for meeting the demand for wood products and protecting a large portion of natural ecosystems, and they agreed that

this should be better conveyed by tree breeders to the general public and to policy makers".

Most tree breeders around the world are convinced that their job is useful. But are they the best messengers ? Wouldn't be better messengers the people that have to be convinced first: seed merchants, nursery managers, forest owners, directors of large forestry companies ? If they are convinced by expected genetic gains they will be in a better position than breeders to convert that into political support.

The message that breeders wish to convey may be summarised as follows.

10.1. Vision

Conventional breeding has a long way to go. This is not just a vow it is a possible outcome of the „2020 Vision": tripling the plantation estate in Australia by 2020 (Anon. 1997 *in* Nikles and Lee 1998). A similar concept is under development in France. Four forest management scenarios have been simulated. They range from a pure protection and recreation role of forests to a pure production role. Each scenario is being analysed to determine how it will face the wood industry needs and requirements in 2020-2030 (Cristofini *et al.* 1998). Each scenario has a different effect on the type of forest genetics which should be developed in the future: from 100% population genetics to 100% breeding.

10.2. Genetic base of varieties

Genetic selection is not genetic erosion (Koski pers. comm.). This can easily be proven by the genetic diversity in clonal varieties. For instance 17 provenances are represented in the 50-clone French Norway spruce variety providing 40% gain on volume and late bud burst (Van de Sype 1994). Bill Libby recommends a minimum of 20 unrelated clones in a clonal variety (Beaulieu 1998). I think he once said 3! Whatever the size of a clonal variety, a sound use may improve its interaction with the environment, including possible pests and diseases; small adjacent monoclonal plots, plots with clone mixtures, withdrawal of varieties after 10-15 years and replacement by new ones. Clonal forestry should be restricted to species with rather short rotation, and in places where possible occurrence of pests and diseases may be handled (Namkoong 1986 *in* Palmberg-Lerche 1993a).

It is suggested to continue the development of propagation techniques (Tibbits *et al.* 1997) which permit capturing the highest genetic gain while maintaining large variability within varieties. When Nikles talks about „family forestry", others probably think of bulk propagated families with high specific combining ability (Nikles 1998). Similarly Baradat suggests the very flexible use of „à la carte" clone-equivalent varieties (same genetic gain as in clonal varieties) produced in seed orchards laid out with polycrossed families originating from progeny-tested „plus trees" and producing an expected volume superiority of 50% compared to current commercial varieties (Baradat and Pastuszka 1992).

All have in mind the greatest possible diversity in forest reproductive material. But everyone should be aware of the need to develop research on mechanisms for maintaining diversity in a changing environment (Koski pers. comm.)

10.3. The right strategy for the right species

Very simply, the size and the type of an improvement strategy should be in relation to the potential economic return of a species (Willan 1985 *in* Palmberg-Lerche 1993a). Provenance testing is the most economical investment (Teissier du Cros 1998) for species that are not among the „top 25" according to FAO's evaluation (Pierre Sigaud pers. comm.). Simultaneously research should be continued on:

- breeding methods for species with high economical return with concepts such as combined short- and long-term strategies (Bisoffi and Gullberg 1996, Kang *et al.* 1998);
- simple improvement strategies merged with gene resource conservation for less utilised species.

10.4. Networking and cooperative endeavours

Whereas large companies or joint ventures in developed countries may afford expensive breeding, developing or small countries may want to invest their effort and money in co-operative research with countries in the same region or by twinning with countries of other regions involved in similar questions. This point is developed by Dieters and Nikles (1997).

Forest genetics and forest tree breeding are sciences that need funds and faith to be present in the 21st Century. But to convince sponsors and clients, breeders need convinced messengers and convincing messages. When not already belonging to the same structure (cooperatives, large companies, joint ventures), partners of the breeding chain will be the best messengers.

Acknowledgement

I am very thankful to all those who have provided me with written information about the situation of forest tree breeding in their country, their region or their organisation : Tom Adams (USA), Philippos Aravanopoulos (Greece), Nuño Borralho (Portugal), Rowland Burdon (New Zealand), Öje Danell (Sweden), Fulvio Ducci (Italy), Tore Ericsson (Sweden), Christel Palmberg-Lerche (FAO for Central and South America, the Caribbeans and Myanmar), Veikko Koski (Finland), Urban Gullberg (Sweden), Howard

Kriebel (USA), Tom Ledig (USA), Owe Martinsson (Sweden) , Garth Nikles (Australia), Rodolfo Salazar (CATIE, Costa Rica), Scott Schlarbaum (USA), Pierre Sigaud (FAO for North China), Tore Skrøppa (Norway), Steve Strauss (USA), Wayne Tibbits (Tasmania) and Marc Villar (France). The fact that each scientist's name is followed by a country does not necessary mean that the answer covered that particular country. Some colleagues have concentrated on a species, a region or a topic.

Literature cited

Baradat, P. and Pastuszka, P., 1992. Le pin maritime. *In* Gallais, A. and Bannerot, H. (eds.): *Amélioration des espèces végétales cultivées.* INRA, Paris, France. 695-709.

Baradat, P., Adams, W.T. and Müller-Starck, G. (eds), 1995. *Population Genetics and Genetics Conservation of Forest Trees.* SPB Academic Publishing. 479 p.

Bastien, J.C. and Roman-Amat, B., 1992. Le douglas. *In* Gallais, A. and Bannerot, H. (eds.): *Amélioration des espèces végétales cultivées.* INRA, Paris, France. 710-719.

Beaulieu, J., Villeneuve, M., and Bousquet J., 1998. Tree Improvement. Its Conntribution to Sustainable Development. Minutes. 26th Meeting of the Canadian Tree Improvement Association. *Dendrome.* 4(2), 4-6.

Bey, C.F., 1996. Report to the Forest Genetics Resources Study Group/NAFC/FAO. National Report – US. Fredericton, New Brunswick, Canada, 17-21 June 1996. 13p.

Bisoffi, S. and Gullberg, U., 1996. Poplar breeding and selection strategies. *In* Stettler, R.F., Bradshaw, H.D., Jr., Heilman, P.E., and Hinckley (eds.): *Biology of Populus and its implications in Management and Conservation.* T.M. NRC Research Press, Ottawa, Ontario, Canada. 139-158.

Bouvarel, P., 1986. L'amélioration génétique des arbres forestiers. Essai d'une histoire. *In* Amélioration génétique des arbres forestiers. *Rev. Forest.* Fr., Numéro spécial 38, 7-11.

Burley, J., Nikles, D.G. (eds), 1973. Tropical Provenance and Progeny Testing and International Cooperation. Proceedings of a joint meeting of IUFRO S2.02.8 Tropical Species Provenances and S2.03.01 Breeding Tropical and Subtropical Species. Nairobi, Kenya, 22-26 October 1973. Commonwealth Forestry Institute, Oxford. 597p.

Burley, J. and Styles, B.T. (eds), 1976. *Tropical Trees. Variation, Breeding and Conservation.* Linnean Society of London, Academic Press. 243p.

Cristofini, B., Lacaze, J.F., Messean, A., and Normandin, D., 1998. Prospective. La forêt, sa filière et leurs lieu au territoire. Synthèse et scénarios. Répercussions sur la recherche. Tome 1. INRA. Paris. France. 232p.

Dieters, M.J. and Nikles, D.G., 1997. The genetic improvement of Carribean pine. Building on a firm foundation. Proc. 24th Biennial Southern Forest Tree Improvement Conference.

Gerhold, H.D., Heybroek, H.M., Patton, R.F., Søgaard, B., Stephan, R. and Weissenberg K. von, 1982. Breeding for balance? An introduction. *Resistance to Diseases and Pests in Forest Trees In* Heybroek, H.M., Stephan, B.R., Weissenberg, K. von (eds.): Proceeding of the Third International Workshop on the Genetics of Host-Parasite Interaction in Forestry. Wageningen, The Netherlands. 14-21 September 1980. ix-xi.

Heybroek, H.M., Stephan, B.R., Weissenberg, K. von (eds), 1982. *Resistance to Diseases and Pests in Forest Trees.* Proceeding of the Third International Workshop on the Genetics of Host-Parasite Interaction in Forestry. Wageningen. The Netherlands. 14-21 September 1980. 503p.

IUFRO, 1998. Division 2. *In:* Structure, Officeholders, Addresses, Telecom Numbers. *IUFRO News*, 27,2,4-6.

Kang, H., Lascoux, M. and Gullberg, U., 1998. Systematic tree breeding. *In* Mandal A.K. and Gibson G.L. (eds.): *Forest Genetics and Tree Breeding.* CBS Publishers & Distributors, New Delhi, 38 - 62.

Ledig, F.T., 1997. Forest Genetics. *In:* McGraw-Hill Encyclopedia of Science and Technology. 8th ed., vol. 7: 412-415.

Ledig, F.T. and Neale, D.B. (Compilers), 1998. World directory of forest geneticists and tree breeders. *Gen. Tech. Rep.* PSW-GTR-170, Pacific Southwest Research Station, Forest Service, USDA, Albany, California, USA, 189p.

Lefèvre, F., Pichot, C. and Pinon, J. 1994. Intra- and interspecific inheritance of some components of resistance to leaf rust in poplar. *Theor. Appl. Genet.* 88. 501-507.

Nikles, D.G., Lee, D.J., 1998. Improved varieties of forest trees available or being developed for commercial planting in Queensland and northern New South Wales. *In* Dyason, R., Dyason, L. and Darsden, R. (eds.): *Plantation and regrowth forestry. A diversity of opportunity.* Proc. Austr. Forest Growers Biennial Conference. Lismore, New South Wales. Australia. 139-154.

Palmberg-Lerche, C., 1992. Conservation of genetic ressources as an integral part of forest management and tree improvement. Proc. International Symposium on Seed Procurement and Regulations in Forest Reproductive Materials in Tropical and Sub-Tropical Countries. KEFRI/GTZ/IUFRO and FAO. Nairobi. Kenya. 4-10 Oct. 1992. 169-180.

Palmberg-Lerche, C., 1993a. Present status of forest plantation and tree improvement in the Americas, with special reference to tropical America. Proc. Pan-American Forestry Congress and the Seventh Brazilian Forestry Congress. Curitiba, Brazil. 19-24 September 1993. Vol. III, 142-149.

Palmberg-Lerche, C., 1993b. Forest Genetic Resources in the History of FAO. Unpublished File Note. Forest Resources Division, Forestry Department, FAO, Rome Italy. 3p.

Palmberg-Lerche, C., 1997. Workshop of international neem network. Personal notes. Forest Ressources Division. FAO, Rome. Italy. 3p.

Patiño Valera, F., 1997. Genetic Ressources of *Swietenia macrophylla* and *Cedrela odorata* in the Neotropics : Priorities for Coordinated Actions. Paper based on a presentation to the International Symposium on Big-leaf Mahogany. Puerto Rico. 13p.

Sigaud, P., 1998. Regional/Sub-Regional Workshops on the Conservation, Management, Sustainable Utilization and Enhancement of Forest Genetic Ressources. I - Sub-Saharan African Region. Identification and Definition of Species Priorities. Draft Document. Forestry Department. Forest Resources Division FAO, Rome, Italy. 13p.

Souvannavong, O., Malagnoux, M. and Palmberg-Lerche, C., 1994. International Cooperation in the Conservation of Mediterranean Forest Genetic Ressources. FAO. Division of Forestry. 7p.

Teissier du Cros, E., 1994. L'amélioration génétique, pourquoi? Génétique et Amélioration des Arbres Forestiers. *Forêt-Entreprise.* 96 (2-3), 15-16.

Teissier du Cros, E., 1998. Genetic improvement reduces loss and accelerates tree growth. International Seminar on Causes and Consequences of Accelerating Tree Growth in Europe. EFI and ECOFOR. Nancy, France, 17-19 May 1998. 13p.

Teissier du Cros, E., Kleinschmit, J., Azoeuf, P., Hoslin, R., 1980. Spiral grain in beech. Variability and heredity. *Silvae Genet.* 29(1), 5-13.

Teissier du Cros, E., Thiébaut, B., 1988. Variability in beech : budding, height growth and tree form. *Ann. Sci. For.* 45 (4), 383-398.

Thielges, B.A. (ed), 1975. *Forest Tree Improvement. The Third Decade.* School of Forestry and Wildlife Management. Louisiana State University. Baton Rouge. 243p.

Tibbits, W.N., Boomsma, D.B. and Jarvis, S., 1997. Distribution, Biology, Genetics and Improvement Programs for *Eucalyptus globulus* and *E. nitens* around the world. 24th Southern Forest Tree Improvement Conference, 9-12 June 1997, Orlando, Florida, USA. 15p.

Toda, R. (ed.), 1974. *Forest Tree Breeding in the World*. Edited and published by the Government Forest Experiment Station, Meguro, Japan. 205p.

Van de Sype, H., 1994. Epicéa commun. L'Amélioration Génétique des Essences Forestières. *Forêt-Entreprise*. 96 (2-3), 52-55.

Villar, M., Lefèvre, F., Bradshaw, H.D., Teissier du Cros, E., 1996. Molecular genetics of rust resistance in poplars (*Melampsora larici-populina* Kleb/*Populus sp.*) by bulked segregant analysis in a 2 x 2 factorial mating design. *Genetics* 143, 531-536.

Tree genomes: what will we understand about them by the year 2020 and how might we use that knowledge?

Ronald R. Sederoff

Introduction

The purpose of this paper is to speculate about the future applications of molecular genetics to the understanding and utilization of tree genomes. Biotechnology of forest trees is a young discipline. The first genetically engineered tree, a glyphosate tolerant hybrid poplar, was produced in 1987 (Fillatti *et al.*, 1987). Since that time, biotechnology of forest trees has incorporated new technology of genetic mapping and has now entered the era of genomics. Much of what follows here is only one person's speculations about scientific progress into a relatively near future. In general, scientists are less effective at predicting the future of science than are writers of fiction, who are less constrained about predictions. The time frame of this paper is to look forward to the year 2020, which is approximately one rotation age for a temperate pine, and as much as four rotations for a tropical hardwood. The purpose of this short review is not to be comprehensive, nor to provide access to key references, but to provide an overview of ideas related to application to forest trees of existing technology in the relatively near future.

I would like to begin this discussion by looking back at my past experience of predicting the future of forest biotechnology. The year 2020 is only 21 years away. I have been gazing into the forest biotechnology crystal ball for about 15 years. In 1984, Tom Ledig and I presented a paper on the potential of forest biotechnology (Sederoff and Ledig, 1984). On the basis of this paper, our record for prediction is mixed. In 1984 we argued that the theoretical and technical basis for the application of recombinant DNA technology was well defined and appropriate for forest trees, and that there was great potential, but that there were some limitations:- (1) lack of information about important processes; (2) absence of technology to transfer genes; and (3) inadequate cell culture systems. We described disease resistance, pest resistance, herbicide resistance, apical dominance, degradation of lignocellulose, improvements of wood properties, drought resistance control of metabolism, nutrient use efficiency, photoperiod response, and photosynthetic efficiency as potential targets for

C. Mátyás (ed.), Forest Genetics and Sustainability, Volume 63, 23–30.

biotechnological manipulation. Many of these predictions have been fulfilled already, and those that have not been achieved would not be thought foolish today. The most important conclusion, however, is that we were not able to imagine the progress of technology and theory that has already occurred. We were far too conservative about what was possible and the rate of progress in molecular biology. Alternatively, we were optimistic about the cost and difficulty of some problems, such as somatic embryogenesis and DNA transfer.

What we missed

We did not consider the control of flowering as a major target because of cytoplasmic male sterility in maize, which was responsible for the southern corn leaf blight epidemic (Laughnan and Gabay, 1978). Furthermore, there was no strategy or known mechanism to create nuclear encoded male sterility at that time. The potential for genomic mapping as a powerful tool for tree breeding was not yet apparent, and the prospects of a genomic science for forest trees was far below the horizon. In 1984, the human genome project was not yet under consideration. It was still a time when the sequence of a single gene was sufficient to merit a graduate thesis, and sequence of a modest virus was the basis for a large multi-investigator project.

What do we anticipate now?

It is now reasonable to expect extensive DNA sequencing and identification of expressed genes for any tree species of long term commercial or scientific interest. In many laboratories, plans are in progress to map all of these expressed genes in some of the most commercially important species, including the pine megagenome, which is 200 times larger than the model plant system, *Arabidopsis thaliana*. With essentially all genes identified by sequence and mapped to physical and genetic location, the remaining essential information becomes gene function. There are currently several tools to determine gene function, including expression analysis, point mutation, induced suppression, transposon tagging, QTL mapping, and comparative sequence analysis.

1. Tools

1.1. Expression analysis with microarrays

The tools to determine the time and location of gene expression have always been valuable indicators of specific functional roles for genes acting in development or in response to specific biotic or abiotic stresses. Assays of the level of gene expression for all expressed genes simultaneously through microarrays or DNA chips looks as if it will become the standard for the industry (Schena *et al.*, 1995). These methods promise to provide information on each gene tested but, more importantly, they will allow detection of coordinated expression, where the interrelationships are not expected or obvious. This methodology will move us into a new era of "holistic" cell biology, with the potential to provide a global view of the extent and significance of epistasis (genetic interactions) at the molecular level (Kauffman, 1993).

1.2. Mutation analysis

Information about function is often inferred from the phenotypes of "loss of function" mutations. In natural populations of most forest trees, mutations are found at high frequencies, presumably due to the balancing forces of mutation and natural selection. In maize and *Arabidopsis* large collections of mutations have been induced or discovered over a period of many years. To identify the genes associated with mutation, collections of tagged libraries have been made so that it is possible to use the phenotype and the location of a mutation to clone the genes of interest. Similarly, it is possible to use transposon libraries to isolate a mutation for a gene where nothing is known but the sequence. It is therefore already possible to find a loss-of-function mutation in a large fraction of all the plant genes of interest in maize or *Arabidopsis*.

1.3. Induced suppression of gene function

Genetic engineering technology has provided yet another approach to suppress gene function through the directed introduction of an entire gene sequence or even a fragment of a specific gene. If the complement of the message is produced, the suppression is due to "antisense", or, alternatively, the expression of the opposite strand is thought to shut down expression due to "sense suppression". Therefore, by genetic engineering, an equivalent of a mutation can be produced to shut down expression of a particular gene. There are limitations of this method because the mechanism of suppression is not yet understood and entire families of related genes can be affected simultaneously.

1.4. Quantitative genetics

Quantitative genetics serves as a means of gene discovery as well. Quantitative trait loci (QTLs) are another class of genetic variants that serve to define new genes that contribute to a complex phenotype. This technology is important in forest trees, because QTLs are common for commercial traits of interest. QTLs are the typical variation observed in natural populations, although they may just be alternative forms of the same genes that will be identified by qualitative genetic methods. Each mapped QTL defines a functional gene or a regulatory locus that is part of the functional genome.

1.5. Comparative sequence analysis

By far the most important tool for understanding the expression of new genes is the comparative approach. Due to the close relationship of all higher plants, the discovery of a function for a gene in any higher plant may extend to all other plants. Many genes are also common between plants and animals, so new functions are discovered by animal molecular biologists that apply to plants as well. The comparative approach will also lead to greater knowledge of the relationship of structure and function. It is extremely important that the function of a gene may be often inferred from the DNA sequence alone. It is possible to infer something about the function for about half of the genes discovered in forest trees, just based on the comparison of their sequence to genes already known and characterized.

2. Additional questions

2.1. Progress in functional genomics might be slow

Even though we can define the objectives and techniques of functional genomics, its rapid progress is not guaranteed. Many times it is not possible to infer a function by mutation. Frequently, the loss of a specific gene activity will have no apparent effect. This result may be due to redundant genes or due to the fact that many genes function only under unusual circumstances. In the model system, baker's yeast, which has been fully sequenced, nearly one third of the approximately proposed 6000 genes have not yet been assigned a function. The use of microarrays to analyze the effects of quantitative and qualitative mutations may help to explain the function and the interactions of the genes that might not be apparent when genes are examined one at a time.

2.2. How many genes are there, and who will own them?

How many genes are there in any specific tree species of interest? At present, the best guess is that any plant may have between 20,000 and 60,000 different genes. To what extent do softwoods and hardwoods have different genes? How many different genes are there in all higher plants? The answer depends on the definition of the function of genes and what makes them different. It might be argued that all plants and animals will have only 100,000 different genes, although the total number of genes defined by different functions could easily exceed a million. This issue is extremely important for intellectual property issues. If all plants share the same basic set of genes, and only a small subset have commercial value, then it will be possible for a small number of companies to establish a disproportionate influence on the future of genetic engineering. This possibility has stimulated much recent investment in genomics. There is some controversy regarding the extent to which control of the technology will be possible. It is of some interest that among the small number of diverse bacteria that have been sequenced, there are large numbers of different genes. What is most important, is that there are a finite number of accessible genes that could be identified within the next 20 years and a number as large as a million is within reach.

2.3. The final piece of the puzzle

A most critical piece of the full picture will be the ability to design proteins, enzymes or other molecules for new functions. If we are able to learn about the relationship of structure and function of proteins, it may be possible to infer structure from sequence, and function from structure. If this ever occurs, then it will really be possible to propose a function that has not existed and to create a molecule to carry out that function. Such molecules will truly be due to molecular design, not so different from those areas of chemistry where new molecules are designed for new materials and products.

3. How will this information be useful for forest trees?

The easy part of the answer to this question is that it will help us to do what we have been trying to do already, much more efficiently. To be more specific, the knowledge should lead us to an engineering phase of genetic modification rather than to a discovery phase. In a discovery phase, we try to figure out what genes do by inserting them and looking for the effects. In an engineering phase, we will design effects and then create them, much as is done in mechanics and microelectronics. Although there will be much to discover and to create, the science of molecular biology will be predictive at a far higher level than biology has ever been. It should be possible to design materials, such as wood, with specific properties, and to design new micro-organisms for specific purposes as well. To some extent we are in an engineering phase already for a few genes such as herbicide resistance or insect resistance. For most biological processes we are still in a discovery phase.

One expectation is that within the next 21 years we will have a considerable amount of information regarding what makes a superior tree. We will learn much about the superior growth properties and general combining ability of superior trees at the molecular level. It is of rather immediate interest to use microarray technology on superior genotypes to define which genes might be over- or under-expressed. Studies of quantitative trait analysis for traits of interest in combination with mapping of expressed genes should lead to a greater understanding of the genetic basis of superior performance.

3.1. Some more specific areas

One obvious area of interest and application is that of biologically based materials. Wood is an industrial raw material and its properties are genetically determined. It is reasonable to ask to what extent the properties of wood, now based on the morphology and composition of the cell walls of xylem in trees, can be modified to create wood with new properties. Recent results on lignin modification suggest we are just beginning to determine the extent of the potential to extend or modify the composition of lignin in wood. The other components of the xylem cells and cell walls including the morphology of the tracheids should be of great interest, to see what kind of changes in content and composition can be made. It may be possible to determine specific changes in fibre length, strength, or inter-fibre bonding, for example. These genotypes could lead to more specialized products or the expanded use of a particular species for an increased number of products.

3.2. Will this knowledge affect what species grow where?

A greater understanding of stress responses controling growth might allow the expansion of the growth range of species, so that rapidly growing species such as *Eucalyptus* might grow in higher latitudes. It seems reasonable that the knowledge of processes for control of growth form, and response to stress, will expand the opportunities to expand both the range of growth of particular species, and to increase the number of useful species as well.

3.3. Wood, agriculture and energy

The properties of wood are properties of the cell wall and a great deal will be learned about cell walls of trees and other plants as well. Commercial properties of plant cell walls extend into every area of forestry and agriculture, including food, forage, textiles, and energy, in addition to the properties of wood. Our energy supply has long depended upon the components of plant cell walls, and it is desirable that this continues because it is renewable and has a relatively low impact on the environment. Much of the world still depends directly on wood for fuel. Modifications of the cell wall will have considerable effects both on growth and fuel efficiency. The properties of cotton and human dietary fibre are two other examples of cell wall properties that are also amenable to improvements through genetic engineering.

3.4. What about tree breeding?

Among the earliest applications of biotechnology in breeding programs will be the use of DNA markers. Marker systems vary, and may still undergo some development, but there is already use of DNA fingerprinting in genotype identification, and for identification of pollen genotypes in controlled crosses and open-pollinated seed orchards. Markers are already useful in predicting performance for traits such as disease resistance and should be extended to predict performance using QTLs through marker aided selection (O'Malley *et al.*, 1996). Increased use of line breeding will provide accelerated improvement, and the incorporation of marker systems should increase the efficiency of selection. The ability to select the trees within families for superior performance should add to the rate of improvement. The use of DNA markers for the genetic dissection of complex traits should lead to the understanding of genetic mechanisms that provide new opportunities for plantation management. Fusiform rust disease is a good case in point, where a complex trait that resisted analysis for decades, responded to DNA marker analysis and was found to be a simple gene-for-gene pathosystem which may be readily managed.

3.5. Some additional issues

Many additional issues deserve discussion, including the use of genetic markers for conservation of genetic diversity in forest trees (Furman *et al.*, 1997), and the potential for use of trees for bioremediation. Public acceptance of genetically engineered plants and animals has been to some extent unpredictable, yet many products, including many in the human diet, have reached a high level of acceptance. How the public will receive the introduction of genetically engineered trees into plantations will be important for the cost of many specific applications. Intellectual property issues continue to be important and could have a significant impact on the wood and paper products industry, following the impact and changes already begun in the agricultural chemical industry.

Literature Cited

Fillatti, J.J., Sellmer, J., McCown, B., Haissig, B. and Comai, L. 1987. *Agrobacterium* mediated transformation and regeneration of *Populus*. *Genetics* 206: 192-199.

Furman, B.J. *et al.* 1997. Analysis of genetic relationships of Central American and Mexican pines using RAPD markers that distinguish species. *Molecular Ecology* 6: 321-331.

Kauffman, S.A. 1993. *The origins of order*. Oxford University Press.

Laughnan, J.R. and Gabay, S. 1978. Nuclear and, cytoplasmic mutations to fertility in male self sterile maize. *In* D.B. Walden (ed.): *Maize Breeding and Genetics*. Wiley and Sons, New York, 427-446.

O'Malley, D.M. *et al.* 1996. Molecular markers, forest genetics and tree breeding. *In* J.P. Gustafson and R.B. Flavell (ed.): *Genomes of plants and animals*. 21st Stadler Genetics Symposium. 87-102.

Schena, M. *et al.* 1995. Quantitative monitoring of gene expression patterns with a complementary DNA microarray. *Science* 270: 467-470.

Sederoff, R.R. and Ledig, F.T. 1984. Increasing forest productivity and value through biotechnology. *Weyerhaeuser Science Symposium* 4: 253-276.

Genetic markers in forest genetics – the tunnel remains dark

Alfred E. Szmidt and Xiao-Ru Wang

Introduction

In the last decade, a diverse array of new molecular tools has become available for various types of genetic studies at the population and higher levels. The rate at which these techniques are being developed far outpaces their competent incorporation into studies of population biology. Furthermore, in many cases, statistical methods for analysing novel genetic data have yet to be formulated (Parker *et al.* 1998). Nevertheless, in many areas of population and evolutionary biology, rapid progress is being made because newly available markers can identify individuals, populations, genetic strains, or closely related species. Thus, it is becoming increasingly important for population biologists to be able to understand and evaluate molecular data, and to know which molecular techniques are appropriate for solving their own research questions. Genetic markers have been subject to several recent reviews (Cruzan 1998, Geburek 1997, Haig 1998, Karp and Edwards 1997, Parker *et al.* 1998, Strauss *et al.* 1992a, Vekemans and Jacquemart 1997). However, these earlier works did not consider certain properties of genetic markers such as genomic origin and coding function or focused only on special applications such as biosystematics (Strauss *et al.* 1992a) or genome mapping (Staub *et al.* 1996, Strauss *et al.* 1992b). In our present contribution, we present a brief review of currently available genetic markers and their applications in various areas of population biology with emphasis on the type of analysed DNA sequences.

1. Allozymes

Examining allozyme variation, which results from changes in protein coding sequences, has been the most common technique in plant population biology (Tanksley 1983). The predominantly codominant character of enzyme variation, low cost and technical simplicity are among the main reasons for the widespread use of this category of markers in population genetic analysis (Wendel and Weeden 1989). Variation detectable by enzyme markers is associated with biparentally inherited nuclear genes coding for enzymatic proteins. Enzyme loci often show considerable degree

C. Mátyás (ed.), Forest Genetics and Sustainability, Volume 63, 31–48.

of polymorphism that makes them particularly useful for studies of gene flow, mating system and effects of drift. However, allozyme studies investigate only a highly restricted part of the variation of enzymatic gene products, and this introduces a clear limit to the resolution in terms of the time-scale generating the observed polymorphic patterns (Strauss *et al.* 1992a). Furthermore, a substantial fraction of allozyme variation is not detected by conventional starch gel electrophoresis (Murphy *et al.* 1996). Another disadvantage of allozyme markers is their low number. Moreover, apparent technical simplicity of enzyme markers is not as great as sometimes suggested due to the lack of standard protocols applicable to different species.

2. DNA markers

In contrast to allozymes, which are products of gene expression, and thus, are not transmitted, DNA sequence information is transmitted among generations (Strauss *et al.* 1992a). Therefore, it is potentially the most accurate source of genetic information. Furthermore, analysis of DNA sequences is particularly versatile for designing various kinds of genetic markers. For instance, by focusing on coding DNA sequences that are strongly related to fitness, it should be possible to assess the genetic variation that is subject to natural selection. Conversely, by focusing on third codon positions and non-coding sequences that are weakly associated with fitness, evolutionary inferences may be nearly free of the constraints imposed by natural selection and are likely to be governed primarily by stochastic rather than selective forces (Parker *et al.* 1998). Other advantages of DNA markers compared to allozymes are the greater variety of scales on which evolutionary processes can be studied, which range from DNA fingerprints to coding regions of nuclear, mitochondrial, and chloroplast genes. In contrast to protocols for analysis of enzyme variation, most DNA protocols are easily standardised, and can be applied to various materials. The physical aspects of DNA also offer several advantages over allozymes. DNA is found in nearly all cells of all organisms and it can be recovered from both living and dead tissue. Furthermore, tissues can be easily stored under field conditions, and only nanograms are needed for analysis (Parker *et al.* 1998).

2.1. Restriction fragment length polymorphism (RFLP) markers

RFLP analyses of single-copy nuclear DNA are used in applications similar to those of allozymes, but with the advantage of a potentially nearly unlimited pool of genetic variants available (Moser and Lee 1994). Depending on the genomic origin of the probe, it is possible to obtain RFLP markers for both nuclear and cytoplasmic genomes showing either biparental or uniparental inheritance respectively. Furthermore, by using probes with known sequence it is possible to develop RFLP markers for both coding and noncoding sequences. Similar to enzyme markers, nuclear RFLPs are typically codominant and display Mendelian inheritance. Unfortunately, the application of RFLP markers in population analysis is constrained by several factors. First, RFLP analysis usually requires large amounts of DNA, making non-destructive sampling difficult. Second, documented homologous probes are scarce, and the use of random probes retrieved from libraries constructed with total DNA digests does not permit to unambiguously determine genomic origin (cytoplasmic or nuclear) of the observed variation. Third, it is not possible to determine the functional significance of genetic variation detected by such probes. Fourth, the technique is expensive, both in effort and in materials (Weatherhead and Montgomerie 1991).

2.2. PCR-based DNA markers

The advent of the polymerase chain reaction (PCR) (Saiki *et al.* 1988) has profoundly improved both the speed and efficiency of detecting all types of DNA variation. Relatively low cost, simple standard extraction and amplification protocols make PCR-based markers applicable to virtually any species (McPherson *et al.* 1991). PCR-based DNA markers involve analysis of DNA fragments amplified with two sequence-specific primers (Mullis *et al.* 1986, Saiki *et al.* 1988) or with a pair of primers of arbitrary sequence (Vos *et al.* 1995, Williams *et al.* 1990). Depending on the primer, PCR-based markers can be either sequence-specific or anonymous and random. Two-primer extension with specific primers usually provides markers of known genomic origin and function of the amplified sequence. On the other hand, arbitrary primers usually reveal variation in randomly selected anonymous regions of the genome. Variation of the amplification products can be analysed with several different detection methods. Amplification products can be separated by size by gel electrophoresis. This approach is typically used to detect variation of fragments amplified with arbitrary primers 0(Isabel *et al.* 1995, Szmidt *et al.* 1996b). It detects size variation among the fragments but is insensitive to sequence differences among them. Such differences can be detected by digestion of individual fragments with restriction enzymes. This approach is usually employed in analysis of fragments amplified with specific primers (Wang *et al.* 1996). The fragments obtained through enzymatic digestion are separated on acrylamide gels and visualised by silver staining (Bassam *et al.* 1991, Caetano-Anolles *et al.* 1991). Alternatively, point mutations within amplified fragments can be detected with physical separation methods such as Single-Strand Conformational Polymorphism (SSCP) (Orita *et al.* 1989); Denaturing

Gradient Gel Electrophoresis (DGGE) (Myers *et al.* 1986); and Temperature Gradient Gel Electrophoresis (TGGE) (Wartell *et al.* 1990).

2.3. Digestion of PCR amplified fragments with restriction enzymes - PCR-RFLP

Perhaps most useful emerging approach based on the PCR technique involves specific amplifications of single-copy target sequences followed by screening of polymorphism through digestion with restriction enzymes (PCR-RFLP) or physical separations such as SSCP (Orita *et al.* 1989). This technique is highly reproducible and generally produces codominant markers (Slabaugh *et al.* 1997, Strand *et al.* 1997). The potential disadvantage of this approach is that it requires prior sequence information on the species investigated. Fortunately, this drawback can be alleviated by designing universal primers using known sequences from various organisms (Demesure *et al.* 1995, Slabaugh *et al.* 1997, Strand *et al.* 1997, Taberlet *et al.* 1991). Specific primers can also be designed using available sequences from complementary (c) DNA libraries. Complementary DNAs contain coding regions of genes and are often homologous to known coding sequences (Blake *et al.* 1996, Inoue *et al.* 1994). The development of markers based on amplification with this type of primers (Sequence-Tagged-Site STS markers) was initiated in order to provide a framework linkage map composed of other markers (Olson *et al.* 1989). However, STS primers can also be used as a source of codominant nuclear markers for population analysis in different species (Tsumura *et al.* 1997).

2.4. Random Amplified Polymorphic DNA – RAPD

The RAPD technique uses short primers of arbitrary sequence to amplify at random from a few anonymous genomic sequences (Williams *et al.* 1990). The method is versatile and has been used in many studies in plant population biology (DeVerno and Mosseler 1997, Furman *et al.* 1997, Legionnet *et al.* 1997, Nesbitt *et al.* 1997, Triest *et al.* 1997, Vicario *et al.* 1995). RAPDs usually detect variation in non-coding regions of nuclear and/or cytoplasmic DNA and show Mendelian inheritance (Lu *et al.* 1995). However, only dominant markers are generated and reliability and reproducibility are sometimes difficult to reach (Lu *et al.* 1995, Lu *et al.* 1997).

2.5. Amplified Fragment Length Polymorphism - AFLP™

AFLP™ is based on the selective amplification of a number of genomic fragments (Vos *et al.* 1995). This is achieved by digesting DNA with two restriction enzymes, followed by ligation of complementary adapters and amplification using primers consisting of the adapter sequences, extended with a variable number of 3' 'selective' nucleotides as primers. The method offers the advantages of the RAPD technique but generates more markers per assay and is more reliable (Breyne *et al.* 1997, Jones *et al.* 1997). However, similar to RAPDs, AFLP markers usually detect variation in anonymous nuclear sequences and show dominant Mendelian inheritance (Lerceteau and Szmidt 1998).

2.6. Microsatellites

The PCR method has been used to amplify highly variable repetitive DNA sequences such as microsatellites, *i.e.* regions of DNA composed of short (< 6 bp) sequences repeated in tandem. Microsatellites or Simple Sequence Repeats (SSR) markers are usually codominant and a very high number of alleles can be detected at some loci (Dow *et al.* 1995, Echt *et al.* 1996, Kostia *et al.* 1995, Powell *et al.* 1996, Smith and Devey 1994). Since the assay is easy to perform and produces very polymorphic markers, microsatellites are a highly attractive tool for studies at the population level (Morgante and Olivieri 1993). Recently, microsatellites have also been identified in chloroplast DNA (Powell *et al.* 1995, Tsumura *et al.* 1996, Vendramin and Ziegenhagen 1997) which represents a great interest as the average level of sequence variation in chloroplast DNA is low (Clegg and Zurawski 1992). A disadvantage of microsatellites is that identifying these regions from a genomic library for a new species can be time-consuming. Furthermore, known primers often cannot amplify the same locus across related taxa unless the microsatellite region is flanked by highly conserved sequences where priming sites are located (Echt *et al.* 1996, Kostia *et al.* 1995).

2.7. Sequencing

The most detailed analysis of DNA variation can be obtained by sequencing the region of interest from different individuals. Until recently, extensive use of DNA sequencing for population studies has not been practical because the DNA fragments to be sequenced for each individual had to be isolated from DNA libraries after identification by Southern blotting and hybridisation. Using PCR, however, specific regions can be targeted for amplification if the sequence of the conserved regions flanking the region of interest is available (Glaubitz and Carlson 1992, Kvarnheden *et al.* 1998, Lu *et al.* 1998). Determining the nucleotide sequence of a DNA fragment is reliably applied to questions at population and taxonomic levels and provides the ultimate resolution for detecting genetic variation (Nei 1987). The versatility of sequencing approaches is provided by focusing on different classes of DNA sequences ranging from conserved chloroplast genes (Clegg and Zurawski 1992) to anonymous nuclear DNA sequences (Burt *et al.* 1994). Sequencing avoids the problems inherent in protein electrophoresis of allelic variants that go undetected owing to redundancy in the genetic code or similar migration distances on a gel. However, DNA sequencing may often involve very laborious and expensive procedures, and has been applied to population-level questions infrequently (Parker *et al.* 1998). Nevertheless, progress is being made in understanding the evolutionary process of DNA sequence variation by sequencing particular DNA fragments such as nuclear, mitochondrial and chloroplast DNA control regions (Glaubitz and Carlson 1992, Lu *et al.* 1998).

3. Applications of genetic markers in population biology

Perhaps the most important application of genetic markers in population biology concerns the study of genetic structure and genetic diversity in forest tree populations. The influence of characteristics such as life form, geographic distribution and mating system on the amount and apportionment of genetic variation has been assessed in many species for allozyme markers (Hamrick *et al.* 1992 and references therein). These studies have contributed a great deal to our understanding of various population processes. However, due to a lack of sufficient allozyme variation, many enticing research directions have been impractical to date, especially those requiring information about genetic relatedness of individuals within populations (Parker *et al.* 1998). Furthermore, taking into account the genomic origin of allozymes these surveys are only instructive as to the relative levels of variation in biparentally transmitted nuclear genes coding for a few enzymatic proteins. The results of studies that have compared morphometric and allozyme variation have not consistently demonstrated positive associations among different types of traits (Hamrick *et al.* 1991). Therefore, it is uncertain whether patterns of allozyme variation can be used to predict patterns of variation in other traits. Current population surveys of nuclear DNA variation typically employ random noncoding regions of nuclear genome (Boscherini *et al.* 1994, Devey *et al.* 1991, Roy *et al.* 1992, Szmidt *et al.* 1996b, Vicario *et al.* 1995). Therefore, they are instructive as to the relative levels of variation in the non-coding portion of the nuclear genome. On the other hand, our knowledge about the variation in other nuclear sequences in trees is restricted to variation in the copy number of RNA genes (Gorman *et al.* 1992, Govindaraju and Cullis 1992, Karvonen *et al.* 1994, Moran *et al.* 1992).

Studies of chloroplast DNA structure in forest trees revealed that in spite of generally conservative character of this genome certain regions are variable in some species (Echt *et al.* 1998, Neale and Sederoff 1988, Parducci and Szmidt 1999, Vendramin *et al.* 1996, Wagner *et al.* 1987, Wang and Szmidt 1994). Recently, similar variable regions have been found in mitochondrial DNA (Sinclair *et al.* 1998). Population analysis of these regions revealed considerable differentiation within and among populations of several species (Dong and Wagner 1993, Strauss *et al.* 1993). Lower, but still substantial differentiation is usually found with respect to chloroplast DNA (Dong and Wagner 1994, Hong *et al.* 1993, Strauss *et al.* 1993, Wagner *et al.* 1987, Wang and Szmidt 1994). These studies provided many valuable insights into the differentiation of cytoplasmic genomes in forest tree populations which until recently were completely inaccessible to genetic analysis. In addition, cytoplasmic markers often serve to identify species and delineate their distribution (Szmidt *et al.* 1996a, Wagner *et al.* 1992, Wang and Szmidt 1994).

3.1. Introgression

Natural hybridisation is an important force affecting the evolution and ecology of plant populations. Cases of hybridisation appear to be relatively frequent in plants, and many taxa have been influenced by hybridisation at some point in their history (Grant 1981). Interspecific hybridisation can have both positive (increased genetic diversity, fitness and adaptability) and negative (loss of genetic diversity, genetic assimilation, outbreeding depression) consequences on population viability (Cruzan 1998). Furthermore, hybrid zones are ideal areas in which to investigate evolutionary processes of speciation. However, reliance upon morphological differences for distinguishing between parental species and their hybrids has shown difficult, even for experts. Taxon-specific markers have contributed to the study of interspecific hybridisation by allowing identification of hybrid individuals, and producing an assessment of the proportional contribution of parental genomes (Bacilieri *et al.* 1996, Bousquet *et al.* 1990, Heinze 1997, Szmidt *et al.* 1993, Wang and Szmidt 1994, Wang *et al.* 1998, Watano *et al.* 1996).

3.2. Gene flow

Gene flow in plants occurs through a combination of pollen and seed dispersal. Seed move genetic material as a result of their physical dispersion from the maternal plant. The contribution of pollen to gene dispersal is affected by pollen movement, seed dispersal, and the frequency of selfing (Ennos 1994). While paternity analyses can be used to estimate effective gene dispersal *via* pollen alone (Adams *et al.* 1992), characterisation of processes contributing to the distribution of genetic variation requires analysis of the separate effects of seeds and pollen on gene movement (Cruzan 1998). Maternally and paternally inherited cytoplasmic markers provide means of tracking seed and pollen dispersal respectively. By contrasting the distribution of these uniparentally inherited cytoplasmic markers with biparentally inherited nuclear markers, estimates of the relative contribution of seed dispersal, pollen dispersal, and the breeding system to patterns of genetic variation can be obtained (Ennos 1994). Thus, studies utilising combined information about population genetic structure from cytoplasmic and nuclear genomes can provide many invaluable insights into the relative contributions of seed and pollen movement to total gene flow (Ennos 1994, McCauley 1995, Szmidt and Wang 1992). New highly polymorphic, paternally inherited chloroplast and biparentally inherited nuclear DNA markers are now available for this purpose (Powell *et al.* 1995, Vendramin *et al.* 1996). However, further progress of studies on gene flow in forest trees is still hampered by the paucity of polymorphic maternally inherited markers (Sinclair *et al.* 1998).

3.3. Mating

Mating pattern is another important factor shaping the amount and distribution of genetic variation. Mating events in plants are governed by the behaviour of pollen vectors and by processes affecting post-pollination siring success (Cruzan 1998). One approach to the description of mating patterns in plant populations involves identification of the paternal parent of seeds collected from known maternal plants (Snow and Lewis 1993). An alternative approach to the study of plant mating uses classifications of mating events to characterise levels of inbreeding and patterns of gene dispersal in populations. These mating system studies usually entail estimation of selfing and outcrossing rates (Muona *et al.* 1990). Genetic markers have provided us with insights into both the patterns of mating occurring in plant populations and the pre- and post-pollination mechanisms that contribute to those patterns (Beaulieu and Simon 1995, Boyle *et al.* 1992, Changtragoon and Finkeldey 1995, Cheliak *et al.* 1985, Muona *et al.* 1990). While allozymes remain the most widely used technique for mating system estimation, AFLP markers may be an attractive alternative due to their high levels of polymorphism (Gaiotto *et al.* 1997).

3.4. Asexual reproduction

Asexual reproduction is relatively common in plant species. It occurs through vegetative spread, production of vegetative propagules, or apomixis to produce varying distributions of identical genotypes within and among populations. Until recently, the lack of markers that can reliably identify all genotypes present in a population has hampered studies of clonal species. In many cases, not all individuals may be separated, and the number of genotypes identified tends to increase when more polymorphic markers are used (Ellstrand and Roose 1987). Recently, these problems have been alleviated by employing hypervariable DNA-based markers, including variable number of tandem repeats (Nybom and Schaal 1990), RAPD (Legionnet *et al.* 1997, Wang 1997, Wang *et al.* 1997), and AFLP markers (Arens *et al.* 1998). The studies on asexual reproduction conducted thus far have been somewhat limited in scope, but this approach holds considerable promise for the assessment of genetic diversity in clonal taxa (Parker *et al.* 1998).

3.5. Phylogeny

Phylogenetic analyses are helpful in understanding relationships among species or higher level comparisons. Furthermore, phylogenetic knowledge serves as the framework for studying the evolution of adaptation at all its levels, ecological through molecular (Strauss *et al.* 1992a). In addition, comparing levels of biodiversity among different groups using phylogenetic measures helps to prioritise species for conservation efforts (Krajewski 1994). Genetic markers have proven to be powerful tools for discerning biosystematic, biogeographic, and phylogenetic relationships among species. In particular, a non-recombining, uniparentally-inherited cpDNA markers have been used successfully to perform intraspecific phylogenies for a large number of forest tree species (Govindaraju *et al.* 1992, Krupkin *et al.* 1996, Sigurgeirsson and Szmidt 1993, Strauss and Doerksen 1990, Tsumura *et al.* 1995, Wang and Szmidt 1993, Wang *et al.* 1998). The available phylogenetic information is now being rapidly complemented with sequence data for various cytoplasmic and nuclear genes (Jeandroz *et al.* 1997, Kajita *et al.* 1998, Laroche *et al.* 1997: Wang *et al.* 1999).

4. The tunnel remains dark

The diversity of genetic markers currently employed in genetic analysis is astounding. Developments in molecular biology and, notably, the spread PCR amplification techniques, have contributed to a greater accessibility and affordability of DNA markers (Vekemans and Jacquemart 1997). By providing many different types of markers for intraspecific analysis, and highly conserved gene sequences for analysis at the interspecific level and above, analysis of DNA variation can provide huge progress in understanding of population biology at all levels of biological organisation (Strauss *et al.* 1992a). However, the development of markers suitable for population analysis has not kept up with progress observed in other areas of forest genetic research such as genome mapping.

Because of the heterogeneous nature of evolutionary determinants of genetic variation, specific genetic markers are needed to study their effects. Depending on the nature of evolutionary determinants, individual markers may or may not respond to them. For instance, in the absence of selection, genetic variation will be affected by mutation, gene flow and drift. These processes will affect all loci, regardless of their function. On the other hand, selective forces are more likely to affect variation of the structural, regulatory and RNA genes than that of noncoding portions of the genome. Furthermore, different coding regions of the genome experience different selective pressures depending on the genetic product and/or the tendency of the DNA to withstand changes in nucleotide sequences (dePamphilis and Palmer 1990, Li *et al.* 1985, Lu *et al.* 1998). For example, strong selection on a DNA sequence will prevent the accumulation of much variation arising by mutation. On the other hand, noncoding regions can accumulate mutational change in a neutral manner, such that only genetic drift influences the level of polymorphism in a population. Therefore, although the rate of mutation may be equal in all parts of the genome, observed mutation rates and subsequent population divergence vary due to different selective pressures and as a result the patterns of variation at selected loci may differ from that of neutral loci (Hattemer 1991).

While one of the primary goals of population genetics has been the study of natural selection, up to now genetic markers have not been particularly useful for this purpose. Allozyme markers have been used to study patterns of selection but this type of application has mostly been limited to highly inbred populations (Ennos 1990). Anonymous character of most currently available DNA markers limits their use in studies of natural selection as it precludes association of the observed levels and patterns of variation with the biological function of the analysed sequences. Since population biologists often need high-resolution genetic markers, the techniques that produce a large number of alleles at a single locus and/or many loci with two or more common alleles are usually regarded as most useful. On the other hand, the coding function and location of the regions identified by the markers are not considered. It should be emphasised however, that apart from providing a polymorphic trait that can be used to trace a character or an individual, molecular techniques can also provide important information about the levels and patterns variation in various types of coding sequences. Studying such sequences can give many important insights, e.g. into various selective forces affecting population genetic structure. In this application, the extent and distribution of polymorphism detected by a given technique becomes the subject of the study and not a desired property of the method. Unfortunately, most recent progress in marker development has focused on the detection of nucleotide sequence variation within introns of a few specific protein-coding genes (Lessa and Applebaum 1993). On the other hand, relatively few studies have focused on the development of primers suitable for population analysis of coding genome regions.

Gene sequences are rapidly accumulating for many commercially and scientifically important forest tree species (*e.g.* Schubert *et al.* 1997 1998). These resources can now be used for the development and use of genetic markers for known genes, thereby augmenting random genetic markers and assessing variation of adaptive traits at the DNA level (Slabaugh *et al.* 1997, Strand *et al.* 1997). Using specific primers can develop such gene-specific markers. Specific primer sequences can be determined from the flanking sequences of the region in the same or closely related species (Slabaugh *et al.* 1997, Strand *et al.* 1997). Once primers are synthesised, PCR products can be sequenced to determine the nature and number of substitution events at the nucleotide level or the degree of allelic diversity at the sequence level. Markers spanning coding sequences are expected to be less polymorphic than noncoding sequences. However, the DNA fragments representing coding sequences can be subjected to a variety of powerful methods for detecting small sequence differences. SSCP (Orita *et al.* 1989) and DGGE (Meyers *et al.* 1986) are two of the more widely used methods. These methods often detect a high frequency of sequence variants even in coding sequences (Liu and Sommer 1994, Slabaugh *et al.* 1997).

While many of the available markers can provide information about levels of genetic relatedness, these approaches are not ideal for characterising levels of genetic structure within and between populations. Descriptions of the population genetic structure and its dynamics are based on analysis of gene and/or genotype frequencies. To test whether such descriptions apply to a living population, experimental data on these frequencies must be obtained. Markers that represent Mendelian loci and have codominant expression are also required for the generation of standard estimates of genetic diversity and gene flow. Several of the available DNA markers (e.g. rDNA, minisatellites) lack clear allelism (*i.e.*, bands cannot be assigned to a particular locus), and with most RAPD and AFLP markers, loci can clearly be identified, but heterozygous genotypes cannot be discriminated. Clearly, many additional codominant markers are needed in order to obtain precise estimates of population differentiation.

The extent of population differentiation is expected to vary among nuclear, chloroplast and mitochondrial sequences (Birky *et al.* 1989, Ennos 1994). As a consequence, depending on their genomic origin genetic markers yield different information about genetic diversity and population differentiation (Tomaru *et al.* 1997, Wang and Szmidt 1994). Therefore, in order to obtain an adequate picture of the pattern and apportionment of genetic variation within and among populations and species, genetic markers representing both cytoplasmic and nuclear genomes should be employed. So far, our knowledge about genetic differentiation for cytoplasmic and nuclear sequences is limited to only a few species. The main limitation is the paucity of polymorphic, codominant nuclear markers and polymorphic mitochondrial markers.

Acknowledgement

This work was supported by grants from the Swedish Council for Agricultural and Forestry Research (SJFR) for the first author, and by an STA fellowship for the second author, sponsored by JISTEC/JRDC, Japan.

Literature cited

Adams W. T., Griffin A. R. and Moran G. F. 1992. Using paternity analysis to measure effective pollen dispersal in plant populations. *American Naturalist* 140: 762-780.

Arens P., Coops H., Jansen J. and Vosman B. 1998. Molecular genetic analysis of black poplar (*Populus nigra* L.) along Dutch rivers. *Molecular Ecology* 7: 11-18.

Bacilieri R., Ducousso A., Petit R. J. and Kremer A. 1996. Mating system and asymmetric hybridization in a mixed stand of European oaks. *Evolution* 50: 900-908.

Bassam B. J., Caetano-Anolles G. and Gresshoff P. M. 1991. Fast and sensitive silver staining of DNA in polyacrylamide gels. *Analytical Biochemistry* 196: 80-83.

Beaulieu J. and Simon J. P. 1995. Mating system in natural populations of eastern white pine in Quebec. *Canadian Journal of Forest Research* 25: 1697-1703.

Birky J., C W, Fuerst P. and Maruyama T. 1989. Organelle gene diversity under migration, mutation, and drift: equilibrium expectations, Approach to equilibrium, effects of heteroplasmic cells, and comparison to nuclear genes. *Genetics* 121: 613-627.

Blake T. K., Kadyrzhanova D., Shepherd K. W., Islam A. K. M. R., Langridge P. L., Mcdonald C. L., Erpelding J., Larson S., Blake N. K. and Talbert L. E. 1996. STS-PCR markers appropriate for wheat-barley introgression. *Theoretical and Applied Genetics* 93: 826-832.

Boscherini G., Morgante M., Rossi P., Vendramin G. and Vicario F. 1994. Detection of DNA polymorphism in *Pinus leucodermis* Ant. using random amplification. *Forest Genetics* 1: 131-137.

Bousquet J., Cheliak W. M., Wang J. and Lalonde M. 1990. Genetic divergence and introgressive hybridization between *Alnus sinuata* and *A. crispa* (Betulaceae). *Plant Systematics and Evolution* 170: 107-124.

Boyle T. J. B., Liengsiri C. and Piewluang C. 1992. Genetic studies in a tropical pine - *Pinus kesiya* III. The mating system in four populations from northern Thailand. *Journal of Tropical Forest Science* 4: 37-44.

Breyne P., Boerjan W., Gerats T., van Montagu M. and van Gysel A. 1997. Applications of AFLP(TM) in plant breeding, molecular biology and genetics. *Belgian Journal of Botany* 129: 107-117.

Burt A., Carter D., White T. and Taylor J. 1994. DNA sequencing with arbitrary primer pairs. *Molecular Ecology* 3: 523-525.

Caetano-Anolles G., Bassam B. J. and Gressholff P. M. 1991. DNA amplification fingerprinting: a strategy for genome analysis. *Plant Molecular Biology Reporter* 9: 294-307.

Changtragoon S. and Finkeldey R. 1995. Patterns of genetic variation and characterization of the mating system of *Pinus merkusii* in Thailand. *Forest Genetics* 2: 87-97.

Cheliak W. M., Pitel J. A. and Murray G. 1985. Population structure and the mating system of white spruce. *Canadian Journal of Forest Research* 15: 301-308.

Clegg, M. T. and Zurawski, G. 1992. Chloroplast DNA and the study of plant phylogeny: present status and future prospects. *In* Soltis P. S., Soltis D. E. and Doyle J. J. (eds.): *Molecular Systematics of Plants*. Chapman and Hall, New York, London.

Cruzan M. B. 1998. Genetic markers in plant evolutionary ecology. *Ecology* 79: 400-412.

Demesure B., Sodzi N. and Petit R. J. 1995. A set of universal primers for amplification of polymorphic non-coding regions of mitochondrial and chloroplast DNA in plants. *Molecular Ecology* 4: 129-131.

dePamphilis C. and Palmer J. 1990. Loss of photosynthetic and chlororespiratory genes from the plastid genome of a parasitic flowering plant. *Nature* 348: 337-339.

DeVerno L. L. and Mosseler A. 1997. Genetic variation in red pine (*Pinus resinosa*) revealed by RAPD and RAPD-RFLP analysis. *Canadian Journal of Forest Research* 27: 1316-1320.

Devey M. E., Jermstad K. D., Tauer C. G. and Neale D. B. 1991. Inheritance of RFLP loci in a loblolly pine 3-generation pedigree. *Theoretical and Applied Genetics* 83: 238-242.

Dong J. and Wagner D. B. 1993. Taxonomic and population differentiation of mitochondrial diversity in *Pinus banksiana* and *Pinus contorta*. *Theoretical and Applied Genetics* 86: 573-578.

Dong J. S. and Wagner D. B. 1994. Paternally inherited chloroplast polymorphism in *Pinus*: Estimation of diversity and population subdivision, and tests of disequilibrium with a maternally inherited mitochondrial polymorphism. *Genetics* 136: 1187-1194.

Dow, B.D., Ashley, M.V., Howe, H.F. 1995. Characterization of highly variable (GA/CT)n microsatellites in the bur oak, *Quercus macrocarpa*. *Theoretical and Applied Genetics* 91: 137-141.

Echt C. S., DeVerno L. L., Anzidei M. and Vendramin G. G. 1998. Chloroplast microsatellites reveal population genetic diversity in red pine, *Pinus resinosa* Ait. *Molecular Ecology* 7: 307-316.

Echt C. S., Maymarquardt P., Hseih M. and Zahorchak R. 1996. Characterization of microsatellite markers in eastern white pine. *Genome* 39: 1102-1108.

Ellstrand N. and Roose M. 1987. Patterns of genotypic diversity in clonal plant species. *American Journal of Botany* 74: 123-131.

Ennos R., Clegg M., Kahler A. and Weir B. and 1990. Detection and measurement of selection: genetic and ecological approaches. *In* Brown A., Clegg M., Kahler A. and Weir B. (eds.): *Plant population genetics, breeding, and genetic resources*. Sinauer, Sunderland, Massachusetts, USA.

Ennos R. 1994. Estimating the relative rates of pollen and seed migration among plant populations. *Heredity* 72: 250-259.

Furman B. J., Grattapaglia D., Dvorak W. S. and O'Malley D. M. 1997. Analysis of genetic relationships of central American and Mexican pines using RAPD markers that distinguish species. *Molecular Ecology* 6: 321-331.

Gaiotto F. A., Bramucci M. and Grattapaglia D. 1997. Estimation of outcrossing rate in a breeding population of *Eucalyptus urophylla* with dominant RAPD and AFLP markers. *Theoretical and Applied Genetics* 95: 842-849.

Geburek T. 1997. Isozymes and DNA markers in gene conservation of forest trees. *Biodiversity and Conservation* 6: 1639-1654.

Glaubitz J. C. and Carlson J. E. 1992. RNA editing in the mitochondria of a conifer. *Current Genetics* 22: 163-165.

Gorman S. W., Teasdale R. D. and Cullis C. A. 1992. Structure and organization of the 5S rRNA genes (5S-DNA) in *Pinus radiata* (Pinaceae). *Plant Systematics and Evolution* 183: 223-234.

Govindaraju D., Lewis P. and Cullis C. 1992. Phylogenetic analysis of pines using ribosomal DNA restriction fragment length polymorphisms. *Plant Systematics and Evolution* 179: 141-153.

Govindaraju D. R. and Cullis C. A. 1992. Ribosomal DNA variation among populations of a *Pinus rigida* Mill (pitch pine) ecosystem. 1. Distribution of copy numbers. *Heredity* 69: 133-140.

Grant V. 1981. *Plant speciation*. Columbia University Press, New York.

Haig S. M. 1998. Molecular contributions to conservation. *Ecology* 79: 413-425.

Hamrick J. L., Godt M. J. W. and Sherman-Broyles S. 1992. Factors influencing levels of genetic diversity in woody plant species. *New Forests* 6: 95-124.

Hamrick, J. L., Godt, M. J. W., Murawski, D. A. and Loveless, M. D. 1991. Correlation between species traits and allozyme diversity: Implications for conservation biology. *In* Falk D. A. and Holsinger K. E. (eds.): *Genetics and conservation of rare plants*. Oxford University Press, Oxford.

Hattemer, H. H. 1991. Measuring genetic variation. *In* Müller-Starck G. and Ziehe M. *Genetic Variation in European Populations of Forest Trees*. Sauerländer's Verlag, Frankfurt am Main.

Heinze B. 1997. A PCR marker for a *Populus deltoides* allele and its use in studying introgression with native European *Populus nigra*. *Belgian Journal of Botany* 129: 123-130.

Hong Y. P., Hipkins V. D. and Strauss S. H. 1993. Chloroplast DNA diversity among trees, populations and species in the California closed-cone pines (*Pinus radiata*, *Pinus muricata* and *Pinus attenuata*). *Genetics* 135: 1187-1196.

Inoue T., Zhong H. S., Miyao A., Ashikawa I., Monna L., Fukuoka S., Miyadera N., Nagamura Y., Kurata N., Sasaki T. and Minobe Y. 1994. Sequence-tagged sites (STSs) as standard landmarkers in the rice genome. *Theoretical and Applied Genetics* 89: 728-734.

Isabel N., Beaulieu J. and Bousquet J. 1995. Complete congruence between gene diversity estimates derived from genotypic data at enzyme and random amplified polymorphic DNA loci in black spruce. *Proc. Natl. Acad. Sci. USA* 92: 6369-6373.

Jeandroz S., Roy A. and Bousquet J. 1997. Phylogeny and phylogeography of the circumpolar genus *Fraxinus* (OLEACEAE) based on internal transcribed spacer sequences of nuclear ribosomal DNA. *Molecular Phylogenetics and Evolution* 7: 241-251.

Jones C. J., Edwards K. J., Castaglione S., Winfield M. O., Sala F., Vandewiel C., Bredemeijer G., Vosman B., Matthes M., Daly A., Brettschneider R., Bettini P., Buiatti M., Maestri E., Malcevschi A., Marmiroli N., Aert R., Volckaert G., Rueda J., Linacero R., Vazquez A. and Karp A. 1997. Reproducibility testing of RAPD, AFLP and SSR markers in plants by a network of European laboratories. *Molecular Breeding* 3: 381-390.

Kajita T., Kamiya K., Nakamura K., Wickneswari W., Yoshimaru H., Tsumura Y., Tachida H. and Yamazaki T. 1998. Molecular phylogeny of Dipterocarpaceae in Southeast Asia based on nucleotide sequence of *matK*, *trnL* intron and *trnL-trnF* intergenic spacer region in chloroplast DNA. *Molecular Phylogenetic Evolution*: 10:202-209

Karp, A. and Edwards, K.J. 1997. DNA markers: a global overview. *In* Caetano-Anolles G. and Gresshoff P. M. (eds.): *DNA Markers*. Wiley-VCH Inc, New York.

Karvonen P., Szmidt A. E. and Savolainen O. 1994. Length variation in the internal transcribed spacers of ribosomal DNA in *Picea abies* and related species. *Theoretical and Applied Genetics* 89: 969-974.

Kostia S., Varvio S. L., Vakkari P. and Pulkkinen P. 1995. Microsatellite sequences in *Pinus sylvestris*. *Genome* 38: 1244-1248.

Krajewski C. 1994. Phylogenetic measures of biodiversity - A comparison and critique. *Biological Conservation* 69: 33-39.

Krupkin A. B., Liston A. and Strauss S. H. 1996. Phylogenetic analysis of the hard pines (*Pinus* subgenus *Pinus*, Pinaceae) from chloroplast DNA restriction site analysis. *American Journal of Botany* 83: 489-498.

Kvarnheden A., Albert V. A. and Engström P. 1998. Molecular evolution of cdc2 pseudogenes in spruce (*Picea*). *Plant Molecular Biology* 36: 767-774.

Laroche J., Li P., Maggia L. and Bousquet J. 1997. Molecular evolution of angiosperm mitochondrial introns and exons. *Proc. Natl. Acad. Sci.* USA 94: 5722-5727.

Legionnet A., Faivre-Rampant P., Villar M. and Lefevre F. 1997. Sexual and asexual reproduction in natural stands of *Populus nigra*. *Botanica Acta* 110: 257-263.

Lerceteau E. and Szmidt A. E. 1998. Properties of AFLP markers in inheritance and genetic diversity studies of *Pinus sylvestris* L. *Heredity: in press.*

Lessa E. and Applebaum G. 1993. Screening techniques for detecting allelic variation in DNA sequences. *Molecular Ecology* 2: 119-129.

Li W., Luo C. and Wu C. 1985. Evolution of DNA sequences. *In* MacIntyre R. (ed.): *Molecular evolutionary genetics*. Plenum Press, New York, USA.

Liu Q. and Sommer S. 1994. Parameters affecting the sensitivities of dideoxy fingerprinting and SSCP. *PCR Methods and Applications* 4: 97-108.

Lu M-Z., Szmidt A. E. and Wang X-R. 1995. Inheritance of RAPD fragments in haploid and diploid tissues of *Pinus sylvestris* (L.). *Heredity* 74: 582-589.

Lu M-Z., Wang X-R. and Szmidt A. E. 1997. Molecular characterisation of RAPDs in *Pinus sylvestris* L. *Forest Genetics* 4: 227-234.

Lu M-Z., Szmidt A. E. and Wang X-R. 1998. RNA editing in gymnosperms and its impact on the evolution of the mitochondrial *coxI* gene. *Plant Molecular Biology* 37: 225-234.

McCauley D. E. 1995. The use of chloroplast DNA polymorphism in studies of gene flow in plants. *Trends in Ecology and Evolution* 10: 198-202.

McPherson, M. J., Quirke, P. and Taylor, G. R. 1991. PCR A practical approach. *In* Rickwood D. and Hames B. D. (eds.): *The practical approach series.* IRL Press at Oxford University Press, Oxford, New York, Tokyo.

Meyers R., Maniatis T. and Lerman L. 1986. Detection and localization of single base changes by denaturing gradient gel electrophoresis. *Methods in Enzymology* 155: 501-527.

Moran G. F., Smith D., Bell J. C. and Appels R. 1992. The 5S-RNA genes in *Pinus radiata* and the spacer region as a probe for relationships between *Pinus* species. *Plant Systematics and Evolution* 183: 209-221.

Morgante M. and Olivieri A. M. 1993. PCR-amplified microsatellites as markers in plant genetics. *The Plant Journal* 3: 175-182.

Moser H. and Lee M. 1994. RFLP variation and genealogical distance, multivariate distance, heterosis, and genetic variance in oats. *Theoretical and Applied Genetics* 87: 947-956.

Mullis K., Faloona F., Scharf S., Saiki R., Horn G. and Erlich H. 1986. Specific enzymatic amplification of DNA *in vitro*: the polymerase chain reaction. *In Cold Spring Harbor Symp. Quant. Biol.*, 263-273.

Muona O., Paule L., Szmidt A. E. and Kärkkäinen K. 1990. Mating system analysis in a central and northern European population of *Picea abies*. *Scandinavian Journal of Forest Research* 5: 97-102.

Murphy, R. W., Sites, J. W., Buth, D. G. and Haufler, C. H. 1996. Proteins: Isozyme electrophoresis. *In* Hillis D., Moritz C. and Mable B. (eds.): *Molecular Systematics.* Sinauer Associates, Sunderland, Massachusetts, USA.

Myers R., Maniatis T. and Lerman L. 1986. Detection and localization of single base changes by denaturing gradient gel electrophoresis. *Methods in Enzymology* 155: 501-527.

Neale D. B. and Sederoff R. R. 1988. Inheritance and evolution of conifer organelle genomes. In: Hanover J. W. and Keathley D. E. (eds.): *Genetic Manipulation of Woody Plants.* Plenum Press, New York and London.

Nei M. 1987. *Molecular Evolutionary Genetics*. Columbia University Press, New York.

Nesbitt K. A., Potts B. M., Vaillancourt R. E. and Reid J. B. 1997. Fingerprinting and pedigree analysis in *Eucalyptus globulus* using RAPDs. *Silvae Genetica* 46: 6-11.

Nybom H. and Schaal B. A. 1990. DNA "fingerprints" applied to paternity analysis in apples (*Malus* x *domestica*). *Theoretical and Applied Genetics* 79: 763-768.

Olson M., Hood L., Cantor D. and Botstein D. 1989. A common language for physical mapping of the human genome. *Science* 245: 1434-1435.

Orita M., Suzuki Y., Sekiya T. and Hayashi K. 1989. Rapid and sensitive detection of point mutations and DNA polymorphisms using the polymerase chain reaction. *Genomics* 5: 874-879.

Parducci L. and Szmidt A. E. 1999. PCR-RFLP analysis of cpDNA in the genus *Abies*. Theoretical and Applied Genetics: *in press*.

Parker P. G., Snow A. A., Schug M. D., Booton G. C. and Fuerst P. A. 1998. What molecules can tell us about populations: Choosing and using a molecular marker. *Ecology* 79: 361-382.

Powell W., Machray G. C. and Provan J. 1996. Polymorphism revealed by simple sequence repeats. *Trend in Plant Science* , Reviews 1: 215-222.

Powell W., Morgante M., McDevitt R., Vendramin G. and Rafalski J. 1995. Polymorphic simple sequence regions in chloroplast genomes: Applications to the population genetics of pines. *Proc. Natl. Acad. Sci.* USA 99: 7759-7763.

Roy A., Frascaria N., Mackay J. and Bousquet J. 1992. Segregating random amplified polymorphic DNAs (RAPDs) in *Betula alleghaniensis. Theoretical and Applied Genetics* 85: 173-180.

Saiki R. K., Gelfand D. H., Stoffel S., Scharf S. J., Higuchi R., Horn G. T., Mullis K. B. and Erlich H. A. 1988. Primer-directed enzymatic amplification of DNA with a thermostable DNA polymerase. *Science* 239: 487-491.

Schubert R., Müller-Starck G., Sandermann Jr H., Ernst D. and Häger K. 1997. The molecular structure and evolutionary relationship of a 16.9 kDa heat shock protein from Norway spruce (*Picea abies* (L.) Karst.). *Forest Genetics* 4: 131-138.

Schubert R., Sperisen C., Müller-Starck G., Lascala S., Ernst D., Sandermann Jr H. and Häger K. P. 1998. The cinnamyl alcohol dehydrogenase gene structure in *Picea abies* (L.) Karst.: genomic sequences, Southern hybridization, genetic analysis and phylogenetic relationships. *Trees* 12: 453-463.

Sigurgeirsson A. and Szmidt A. E. 1993. Phylogenetic and biogeographic implications of chloroplast DNA variation in *Picea. Nordic Journal of Botany* 13: 233-246.

Sinclair W. T., Morman J. D. and Ennos R. A. 1998. Multiple origins for Scots pine (*Pinus sylvestris* L) in Scotland: evidence from mitochondrial DNA variation. *Heredity* 80: 233-240.

Slabaugh M. B., Huestis G. M., Leonard J., Holloway J. L., Rosato C., Hongtrakul V., Martini N., Toepfer R., Voetz M., Schell J. and Knapp S. J. 1997. Sequence-based genetic markers for genes and gene families: Single-strand conformational polymorphisms for the fatty acid synthesis genes of *Cuphea. Theoretical and Applied Genetics* 94: 400-408.

Smith D. N. and Devey M. E. 1994. Occurrence and inheritance of microsatellites in *Pinus radiata*. *Genome* 37: 977-983.

Snow A. A. and Lewis P. O. 1993. Reproductive traits and male fertility in plants - Empirical approaches. *Annual Review of Ecology and Systematics* 24: 331-351.

Staub J. E., Serquen F. C. and Gupta M. 1996. Genetic markers, map construction, and their application in plant breeding. *Hortscience* 31: 729-741.

Strand A. E., Leebensmack J. and Milligan B. G. 1997. Nuclear DNA-based markers for plant evolutionary biology. *Molecular Ecology* 6: 113-118.

Strauss S. H. and Doerksen A. H. 1990. Restriction fragment analysis of pine phylogeny. *Evolution* 44: 1081-1096.

Strauss S., Bousquet J., Hipkins V. and Hong Y.-P. 1992a. Biochemical and molecular genetic markers in biosystematic studies of forest trees. *New Forests* 6: 125-158.

Strauss S., Lande R. and Namkoong G. 1992b. Limitations of molecular-marker-aided selection in forest tree breeding. *Canadian Journal of Forest Research* 22: 1050-1061.

Strauss S. H., Hong Y. P. and Hipkins V. D. 1993. High levels of population differentiation for mitochondrial DNA haplotypes in *Pinus radiata, muricata*, and *attenuata*. *Theoretical and Applied Genetics* 86: 605-611.

Szmidt, A. E. and Wang, X-R. 1992. DNA markers in forest genetics. *In* Müller-Starck G. and Ziehe M. (eds.): *Genetic Variation in Forest Tree Populations in Europe*. Sauerländer's Verlag, Frankfurt am Main.

Szmidt A. E., Wang X-R. and Changtragoon S. 1996a. Contrasting patterns of genetic diversity in two tropical pines: *Pinus kesiya* (Royle ex Gordon) and *P. merkusii* (Jungh et De Vriese). *Theoretical and Applied Genetics* 92: 436-441.

Szmidt A. E., Wang X-R. and Lu M-Z. 1996b. Empirical assessment of allozyme and RAPD variation in *Pinus sylvestris* (L) using haploid tissue analysis. *Heredity* 76: 412-420.

Taberlet P., Gielly L., Pautou G. and Bouvet J. 1991. Universal primers for amplification of three non-coding regions of chloroplast DNA. *Plant Molecular Biology* 17: 1105-1109.

Tanksley S. D. 1983. Molecular markers in plant breeding. *Plant Molecular Biology Reporter* 1: 3-8.

Tomaru N., Takahashi M., Tsumura Y., Takahashi M. and Ohba K. 1997. Intraspecific variation and phylogeographic patterns of *Fagus crenata* (Fagaceae) mitochondrial DNA. *American Journal of Botany* 85: 629-636.

Triest L., deGreef B., deBondt R., van den Bossche D. and Dhaeseleer M. 1997. Use of RAPD markers to estimate hybridization in *Salix alba* and *Salix fragilis*. *Belgian Journal of Botany* 129: 140-148.

Tsumura Y., Yoshimura K., Tomaru N. and Ohba K. 1995. Molecular phylogeny of conifers using RFLP analysis of PCR-amplified specific chloroplast genes. *Theoretical and Applied Genetics* 91: 1222-1236.

Tsumura Y., Ohba K. and Strauss S. H. 1996. Diversity and inheritance of inter-simple sequence repeat polymorphisms in Douglas-fir (*Pseudotsuga menziesii*) and sugi (*Cryptomeria japonica*). *Theoretical and Applied Genetics* 92: 40-45.

Tsumura Y., Suyama Y., Yoshimura K., Shirato N. and Mukai Y. 1997. Sequence-tagged-sites (STSs) of cDNA clones in *Cryptomeria japonica* and their evaluation as molecular markers in conifers. *Theoretical and Applied Genetics* 94: 764-772.

Vekemans X. and Jacquemart A. L. 1997. Perspectives on the use of molecular markers in plant population biology. *Belgian Journal of Botany* 129: 91-100.

Vendramin G. G., Lelli L., Rossi P. and Morgante M. 1996. A set of primers for the amplification of 20 chloroplast microsatellites in Pinaceae. *Molecular Ecology* 5: 595-598.

Vendramin G. G. and Ziegenhagen B. 1997. Characterisation and inheritance of polymorphic plastid microsatellites in *Abies*. *Genome* 40: 857-864.

Vicario F., Vendramin G. G., Rossi P., Liò P. and Giannini R. 1995. Allozyme, chloroplast DNA and RAPD markers for determining genetic relationships between *Abies alba* and the relic population of *Abies nebrodensis*. *Theoretical and Applied Genetics* 90: 1012-1018.

Vos P., Hogers R., Bleeker M., Reijans M., Vandelee T., Hornes M., Frijters A., Pot J., Peleman J., Kuiper M. and Zabeau M. 1995. AFLP: A new technique for DNA fingerprinting. *Nucleic Acids Research* 23: 4407-4414.

Wagner D. B., Furnier G. R., Saghai-Maroof M. A., Williams S. M., Dancik B. P. and Allard R. W. 1987. Chloroplast DNA polymorphisms in lodgepole and jack pines and their hybrids. *Proc. Natl. Acad. Sci.* USA 84: 2097-2100.

Wagner D. B., Nance W. L., Nelson C. D., Li T., Patel R. N. and Govindaraju D. R. 1992. Taxonomic patterns and inheritance of chloroplast DNA variation in a survey of *Pinus echinata, Pinus elliotti, Pinus palustris*, and *Pinus taeda*. *Canadian Journal of Forest Research* 22: 683-689.

Wang X-R. and Szmidt A. E. 1993. Chloroplast DNA-based phylogeny of Asian *Pinus* species (*Pinaceae*). *Plant Systematics and Evolution* 188: 197-211.

Wang X-R. and Szmidt A. E. 1994. Hybridization and chloroplast DNA variation in a *Pinus* species complex from Asia. *Evolution* 48: 1020-1031.

Wang X-R., Szmidt A. E. and Lu M.-Z. 1996. Genetic evidence for the presence of cytoplasmic DNA in pollen and gametophytes and maternal inheritance of mitochondrial DNA in *Pinus*. *Forest Genetics* 3: 37-44.

Wang X-R. 1997. Genetic variability in a canker pathogen fungus *Gremmeniella abietina* (Lagerb.) Morelet. 1. Contribution of sexual *vs*. asexual reproduction. *Mycological Research* 1202-1206.

Wang X-R., Ennos R. A., Szmidt A. E. and Hansson P. 1997. Genetic variability in the canker pathogen fungus, *Gremmeniella abietina* (Lagerb.) Morelet. 2. Fine scale investigation of the population genetic structure. *Canadian Journal of Botany* 75: 1460-1469.

Wang X.-R., Szmidt A. E. and Nghia N.-H. 1998. The phylogenetic position of the endemic flat-needle pine *Pinus krempfii* (Pinaceae) from Vietnam, based on PCR-RFLP analysis of chloroplast DNA. *Plant Systematics and Evolution: in press.*

Wang X.-R., Tsumura Y., Yoshimaru H., Nagasaka K. and Szmidt A. 1999. Phylogenetic relationships of Eurasian pines (*Pinus*, Pinaceae) based on chloroplast *rbc*L, *mat*K, *rpl*20-*rps*18 spacer and *trn*V intron sequences. *American Journal of Botany: in press*.

Wartell R., Hosseini S. and Moran C. 1990. Detecting base pair substitutions by temperature gradient gel electrophoresis. *Nucleic Acid Research* 18: 2699-2705.

Watano Y., Imazu M. and Shimizu T. 1996. Spatial distribution of cpDNA and mtDNA haplotypes in a hybrid zone between *Pinus pumila* and *P. parviflora* var *pentaphylla* (Pinaceae). *Journal of Plant Research* 109: 403-408.

Weatherhead P. and Montgomerie R. 1991. Good news and bad news about DNA fingerprinting. *TREE* 6: 173-174.

Wendel J. F. and Weeden N. F. 1989. Visualization and interpretation of plant isozymes. *In* Soltis D. E. and Soltis P. S. (eds.): *Isozymes in plant biology*. Chapman and Hall, London.

Williams J. G. K., Kubelik A. R., Livak K. J., Rafalski J. A. and Tingey S. V. 1990. DNA polymorphisms amplified by arbitrary primers are useful as genetic markers. *Nucleic Acids Research* 18: 6531-6535.

3. ADAPTIVE POTENTIAL OF POPULATIONS

Patterns of adaptive genetic variation in forest tree species; the reproductive enviroment as an evolutionary force in *Picea abies*

Tore Skrøppa and Øystein Johnsen

Introduction

Common garden plantations have provided data for studies of the variability in quantitative adaptive traits among and within wild populations of forest tree species. The distribution of variability among and within populations, its patterns across the landscape and relationships to environmental variables have been used to interpret the ecological basis for genetic differentiation and evolution of the species (Rehfeldt 1988, 1989,1995; Campbell *et al.* 1989). The studies have shown that climatic adaptation seems to be the most important component in the evolutionary process of temperate and boreal conifers. Most species show substantial variability in adaptive traits both within and among populations, even at the climatic margin of their range (Savolainen 1997). The main evolutionary forces influencing the variability patterns are natural selection, genetic drift, gene flow, mutations and phenotypic plasticity (Eriksson 1998a). The many types of variability patterns shown by different species and their degree of adaptedness to the sites suggest that species occupying the same environment may be influenced by different evolutionary forces, see examples discussed by Eriksson (1998b). Several selective forces may even operate on the same population, some simultaneously and some at different life stages or during different seasons of the year (Eriksson 1998a). The variability patterns are therefore most likely under the influence of several evolutionary forces interacting in rather complex ways. This conclusion is supported by the rather high levels of genetic variation in survival and annual growth rhythm traits maintained in the boreal forest (Savolainen 1997), in spite of a strong directional selection against the traits.

In this presentation we will introduce another phenomenon, which may contribute to the interpretation of the observed variability patterns of these traits, namely the role of the environmental conditions during the generative

C. Mátyás (ed.), Forest Genetics and Sustainability, Volume 63, 49–58.

reproductive process in Norway spruce. Skrøppa and Johnsen (1994) and Johnsen and Skrøppa (1996) have earlier presented reviews of this research.

1. Results with *Picea abies*

1.1. Origin and immigration history in Norway

Picea abies is a relatively young species in the Scandinavian Peninsula. It migrated from the Russian refugium, advanced through the forest area at the present border between Russian and Finland and passed through Finland and northern Sweden into Norway in the period 3500-500 BC (Moe 1970; Schmidt-Vogt 1977). Some new findings have suggested an earlier establishment of the species in some areas (Kullmann 1995). The border area between Sweden and Central Norway was crossed approx. 500 BC. In the southern part of Norway the earliest immigration took place in the area between the present border between Norway and Sweden in the years 400-200 BC. Eight hundred years later the species was established in the lowland south-eastern area. The migration into the valleys to its present altitudinal boundaries was not completed until the period AD 1000-1300. The species must thus generally be considered to be young in Norway and has had a short time to adapt to local climatic conditions, taking the long generation interval into account.

1.2. Variability patterns

In spite of the few generations of the species in Norway and an extensive gene flow among populations (Koski 1970), a substantial variation has been observed among populations for traits related to climatic adaptation. In an artificial test of autumn frost hardiness of one-year-old seedlings (Dæhlen *et al.* 1995) a clinal variation among populations was observed, and the latitude and altitude of the natural stands explained 88% of the variability. For a number of traits known to have adaptive significance similar clinal trends have been found. However, a high variability among families within populations has been observed for the same traits, even for populations at the climatic margin of the species. The range of variation in autumn frost hardiness among 10 families sampled from the same population corresponds to a provenance variation of 400 m in altitude or 3-4° of latitude (Skrøppa 1991, Johnsen and Skrøppa unpublished). The cessation of shoot elongation at the age of six years of populations from altitude 300 m and 500 m and only 15 km apart differed by six days (Skrøppa 1982), while the range of variation among families from the same populations was ten days. The genetic variation is chiefly additive (Eriksson *et al.* 1978; Ekberg *et al.* 1991; Skrøppa unpublished), and most traits have a strong and variable inbreeding depression (Skrøppa 1996). Different relationship patterns between traits have been observed for provenances and families (Ekberg *et al.* 1994; Skrøppa *et al.* 1999). For provenances, rather high correlation coefficients have been estimated between the annual growth rhythm traits and hardiness and total growth, while such relationships have been weak or non-existing for families within populations. At the provenance level, the duration of the shoot growth period explained 80% of the variability in shoot growth, while for families the duration and the rate of growth each explained 50% (Skrøppa and Magnussen 1993; Skrøppa unpublished). For provenances, the basic wood density at age 27 years was strongly correlated to growth rhythm traits measured at age 6-7 years, whereas similar relationships were not present between families grown in the same tests (Skrøppa *et al.* 1999). These results indicate that different types of mechanisms may influence the genetic variability within and among populations.

1.3. Reproduction outside native environment

In Norway, spruce seed orchards have been located in warm climates in order to enhance seed production. For the northern and central part of the country (lat. 63-67°N) a seed orchard is situated at lat. 58°N, and seeds for the high altitudes are produced in an orchard near the sea level. The parental clones in these orchards are phenotypically selected trees in native stands in the same area where the seedlings from the orchard seed are to be planted. Both orchards are located outside the natural spruce area and well isolated from outside pollen sources. They were expected to give rise to seedlings adapted to the native areas of the parents. It was soon observed, however, that the seedlings from the southern orchard exhibited a phenology different from that of seedlings of northern autochthonous forest stands (Bjørnstad 1981). In experiments comparing controlled cross families from the seed orchard and families from open pollination of the same plus trees in the native stands (Bjørnstad 1981; Johnsen 1989a, 1989b), the seed orchard families were different from their northern half-sibs in many traits. They flushed later in the spring, terminated leader growth later in the summer, had higher frequencies of lammas growth, lignified the annual ring later in the autumn and a had slower development of autumn frost hardiness. These differences were persistent after a clonal propagation by rooted cuttings (Johnsen 1989a). Observations of phenology traits in progeny tests showed that the seed orchard families continued to perform like a more southern provenance 17 years from seed (Edvardsen *et al.* 1996). The hypothesis that the observed effects were due to the phenotypic plus tree selection could be rejected as families of plus trees in natural stands did not have a phenology different from non-selected trees in the same stands (Johnsen and Østreng 1994). When transfers where made at the same latitude from high to low altitudes where sexual reproduction took place, similar effects on the progeny were observed (Johnsen 1988; Skrøppa 1994).

Practical implications of these effects have been observed both in field trials and in plantations. They have been similar to experiences after provenance transfers with Norway spruce. The seed orchard progenies have generally shown superior height growth, they have suffered in early autumn frost events and have benefited in late spring frost due to a later bud flushing compared to trees from the native provenance (Johnsen *et al.* 1989; Skrøppa unpublished).

Further results were obtained after tests with controlled cross families generated in contrasting environments. Identical full-sib families were produced at latitudes 62 and 66°N in Finland (Skrøppa *et al.* 1994) and inside a heated greenhouse and in a nearby seed orchard in Norway (Johnsen *et al.* 1995). In both cases the progenies that were produced under the warm spring conditions were less frost hardy than their full-sibs from the colder conditions. The observed results lead to the formulation of the hypothesis that phenological traits of Norway spruce progenies are influenced also by the climatic conditions during sexual reproduction, in addition to the genetic constitution of their parents.

To test the potential effects of differential selection during male meiosis and pollen maturation, pollen was produced in a phytotron under short and long days and high and low temperature. However, no evidence could be found for an altered autumn frost hardiness of the progeny obtained after controlled crosses with pollen from these treatments (Johnsen *et al.* 1996). In another experiment crosses were made inside a heated greenhouse in early spring (late March, day length 14 h) and were repeated both inside and outside the greenhouse in May (day length 18 h) (Johnsen *et al.* 1996). The progenies reproduced in early spring were less frost hardy in the autumn than those produced in May. Newer experiments have given some hints that the period before pollination (female meiosis) seems to be an important sensitive stage of the reproductive process, and that photoperiod might be a triggering factor in addition to temperature. Long day during seed production seems to give more hardy seedlings compared to siblings reproduced under short days. Future experiments, however, are needed to confirm these observations (Johnsen, Dæhlen and Skrøppa, unpublished).

If the timing of flowering in the spring influences progeny traits, then phenotypic performance should depend on early versus late flowering. This has indeed been observed for seedlings from bulked seed lots from two seed orchards, from a cool spring with late flowering (1987) and a warm spring with early flowering (1989). Seedling from the cool seed year formed terminal buds at shorter nights in growth chambers and formed buds earlier in the autumn under natural days in nurseries than did seedlings from the warm year (Kohmann and Johnsen 1994). The mean difference between the two years corresponds to a provenance difference of two degrees of latitude.

Norway spruce provenances have been extensively transferred from southern to northern latitudes. Seeds have been collected in stands in West

Norway (lat. 61°N) and in Central Norway (lat. 64°N) being established with seedlings of origin Harz, Germany (lat. 52°N). Large differences in the timing of bud-set were observed between progenies from the northern Harz stands and the original Harz provenance, but not between those of the northern Harz stand and the native northern provenances (Skrøppa and Kohmann 1997). The progenies from the northern Harz stands have a similar phenotypic response to photoperiod as the local northern provenance, an effect larger that could be explained by natural selection and pollen migration from local stands.

1.4. Hypotheses for explanation

When seeds are produced under warm non-native conditions, then an increased heat sum in the spring induces an earlier start of the reproductive process at a shorter photoperiod than in the native environment further north or at higher altitudes. Transfers from north to south and from a high to a lower altitude at the same latitude can result in a two hours' change in photoperiod at the time of pollination. Normally that takes place in the first half of May at low altitudes in southern Norway compared to four weeks later at high altitudes, or even more, further north. Our hypothesis is therefore that temperature, photoperiod or an interaction between the two factors, function as the environmental trigger influencing the progenies by a mechanism not yet described, operating during the reproductive process.

The reproductive cycle offers possibilities for selection both during the haploid and the diploid phase. The hypothesis of a gametophytic selection during male meiosis and pollen maturation (meiotic drive) was rejected in two independent experiments. Thus, the process with the greatest potential for selection, due to the enormous amount of pollen grains produced, does not seem to be involved. There are further possibilities for selection in the female flower during tube growth, fertilisation, megaspore degeneration and postmeiotic embryo competition. However, only 3-7 pollen grains can be accommodated in the pollen chamber, and 2-4 embryos are initially present in the developing seed (Owens and Blake 1985). The potential selection intensity of a directional selection at each of these stages is low, and it has to go in the same direction at each consecutive stage to account for the changes observed. We therefore tend to conclude that the observed results are most likely not due to a directional selection operating during the reproductive cycle.

A second mechanism may be that the influencing environmental factor at some reproductive stage in the female cones regulates the level of gene expression in the progenies and that this potential gene regulation has a long lasting effect. Genomic imprinting is increasingly being accepted as a fundamental process determining, in ways not predicted by the laws of Mendelian inheritance, whether a particular gene should be expressed or not

(Matzke and Matzke 1993). However, we can only speculate that this is a possible cause of the observed influences of the maternal reproductive environment on the phenotypic expression of adaptive traits of Norway spruce progenies.

2. Results with other species

Effects of the crossing environment on the next generation have been observed in Scots pine (*Pinus sylvestris*), but to a lesser degree than in Norway spruce (Lindgren and Wang 1986; Dormling and Johnsen 1992; Andersson 1994; Lindgren and Wei 1994). In white spruce (*Picea glauca*), Bigras and Beaulieu (1997) made controlled crosses in a greenhouse in January and repeated the same crosses on field-grown grafts. They found that the families from the greenhouse crosses were less frost tolerant in the autumn than families from field crosses. With the same species, Stoehr *et al.* (1998) made identical controlled pollinations, using a pollen-mix, at two locations, at lat. 53°45'N and lat. 50°15'N. The seed location significantly affected germination percent and rate of germination, number of needle primordia, greenhouse and field heights and autumn frost hardiness. The height growth differences decreased after two growing seasons in the field. With *Larix spp.* controlled cross families produced in a greenhouse and outdoor adjacent to the greenhouse showed significant growth differences, and the phenotypes of the indoor crosses tended to resemble more southern ecotypes (Greenwood and Hutchison 1996, Hutchison *et al.* 1996). In these crosses, a segregation distortion at the chlorophyll *cab*-3 locus as a function of crossing environment was observed.

3. Land races of introduced species

The more northern-like performance of seedlings from seeds harvested in Norway spruce stands established with provenances transferred from a southern to northern latitude suggests a fast formation of land races that phenotypically are close to the local populations. It has been observed both with this and other conifer species that seedlings from seed harvested in planted stands with transferred provenances perform differently from seedlings from the original provenance. In Denmark, seedlings from seeds harvested in first generation stands of *Picea sitchensis* have better survival and less climatic damage that seedlings of the same provenances from direct import from North America (Nielsen 1994). Similar experiences are made with *Abies normaniana* in Denmark (Nielsen, personal comm.). In Scotland, seedlings from first generation of stands of *Abies grandis* perform far better than direct import of the same provenance (Ennos *et al.* 1998). In Germany,

Douglas fir seedlings from German stands have performed better than the other provenances tested (Kleinschmit *et al.* 1974). Certainly, these differences can be explained by natural selection, but may partly also be influenced by the climatic conditions at the site of seed production.

4. Implications

Regardless of the type of mechanisms involved, the observed influences of the maternal crossing environment on the phenotypic variability of the progeny have important implications. Results from both field tests and practical plantings have demonstrated that the change in phenotype, in particular of the annual growth rhythm, may influence both survival and other fitness traits (Skrøppa unpublished). The consequence may either be positive or negative for the plantation, depending on how climatic extremes are related to the timing of growth start, its cessation and development of autumn frost hardiness. This factor may therefore be an important component in the climatic adaptation of Norway spruce. The changes in the phenology traits of the progeny will be in the same direction as that of natural selection. Depending on the type of genetic control involved, this force may contribute to the maintenance of a high genetic variability within populations. Its effect on fitness may be similar to that of phenotypic plasticity (Mátyás 1996; Eriksson 1998a), disguising the genotypic value and maintaining a larger portion of the genotypes even under a strong selection pressure. It is, however, different in the respect that it operates through the phenotypes of the next generation. The observed effects can also be a contributing explanation of the strong clinal variation relative to latitude and altitude found for traits that describe the annual growth cycle in *Picea abies*. They must therefore be taken into account when interpreting provenance differences in these and related traits. Similarly, they may also contribute to the understanding of why trait relationships at the provenance level and among families within populations may be so different, due to different causal factors influencing the phenotypic variability at the two genetic levels.

The influence of the crossing environment on the phenotypes of the next generation should be taken into account both in tree breeding and in the conservation of forest genetic resources. It can directly influence the phenology of seedlings from the seed orchard crop after a transfer of the parents to a warmer climate, as observed in Norway. Considerations should therefore be made of both the location of seed orchards and where controlled crosses for the generation of breeding material are made. For the conservation activities, its importance will depend on how the mechanism operates and its genetic control, which are yet not known. It should also be accounted for in the models predicting the effects of global warming on both natural and planted boreal forests.

The observed effects of the reproductive environment on the phenotypes of the next generation certainly must have influenced the evolution of the boreal *Picea abies* forests. It should therefore be considered as an additional evolutionary force to mutation, genetic drift, natural selection, gene flow and phenotypic plasticity. It may be one of the reasons for the apparent high adaptedness of this species under harsh boreal conditions.

Acknowledgement

The Nordic Forest Research Co-operation Committee has supported this research.

Literature cited

Andersson, B. 1994. After-effect of maternal environment on autumn frost hardiness in *Pinus sylvestris* seedlings in relation to cultivation techniques. *Tree Phys*. 14: 13-322.

Bigras, F. and J. Bonlieu, 1997. Influence of maternal environment on frost tolerance of *Picea glauca* progenies. *In* Proceedings of the twenty-sixth meeting of the Canadian Tree Improvement Association. Part 2: 91.

Bjørnstad, Å., 1981. Photoperiodical after-effect of parent plant environment in Norway spruce (*Picea abies* (L.) Karst) seedlings. *Medd. Nor. Inst. Skogforsk*. 36:1-30.

Campbell, R. K., W. A. Pawuk and A. S. Harris, 1989. Microgeographic genetic variation of Sitka spruce in southeastern Alaska. *Can. J. For. Res*. 19:1004-1013.

Dormling, I. and Ø. Johnsen, 1992. Effects of the parental environment on full-sib families of *Pinus sylvestris*. *Can. J. For. Res*. 22:88-100.

Dæhlen, A.G., Ø. Johnsen and K. Kohmann, 1995. Autumn frost hardiness in young seedlings of Norway spruce from Norwegian provenances and seed orchards. (Norwegian with English summary). *Rapp. Skogforsk*. 1/95:1-24.

Edvardsen, Ø. M., Ø. Johnsen and J. Dietrichson, 1996. Growth rhythm and frost hardiness in northern progeny trials with plants from Lyngdal seed orchard. (Norwegian with English summary). *Rapp. Skogforsk*. 9/96:1-9.

Ekberg, I., G. Eriksson and C. Nilsson, 1991. Consistancy of phenology and growth of intra- and interprovenance families of *Picea abies*. *Scand. J. For. Res*. 6:323-333.

Ekberg, I., G. Eriksson, G. Namkoong, C. Nilsson and L. Norell, 1994. Genetic correlations for growth rhythm and growth capacity at ages 3-5 years in provenance hybrids of *Picea abies*. *Scand. J. For. Res*. 9:25-33.

Ennos, R., R. Worrell and D. C. Malcolm, 1998. The genetic management of native species in Scotland. *Forestry* 71:1-23.

Eriksson G., 1998a. Evolutionary forces influencing variation among populations of *Pinus sylvestris*. *Silva Fennica* 32:173-184.

Eriksson, G., 1998b. Sampling for genetic resources populations in the absence of genetic knowledge. *In* Turok, J., E. Collin, B. Demesure, G,. Eriksson, J. Kleinschmit, M. Rusanen and R. Stephan (compilers): Noble Hardwood Network. Report of the second meeting, 22-25 March 1997. Lourizàn, Spain. International Plant Genetic Resources Institute, Rome, Italy. 61-75.

Eriksson, G., I. Ekberg, I. Dormling, B. Matern and D. von Wettstein, 1978. Inheritance of bud-set and bud-flushing in *Picea abies* (L.) Karst. *Theor. Appl. Genet.* 52:3-19.

Greenwood, M. S. and K. W. Hutchison, 1996. Genetic aftereffetcs of increased temperature in *Larix*. *In* J. Hom, R.Birdsey and K. O'Brian (eds.): Proceedings 1995 Meeting of the Northern Global Change Program 14-16 March 1995. 56-62.

Hutchison K., M. Greenwood, Ø. Johnsen and T. Skrøppa, 1996. Genetic aftereffects of environmental stress in conifers. Abstract from a Scientific Regional Information Exchange Group Workshop, Texas A and M University, Houston.

Johnsen, Ø., 1988. After-effects on progenies from orchard clones moved to non-native environments. *In* J. Worrall, J. Loo-Dinkins and D.T. Lester (eds.): *Physiology and genetics of reforestation*. Proc. of the Tenth North American Forest Biology Workshop, Vancouver, B.C. July 20 - 22, 1988: 1-11.

Johnsen, Ø., 1989a. Phenotypic changes in progenies of northern clones of *Picea abies* (L.) Karst. grown in a southern seed orchard. I. Frost hardiness in a phytotron experiment. *Scan. J. For. Res.* 4: 317-330.

Johnsen, Ø., 1989b. Phenotypic changes in progenies of northern clones of *Picea abies* (L.) Karst. grown in a southern seed orchard. II. Seasonal growth rhythm and height in field trials. *Scan. J. For. Res.* 4:331-341.

Johnsen, Ø. and G. Østreng, 1994. Effects of plus tree selection and seed orchard environment on progenies of *Picea abies*. *Can J. For. Res.* 24:32-38.

Johnsen, Ø. and T. Skrøppa, 1996. Adaptive properties of *Picea abies* progenies are influenced by environmental signals during sexual reproduction. *Euphytica* 92:67-71.

Johnsen, Ø., J. Dietrichson and G. Skaret, 1989. Phenotypic changes in progenies of northern clones of *Picea abies* (L.) Karst. grown in a southern seed orchard. III. Climatic damage in a progeny trial. *Scan. J. For. Res.* 4:343-350.

Johnsen, Ø., T. Skrøppa, G. Haug, I. Apeland and G. Østreng, 1995. Sexual reproduction in a greenhouse and reduced autumn frost hardiness of *Picea abies* progenies. *Tree Physiol.* 15: 551-555.

Johnsen, Ø., T. Skrøppa, O. Junttila and O. G. Dæhlen, 1996. Influence of the female flowering environment on autumn frost-hardiness of *Picea abies* progenies. *Theor. Appl. Genet.* 92:797-802.

Kohmann, K and Ø. Johnsen, 1994. The timing of bud-set in seedlings of *Picea abies* from seed crops of a cool versus a warm summer. *Silvae Genet.* 43:328-332.

Kleinschmit, J., J. Racz, H. Weissgerber, W. Dietze, H. Dieterich and R. Dimpflmeier,1974. Ergebnisse aus dem internationalen Douglasien-herkunftversuch von 1970 in der Bundesrepublik Deutschland. *Silvae Genet.* 23:167-176.

Koski, V. 1970. A study of pollen dispersal as a mechanism of gene flow in conifers. *Comm. Inst. For. Fenn.* 70.4:1-78.

Kullmann, L. 1995. New and firm evidence for Mid-Holocene appearance of *Picea abies* in the Scandes Mountains, Sweden. *Journal of Ecology* 83:439-447.

Lindgren, D. and Q. Wang, 1986. Are genetic results influenced by the environment during seed maturation ? *In* Proc. IUFRO Conference on Breeding Theory, Progeny Testing and Seed Orchards. Williamsburg, Virginia, USA. 192-199.

Lindgren, D. and R.-P. Wei, 1994. Effects of maternal environment on mortality and growth in young *Pinus sylvestris* in field trials. *Tree Phys.* 14:323-327.

Mátyás, C., 1996. Climatic adaptation of trees: rediscovering provenance tests. *Euphytica* 92:45-54.

Matzke, M. and A. J. M. Matzke, 1993. Genomic imprinting in plants: Parental effects and transinactivation phenomena. *Ann. Rev. Plant. Phys. Plant. Mol. Biol.* 44:53-76.

Moe, D. 1970. The post-glacial immigration of *Picea abies* into Fennoscandia. *Bot. Notiser* 123: 61-66.

Nielsen, U. B., 1994. Genetic variation in Sitka Spruce (*Picea sitchensis* (Bong.) Carr.) with respect to height growth, stem form and frost hardiness - investigated on the basis of Danish provenance, progeny and clonal field experiments. (Danish with Eng. summary). *Forskingsserien*. Danish Landscape and Forest Research Institute, Hørsholm, Denmark. No 11:1-330.

Owens, J. N. and M. D. Blake, 1985. Forest tree seed production. Information Report PI-X-53 Petawawa National Forestry Institute, Chalk River, Ontario.

Rehfeldt, G. 1988. Ecological genetics of *Pinus contorta* from the Rocky Mountains (USA): a synthesis. *Silvae Genet*. 37:131-135.

Rehfeldt, G., 1989. Ecological adaptations in Douglas-fir (*Pseudotsuga menziesii* var glauca): a synthesis. *For. Ecol. Manage*. 28:202-215.

Rehfeldt, G., 1995. Genetic variation, climate models and the ecological genetics of *Larix occidentalis*. *For. Ecol. Manage*. 78:21-37.

Savolainen, O., 1997. Pines beyond the polar circle. *In* P.M.A. Tigerstedt (ed.): *Adaptation in plant breeding*. Kluwer Academic Publishers. 153-159.

Schmidt-Vogt, H. 1977. Die Fichte. Band I.Taxonomie - Verbreitung - Morphologie - Ökologie -Waldgensellschaften. Verlag Paul Parey, Hamburg und Berlin.

Skrøppa, T., 1982. Genetic variation in growth rhythm characteristics within and between natural populations of Norway spruce. A preliminary report. *Silva Fennica*. 16:160-167.

Skrøppa, T., 1991. Within-population variation in autumn frost hardiness and its relationships to bud-set and height growth in *Picea abies*. *Scand. J. For. Res*. 6:353-363.

Skrøppa, T., 1994. Growth rhythm and hardiness of *Picea abies* progenies of high altitude parents from seeds produced at low elevation. *Silvae Genet*. 43:95-100.

Skrøppa, T., 1996. Diallel crosses in Norway spruce II. Performance and inbreeding depression of selfed families. *Forest Genetics* 3:69-79.

Skrøppa, T. and S. Magnussen, 1993. Provenance variation in shoot elongation parameters in Norway spruce. *Silvae Genet*. 42:111-120.

Skrøppa, T. and Ø. Johnsen, 1994. The genetic response of plant population to a changing environment: the case for non-Mendelian processes. *In* T.J.B. Boyle and C.E.B.Boyle (eds.): Biodiversity, Temperate Ecosystems, and Global Change. NATO ASI Series, Vol. I 20 Springer-Verlag Berlin Heidelberg. 183-199.

Skrøppa. T. and K. Kohmann, 1997. Adaptation to local conditions after one generation in Norway spruce. *Forest Genetics* 4:171-177.

Skrøppa, T., G. Hylen and J. Dietrichson, 1999. Relationships between wood density and juvenile heights and growth rhythm traits for Norway spruce provenances and families. (Submitted).

Skrøppa, T., T. Nikkanen, S. Routsalainen, and Ø. Johnsen, 1994. Effects of sexual reproduction at different latitudes on performance of the progeny of *Picea abies*. *Silvae Genet*. 43:298-304.

Stoehr, M. U., S. J. L'Hirondelle, W. D. Binder and J. E. Webber, 1998. Parental environmental aftereffects on germination, growth and adaptive traits in selected white spruce families. *Can. J. For. Res*. 28:418-426.

Population differentiation for adaptive traits and their underlying loci in forest trees: theoretical predictions and experimental results

Antoine Kremer, Valérie Le Corre and Stéphanie Mariette

Introduction

Population differentiation has been investigated in forest trees since provenance tests were established. A vast amount of results has accumulated in numerous reports and articles about intraspecific variation, that have been summarized in textbooks about forest genetics (Wright 1976). Provenance differences exist for almost any adaptive trait that has been measured in provenance test and for almost any species. These results contrast markedly with data based on biochemical markers as isozymes. As shown by the literature review by Hamrick *et al.* (1992), forest trees usually exhibit extremely low levels of differentiation for isozymes. Results derived from isozyme surveys are confirmed by other molecular markers, with the exception of organelle markers which follow usually uniparental inheritance. The discrepancy of differentiation between adaptive traits and molecular markers is attributed to the fact that different evolutionary forces affect these traits and that they are controlled by different sets of loci likely to be unlinked.

Differentiation for adaptive traits is measured by estimating the variance due to population differences for the metric trait investigated; differentiation for molecular markers is estimated by the variance of allele frequencies among populations. Assuming that the loci that control the adaptive trait and their alleles are known, we attempt to calculate and compare differentiation for the adaptive trait and for the allele frequencies at the loci controlling the adaptive trait (LCAT). The objective of these investigations is to test the hypothesis that differentiation for phenotypic traits could be predicted by differentiation for molecular markers, provided that the latter is calculated for loci controlling the metric trait under investigation. If this hypothesis is verified, than molecular screening could be an alternative to provenance testing. Of course, the basic assumption requires that the LCATs are known and their alleles as well. Progress in molecular genetics and marker techniques is continuous, and more QTLs involved in phenotypic traits, including adaptive traits, are constantly been identified. Although the assumption of known LCATs may be unrealistic today, this situation may

C. Mátyás (ed.), Forest Genetics and Sustainability, Volume 63, 59–74.

rapidly change in the near future, when alleles at the LCATs will be identified.

In the present paper, we will compare differentiation for adaptive traits and LCATs, both on an experimental and theoretical basis. A theoretical comparison will first be made inspired from earlier investigations in quantitative genetics in a simple monolocus system. It will be followed by a comparison between differentiation values obtained by simulation methods where complex multilocus structures can be artificially created. Finally, theoretical predictions will be compared with the scarce experimental results obtained so far.

1. Measures of genetic differentiation for adaptive traits and LCATs

The comparison of differentiation for both types of traits is straightforward in an unrealistic simple situation, when the adaptive trait is controlled by a single and biallelic locus and when it is not affected by selection. Assuming that a given adaptive trait is controlled by only one LCAT with only two alleles (a and b), the variation of allele *a* can be partitioned according to the ANOVA model of Weir (1990).

$$Y_{ijk} = \mu + Pl_i + Zl_{ij} + Gl_{ijk} \qquad \textbf{(1)}$$

Where Y_{ijk} corresponds to the observation of allele k in an individual j belonging to a population i, and Pl_i, Zl_{ij} and Gl_{ijk} represent the random effects due to population, individual (zygote) within-population and gene within zygote. Y_{ijk} equals 1 if the allele k is a and 0 if allele k is b.

Fst of allele *a* can then be calculated as:

$$Fst = \frac{V_{Pl}}{V_{Pl} + V_{Zl} + V_{Gl}} \qquad \textbf{(2)}$$

Similarly, one could subdivide the genotypic value of the adaptive trait into allelic effects. Assume that X_{ijk} is the component of the genotypic value of the LCAT due to allele k belonging to tree i in population j:

$$X_{ijk} = \mu + Pt_i + Zt_{ij} + Gt_{ijk} \qquad \textbf{(3)}$$

where Pt_i , Zt_{ij} and Gt_{ijk} correspond to the same effects as Pl_i , Zl_{ij} and Gl_{ijk} in model (1) except that they refer to the gene effects. Gt_{ijk} , for example, is the additive effect of allele k in tree j.

As for allele frequencies, differentiation for the adaptive trait can then be written as Fst_q (Kremer et al. 1997). To simplify the writing and to avoid confusion with Fst we prefer to use the terminology introduce by Spitze (1993) and rename Fst_q by Qst.

$$Fst_q = Qst = \frac{V_{Pt}}{V_{Pt} + V_{Zt} + V_{Gt}} \quad \textbf{(4)}$$

However adaptive traits can not be measured at the gametic level as marker loci where the two alleles can be identified by electrophoretic methods. For phenotypic traits only the zygotic value $X_{ij.}$ ($X_{ij.} = X_{ij1} + X_{ij2}$) can be measured, which includes the value of the two allelic effects of a diploid individual. Actually the model that is then used is the following, assuming additivity of gene effects:

$$X_{ij.} = m + P_i + Z_{ij} + A_{ij} \quad \textbf{(5)}$$

Where

$$P_i = 2Pt_i \text{ and } V_{Pt} = \frac{V_P}{4} \quad \textbf{(6)}$$

$$Z_{ij} = 2Zt_{ij} \textit{ and } V_{Zt} = \frac{V_Z}{4} \quad \textbf{(7)}$$

$$A_{ij} = Gt_{ij1} + Gt_{ij2} \textit{ and } V_{Gt} = \frac{V_A}{2} \quad \textbf{(8)}$$

A_{ij} is the additive value of individual ij at the locus, P_i is the population value and Z_{ij} is the zygotic value. As a result V_A is the additive variance for a given population.

Modifying (4) according to (6), (7) and (8), Qst becomes:

$$Qst = \frac{V_P}{V_P + V_Z + 2V_A} \quad \textbf{(9)}$$

Assuming that each population is in Hardy-Weinberg equilibrium (V_Z =0), *Qst* has now the more familiar expression:

$$Qst = \frac{V_P}{V_P + 2V_A} \tag{10}$$

Assuming, that the different subpopulations originated from an ancestor population, covariances between genetic values of individuals in model (5) can be calculated , and their expectations can be derived in the form of linear function of coancestry parameters and components of the phenotypic variance in the whole population (Tachida and Cockerham 1987; Kempthorne, 1957). Assuming that *Va* is the additive variance in the total ancestor population the expectations of the covariances are:

$$\begin{aligned} Cov(X_{ij.}, X_{ij'.}) &= V_P \\ Cov(X_{ij.}, X_{ij'.}) &= 2Fst\,Va \end{aligned} \tag{11}$$

$$\begin{aligned} V(X_{ij.}) &= V_P + V_A \\ V(X_{ij.}) &= (1 + Fit)\,Va \end{aligned} \tag{12}$$

Where *Fst* and *Fit* are the two F-statistics at the LCAT (also coancestry coefficients in this case). The results in (11) and (12) were found in different ways by Wright (1965), Cockerham (1969) and Falconer (1960). The expectation of the covariances in (11) and (12) have also a component comprising the dominance variance at the whole population scale. We will assume here that dominance is absent.

As a result the differentiation for the adaptive trait is:

$$Qst = \frac{Fst}{(1 + Fit - Fst)} \tag{13}$$

If we assume that the populations are in HW equilibrium, than $Fit = Fst$ and

$$Qst = Fst \tag{14}$$

In the simplest case of a single neutral biallelic locus with only additive effects, and populations in HW equilibrium, the expectation of *Qst* is therefore the value of *Fst* (Lande 1992). This is unlikely to be the case for major adaptive traits that are usually non neutral and that depend most likely on many loci and several alleles at a given locus.

When multiple loci are involved in adaptive traits, the differentiation values at the LCATs can be written as (Weir and Cockerham, 1984)

$$Fst = \frac{\Sigma\Sigma V_{Pl}}{\Sigma\Sigma(V_{Pl} + V_{Zl} + V_{Gl})} \quad \textbf{(15)}$$

Where the double summation indicates sums over alleles and loci.

Similarly for the metric adaptive trait the differentiation is:

$$Qst = \frac{\Sigma V_{P_i} + \Sigma\Sigma Cov(P_i, P_{i'})}{\Sigma V_{P_i} + \Sigma\Sigma Cov(P_i, P_{i'}) + 2\Sigma V_{A_i} + 2\Sigma\Sigma Cov(A_i, A_{i'})} \quad \textbf{(16)}$$

Where P_i, $P_{i'}$, and A_i, $A_{i'}$ are the population and additive values corresponding to locus *i* and *i'*. The sum signs indicate here sums over loci.

Because the value of a metric trait depending on many loci is the sum of all values at single loci, covariances between loci have to be introduced in the expression of the variances of the trait in case these loci show gametic disequilibria. In the expression of *Qst*, two kinds of disequilibria appear:

- Disequilibria at the between (*Db*) population level (covariances between P_i)
- Disequilibria at the within (*Dw*) population level (covariances between A_i)

Based on the importance of disequilibria, three different situations can be considered:

1. if there are no disequilibria at any level, the expected value of *Qst* becomes similar to *Fst;*
2. if there is an anisotropy between the two sources of disequilibrium, *Db* being higher than *Dw,* then it is likely that *Qst* will be higher then *Fst;*
3. if there is an anisotropy between the two sources of disequilibrium, *Db* being lower than *Dw,* then it is likely that *Qst* will be lower then *Fst;*
4. if *Db* and *Dw* are equal (isotropy of disequilibria) that it is likely than *Qst* and *Fst* are of similar magnitude.

What situation is most likely to be the case in forest trees? Disequilibria can be created by a number of different evolutionary processes including epistatic selection (Hedrick, 1985), Wahlund effects (Feldman and Christiansen, 1975) or genetic drift (Ohta 1982, a,b). In addition, historical factors, such as separation of the species in different refugial zones may also generate disequilibrium (Kremer and Zanetto, 1997). Because forest trees are generally allogamous, having large population sizes and reproducing in panmixia, within-population disequilibria (*Dw*) is generally extremely low (Muona,1982). However in the case of diversifying selection, e.g. selection towards different optimal phenotypes in different environments, favorable alleles at different loci will show a tendency to be associated. As a result *Db* will continuously increase. Many forest trees exhibit clinal geographic variation along an altitudinal, latitudinal or longitudinal gradients (Wright, 1976), indicating that natural selection acts in the form of diversifying selection where optimal phenotypes that are favored vary along these gradients. If there is stabilization selection within a given environment, e.g. that selection will favor trees exhibiting the mean phenotypic value of that environment - which is the optimal value for that environment- selection will not build up disequilibrium within a population. In summary, the association of diversifying selection across environments, with stabilizing selection within an environment, may result in a strong anisotropy between *Dw* and *Db*.

2. Differentiation of adaptive traits and gene loci in simulation studies

Because adaptive traits are likely to be built up as multilocus systems, analytical calculation of *Qst* and *Fst* at the LCATs are extremely complex and can only be simplified when crude approximations are made as discussed before (neutral, monolocus, biallelic locus and additive effects). However, empirical and numerical calculations can be made through simulations in order to compare differentiation for molecular and adaptive traits in scenarios where selection and multiple loci are involved (Latta, 1998). The main aim is to test the theoretical predictions *Qst* and *Fst* that were outlined before in more real situations. In the simulations, we introduced the multilocus structure of an adaptive trait, by assuming that it was controlled by a large number of LCATs distributed randomly throughout the genome, and that selection pressures were affecting the phenotypic value of the adaptive trait.

The simulation model METAPOP that is used in this study has been described elsewhere (Le Corre et al., 1997), where it was applied for studying the effects of colonization dynamics on the genetic diversity in forest trees. It has been completed since in order to introduce quantitative traits. Basically it can be subdivided in three major parts:

1. *Simulation of a genome for each tree*. Marker loci and LCATs can be distributed on a set of linkage groups; mutation rates can be defined for each type of locus. The phenotypic value of the adaptive trait of a given tree is constructed according to classical quantitative theory, as the sum of genetic value and an environmental value (Falconer, 1960). Interaction effects (dominance and epistasis) are assumed to be absent. The additive value of a genotype is calculated after summing the allelic effects over all

loci. Each LCAT was assumed to participate equally to the phenotypic variation of the adaptive trait.

2. *Simulation of a metapopulation*. A set of populations is created, that exchange genes through pollen and seed flow. Each population is of the same size, and gene exchange is defined in a matrix, where pollen and seed migration rates between all pairs of populations are indicated. This matrix allows to generate any type of multipopulation gene exchange model, including the island and stepping stone model.
3. *Simulation of an evolutionary scenario*. Sexual reproduction is modeled by defining the mating system and the mating laws, population dynamics by indicating a population growth model. Selection acting on the adaptive trait is modeled in two different ways:
   - At the within-population level, stabilizing selection towards a local optimum being Z_i is chosen as the selection mechanism for a given population i. In other words, the fitness $W(Z)$ of an individual having phenotypic value Z for the adaptive trait is given by:

$$W(Z) = \exp\left[\frac{-(Z - Z_i)^2}{2\omega^2}\right] \quad \textbf{(17)}$$

     where ω^2 is the strength of selection according to Turelli (1984).
   - At the between-population level, diversifying selection is assumed. In the simulation model, diversifying selection is introduced by defining a separate optimal Z_i for each population. The higher the difference between the optimal values, the stronger the diversifying selection.

If there is diversifying selection across environments associated to stabilizing selection within environments, than strong between-population disequilibria will be generated, whereas within-population disequilibria will remain low. As a result a strong discrepancy between *Qst* and *Fst* is expected under those circumstances.

The simulation model METAPOP was then applied to a case study, likely to correspond to a large number of forest trees and more specifically to oaks. The values of the different parameters in the simulation were chosen according to observations and data available in oaks. The adaptive trait was supposed to be controlled by 50 LCATs distributed randomly on 10 linkage groups. The contribution of each LCAT to the additive variance of the adaptive trait was supposed to be equal. Reported values of the number of QTLs are usually much lower, but they were detected only in single families. Hence, we assumed that the number of QTLs contributing to a quantitative trait is much higher on a population basis. Mutation rates at LCATs was 10^{-5}, and the heritability of the adaptive trait at the beginning of the simulations was assumed to be 0.5. We considered a set of 9 populations distributed spatially on a square grid system (3×3), each population being composed of 1000 trees. These populations exchanges genes by pollen and seed according to the island model with following migration rates: 10^{-2} and 10^{-4}. These values correspond to those inferred *a posteriori* from *Fst* values in gene diversity studies (Kremer *et al.*, 1998). The species was considered to be fully outbreeding, and matings were assumed to be random (except self-fertilization). Stabilizing selection was acting within each population, but the optimal phenotypic values were different in each population, thus creating diversifying selection across environments. We assumed that there was a linear relationship between the optimal phenotypic value and one axis of the grid where the populations were distributed. In the three „southern" populations, the optimal phenotypic value was fixed to -0.5, whereas in the three „central" and three „northern" populations, optimal phenotypic values were fixed respectively to 0 and +0.5. This linear relationship is likely to be representative to the clinal geographic variation observed in provenance tests of numerous forest trees (Wright, 1976).

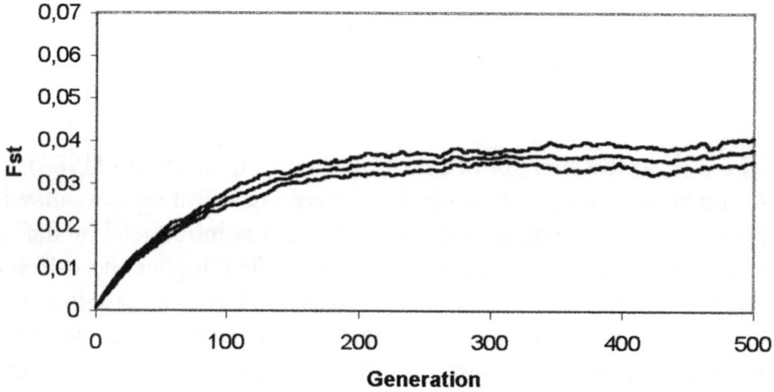

Figure 1. Evolution of *Fst* for loci controlling a neutral adaptive trait (LCATs). *Fst* was calculated over all 50 LCATs (see equation 15). The middle curve represents the mean value of *Fst* over 10 repetitions of simulations. The top curve is the mean value plus one standard deviation. The bottom curve is the mean value minus one standard deviation.

The simulation started by considering a single population of 9000 trees undergoing random mating during 20 000 generations without selection. This step was achieved in order to obtain a starting point, where equilibrium is achieved for the additive variance and the heterozygosity. Then the main initial population was split in 9 separate populations by random choice of 1000 individuals with replacement. *Qst* for the adaptive trait (equation 16) and *Fst* for the 50 LCATs (equation 15) were calculated at each generation during 500 successive generations.

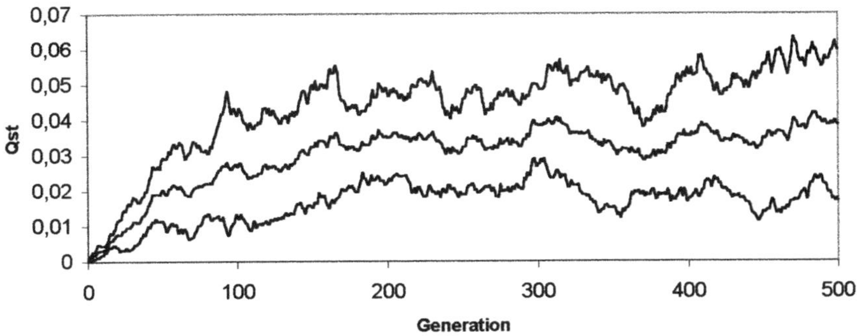

Figure 2. Evaluation of *Qst* for a neutral adaptive trait. *Qst* was calculated according to equation 16. The middle curve represents the mean value of *Qst* over the 10 representations of simulations. For explanation see Figure 1.

Simulations were run in two different scenarios: first, assuming that the adaptive trait controlled by the 50 loci was neutral (Figure 1-2) and second, assuming that diversifying selection was acting as indicated above (Figure 3-4). For each scenario simulations were repeated 10 times to assess the variation of *Fst* and *Qst*.

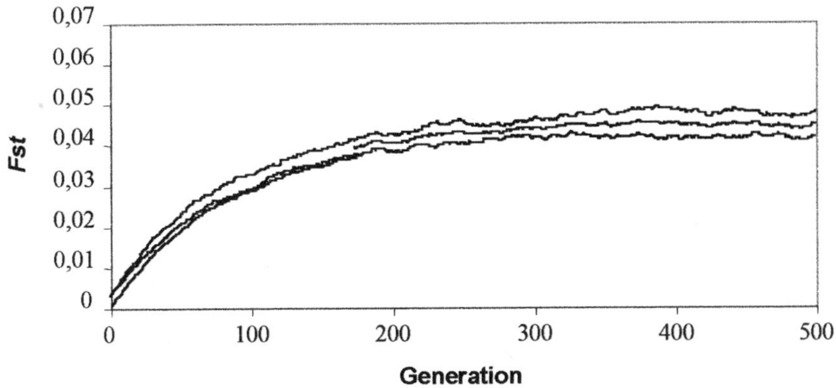

Figure 3. Evolution of *Fst* for loci controlling an adaptive trait (LCATs) undergoing diversifying selection. *Fst* was calculated over all 50 LCATs (see equation 15). For explanation see Figure 1.

The results of the simulations confirm the theoretical expectations predicting that *Qst* is equal in expectation to *Fst* (equation 14 and Figure 1 and 2) if the adaptive trait is neutral. The mean value of *Fst* and *Qst* over the

10 repetitions after 500 generations was respectively 0.0381 and 0.0385. There is however a much larger variation among the repetitions for *Qst* than for *Fst*.

When diversifying selection is occurring, a strong discrepancy appears between *Qst* and *Fst* (Figure 3 and 4) The differentiation of the adaptive trait is rapidly increasing from generation 0 and 150 and then finally attaining 0.2051 at generation 500. At the same time *Fst* is increasing much slower and reaches 0.0428 at generation 500. As a result important differentiation can exist for an adaptive trait whereas the allele frequencies of the LCATs show only minor population differences. Furthermore when *Fst* of the 50 LCATs in the selection scenario (Figure 3) is compared to *Fst* in the neutral scenario (Figure 1), their variation is parallel and they attain nearly the same asymptotic value. Surprisingly, allele frequencies at LCATs controlling an adaptive trait submitted to selection are only slightly more differentiated than those of neutral loci.

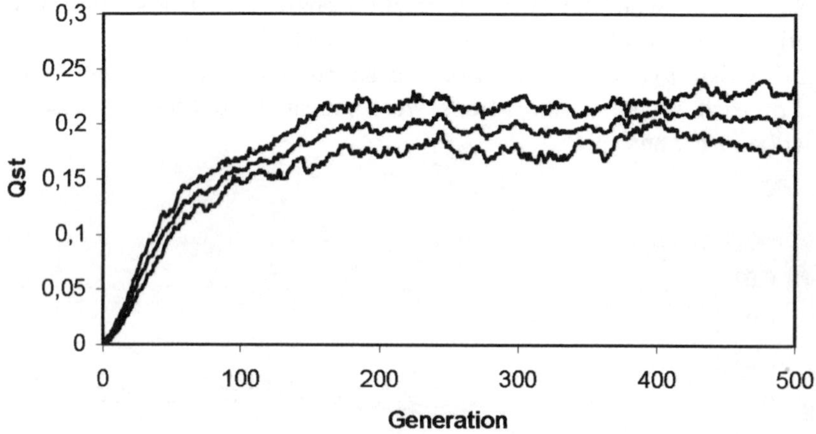

Figure 4. Evolution of *Qst* for an adaptive trait undergoing diversifying selection. *Qst* was calculated according to equation 16. For explanation see Figure 1.

Interestingly, Latta (1998) found similar results in his simulations. There were only a few major differences between his and our input parameters in the selection scenario. The difference in the number of LCATs was large (20 vs. 50). The number of populations was nearly the same (10 vs. 9), but populations size was different (200 vs. 1000). Migration rate was higher in his model (10^{-1} vs. 10^{-2}), and Latta considered also different levels of diversifying selection. A major trend that was shown in his study was the increase of the discrepancy between *Qst* and *Fst* as diversifying selection became stronger. In other words, as more contrasting phenotypes were favored in different environments, *Qst* increased more rapidly than *Fst*. The discrepancy between *Qst* and *Fst* can therefore be interpreted as the result of the combined effect of the multilocus structure of the adaptive trait and selection. More specifically, diversifying selection generates between-population disequilibria at different LCATs, that are not due to physical linkage. These disequilibria inflate the between-population covariance components of *Qst* (see equation 16). However, because random mating occurs within each population the within-population disequilibrium is not increasing. These results also show how little allele frequencies vary at the LCATs despite strong diversifying selection.

3. Comparative experimental results of *Qst* and *Fst*

Although massive data of provenance tests are available from numerous reports on intraspecific variation in forest trees, only scarce comparisons were made between differentiation of adaptive traits and molecular markers. The most likely reason for the lack of comparative data is the fact that the comparison on theoretical grounds became available only recently (Spitze, 1993; Kremer et al., 1996) despite earlier attempts (Falconer, 1960; Cockerham, 1969). In addition, population differentiation was usually investigated in two different disciplines: molecular differentiation in population genetics, differentiation of metric traits in quantitative genetics. Because so far neither loci nor alleles at specific LCATs are known in forest trees, there are no reports about *Qst* and *Fst* at the same loci. Our comparison is therefore restricted to the differentiation of quantitative traits and differentiation at isozymes that are not involved in the expression of those traits.

A major constraint to the calculation of genetic differentiation for quantitative traits is the need to obtain data from combined provenance-progeny tests (see equation 10), where additive variances can be calculated in different populations. If such experiments are available, the number of populations is rather limited in comparison to gene diversity surveys based on isozymes or DNA markers. An alternative way is to use data from provenance tests, assuming that heritability values are known (Kremer et al., 1997); in this case additive variances can be derived from phenotypic variances. But only crude estimates can be obtained from such approximations. In some cases, differentiation was computed from phenotypic data of provenance test by calculating only phenotypic variances; in this case V_A in equation 10 is replaced by the mean phenotypic variance over all populations (Kremer, 1994), thus underestimating *Qst*. There are three reported studies that compared *Qst* and *Fst* on the same sample of populations (see Table 1). These three studies illustrate the difficulties in estimating *Qst* on forest trees, due to the experimental constraints for estimating the additive variance in each population. In the oak study, because heritability values were unknown, different *a priori* values were chosen that encompass the range of variation of heritability estimates reported in the literature (Cornelius, 1994). Each study used a different method for calculating the within-population variance component necessarily for estimating *Qst*. Despite of these difficulties, the results are quite concordant across the three studies (Table 2). Although the metric traits are not necessarily adaptive traits, the difference between *Qst* and *Fst* is large in each study by a factor varying between 10 to 20. Among the 12 phenotypic traits that were assessed in the 3 studies, only one (branch angle in *Pinus contorta*) exhibited lower differentiation than isozymes. In all other pairwise comparisons (isozyme locus with phenotypic trait), *Qst* is consistently higher than *Fst*.

Table 1. Sampling strategies in comparative studies of *Qst* and *Fst*

	Study 1	**Study 2**	**Study 3**
Species	*Pinus contorta* ssp. *latifolia*	*Salix viminalis*	*Quercus petraea*
Number of populations for estimating *Fst*	5	14	100
Number of populations for estimating *Qst*	5 (same pops as for *Fst*)	14 (same pops as for *Fst*)	69 (subset of the 100)
Number of trees/pop for *Fst*	?	7 to 158	120
Within population variance component used for calculating *Qst* (see equation 10)	Additive variance (18 to 27 open pollinated progenies/pop)	Genotypic variance (each sampled tree was vegetatively propagated)	Additive variance (inferred from phenotypic variance assuming given heritability values)
Reference	Yang *et al.* (1996)	Lascoux *et al.* (1996)	Kremer *et al.* (1998) Zanetto and Kremer (1995)

Interestingly the discrepancy between *Qst* and *Fst* observed on experimental data (Table 2) is of the same magnitude than the one obtained in simulation studies including ours (Figure 1-2) and those of Latta (1998). Given that *Fst* was measured on the LCATs contributing to the adaptive trait in the simulations (Figure 1-2 and 3-4), whereas in the observed data isozyme loci were most likely not involved in any metric trait assessed, the concordance in the discrepancy may be considered as a coincidence.

Table 2. Comparative results of *Qst* and *Fst*

Locus	Fst	Trait	Heritability	Qst
Study 1 (*Pinus contorta ssp. latifolia*, Yang et al. (1996))				
Isozymes		Phenotypic traits		
Locus	Fst	Trait		Qst
Aat-1	0.0077	Specific gravity		0.133
Aat-2	0.0097	Diameter		0.166
Aco	0.0114	Height		0.195
Adh	0.0032	Branch angle		0.006
Aph	0.0194	Branch diameter		0.057
Dia-2	0.0118	Branch length		0.161
Dia-3	0.0062			
Gdh	0.0002			
G6p	0.0648			
Idh	0.0001			
Mdh-1	0.0038			
Mdh-2	0.0026			
Mdh-3	0.0139			
Mdh-4	0.0027			
Me	0.0167			
Pgi	0.0039			
Pgm	0.0110			
6Pg-1	0.0207			
6Pg-2	0.0095			
Overall	0.0190			
Study 2 (*Salix viminalis*, Lascoux et al. (1996))				
Aat1	0.038	Number of shoots in season 2		0.229
Acon	0.029			
Mdh1	0.061	Number of shoots in season 4		0.077
Mdh3	0.011			
Pghd2	0.036	Number of shoots in season 5		0.065
Pgi1	0.055			
Overall	0.041	Diameter season 5		0.418
Study 3 (*Quercus petraea*, Kremer et al. (1998), Zanetto and Kremer, 1995)				
Locus	Fst	Trait	Heritability	Qst
Aap-A	0.026	Height age 7	0.05	0.22
Acp-C	0.028		0.10	0.12
Dia-A	0.021		0.15	0.08
Got-B	0.016		0.20	0.06
Lap-A	0.036			
Mr-A	0.022	Bud burst season 6	0.10	0.69
Pgi-B	0.018		0.20	0.53
Pgm-A	0.031		0.30	0.43
Overall	0.025		0.40	0.36

However, the simulations also showed that selection pressures do not significantly affect differences in allelic frequencies at the LCATs, suggesting that LCATs behave almost as neutral markers. As a result, differentiation at LCATs are expected to be of the same magnitude as any other neutral marker including isozymes.

4. Conclusion

The theoretical comparison of differentiation of adaptive traits with molecular markers has indicated genetic causes that may explain the strong discrepancy observed in experimental results between both measures of population differences. Theoretical comparisons were restricted to extreme, simplified situations. However, these causes were further confirmed by simulations more likely to reflect real situations, and are related to the combined effect of the multilocus structure of adaptive traits and selection. Diversifying selection is creating disequilibria at the between-population level between loci controlling adaptive traits, thus inflating population differences. Conversely, disequilibrium at the within-population level is not increasing due to random mating. Because forest environments can be extremely variable, strong diversifying selection is expected to occur, on the other hand, most forest tree population are nearly random mating. As a result, a strong anisotropy of gametic disequilibria for LCATs is expected in forest trees between the two sources of disequilibria: the between- and within-population component. As was shown earlier (Kremer et al., 1997), anisotropy of disequilibria is a major cause of population differentiation in multilocus systems. These results need to be further evaluated in other contexts with different genetic controls of adaptive traits (number of loci, different contribution of each locus to the adaptive trait, dominance and epistatic effects) and different evolutionary scenarios.

Besides the theoretical interpretation of differentiation, this study leads also to practical issues. Even if LCATs can be identified by QTL detection in segregating families and alleles at LCATs are expected to be identified in the future, molecular assessments of differentiation seems not likely to be a relevant alternative to provenance testing. However, there is still place for further theoretical investigations before drawing a definite conclusion on this issue. As indicated by our methods, Qst is basically a multilocus measure of differentiation (equation 16), whereas Fst remains an average monolocus measure (equation 15). Further investigations should address the measurement of differentiation of multilocus gametic frequencies in order to obtain a true multilocus Fst. Such attempts have already been made (Kremer et al., 1997), but various alternatives were suggested that do not lead to identical results.

At last, based on the results of our simulations, conclusions drawn from earlier isozyme studies can be revisited with a different interpretation. If isozyme loci are involved in biochemical pathways controlled by several loci, then selection acting on the end product of the pathway may not significantly affect the population differences for allelic frequencies at the isozyme loci. As a result, isozymes may not be as neutral as usually accepted, but will still express low population differentiation.

Literature cited

Cockerham, C.C. 1969. Variances of gene frequencies. *Evolution 23: 72 – 84.*

Cornelius, J. 1994. Heritabilities and additive genetic coefficients of variation in forest trees. *Canadian Journal of Forest Research 24: 372-379.*

Falconer, D.S. 1960. Introduction to quantitative genetics. *Oliver and Boyd., Edinburgh and London.*

Feldman, M.W. and F.B. Christiansen 1975. The effect of population subdivision on two loci without selection. *Genet. Res. Camb. 24:* 151 - 162.

Hamrick, J.J., Godt, M.J. and S.L. Sherman-Broyles 1992. Factors influencing levels of genetic diversity in woody plant species. *New For. 6: 95-124.*

Hedrick, P.W.S. 1985. Genetics of populations. Jones and Bartlett Publishers, Boston, MA., USA.

Kempthorne, O. 1957. An introduction to genetic statistics. *John Wiley, New York.*

Kremer, A. 1994. Diversité génétique et variabilité des caractères phénotypiques chez les arbres forestiers. *Génétique Sélection Evolution 26*: *Suppl1: 105s-123s.*

Kremer, A., Zanetto, A. and A. Ducousso 1997. Multilocus and multitrait measures of differentiation for gene markers and phenotypic traits. *Genetics 145: 1229-1241.*

Kremer, A. and A. Zanetto 1997. Geographical structure of gene diversity in *Quercus petraea* (Matt.) Liebl. II: Multilocus patterns of variation. *Heredity* **78**: *476-489.*

Kremer, A., Petit, R.J. and A. Ducousso 1998. Structure of gene diversity, gene flow and gene conservation in *Quercus petraea*. *In* J. Turok, A. Kremer and S. De Vries (compilers): Proceedings on the first EUFORGEN meeting on social broadleaves. IPGRI. 133-144.

Lande, R. 1992. Neutral theory of quantitative genetic variance in an island model with local extinction and colonization. *Evolution 42: 455-466.*

Lascoux, M., Thorsen, J. and U. Gullberg 1996. Population structure of a riparian willow species , *Salix viminalis* L. *Genet. Res. Gamb. 68:45-54*

Latta, R.G. 1998. Differentiation of allelic frequencies at quantitative trait loci affecting locally adaptive traits. *American Naturalist 151: 283-292.*

Le Corre, V., Machon, N., Petit, R.J. and A. Kremer 1997. Colonization with long-distance seed dispersal and genetic structure of maternally inherited genes in forest trees: a silmulation study. *Genet. Res., Camb. 68: 117-125.*

Muona, O. 1982. Potential causes for multilocus structure in predominantly outcrossing populations. *Silva Fennica 16*: *107-114.*

Ohta, T. 1982a. Linkage disequilibrium with the island model. *Genetics* 101*: 139 - 155.*

Ohta, T. 1982b. Linkage disequilibrium due to random genetic drift in finite subdivided populations. *Proc. Natl. Acad. Sci. 79: 1940 - 1944.*

Spitze, K. 1993. Population structure in *Daphnia obtusa*: Quantitative genetic and allozymic variation. *Genetics 135: 367-374.*

Tachida, H. and C.C. Cocherham 1987. Quantitative genetic variation in an ecological setting. *Theoretical Population Biology 32: 393-429.*

Turelli, M., 1984. Heritable genetic variation via mutation-selection balance: Lerch's zeta meets the abdominal bristle. *Theoretical Population Biology 25*: *138-193.*

Weir B.S., 1990. *Genetic data analysis*. Sinauer Associates, Sunderland,. MA.

Weir, B.S. and C.C. Cockerham 1984. Estimating F-statistics for the analysis of population structure. *Evolution 38: 1358 – 1370.*

Wright, J. 1976. *Introduction to forest genetics.* Academic press. New York.

Wright, S. 1968. *Evolution and the genetics of populations.* Vol 2. The theory of gene frequencies. University of Chicago Press, Chicago.

Wright, S. 1965. The interpretation of population structure by *F*-statistics with special regard of mating. *Evolution 19: 395-420.*

Yang, R.C. and F.C. Yeh 1993. Multilocus structure in *Pinus contorta* Dougl. *Theor. Appl. Genet. 87: 568 - 576.*

Yang Rong-Chai, Yeh, F.C. and A.D. Yanchuk 1996. A comparison of isozyme and quantitative genetic variation in *Pinus contorta* ssp *latifolia* by Fst. *Genetics 142: 1045-1052*

Zanetto, A. and Kremer A. 1995. Geographical structure of gene diversity in *Quercus petraea* (Matt.) Liebl. I. Monolocus patterns of variation. *Heredity 75: 506-517.*

Genetic structures as indicators for adaptation and adaptational potentials

Martin Ziehe, Hans H. Hattemer, Rommy Müller-Starck
and Gerhard Müller-Starck

Introduction

Forests have become increasingly affected by human influence and forest genetics has gained substantial importance in view of the recently detected genetic implications of environmental changes. Conclusions drawn from genetic structures and their dynamics can help to evaluate the genetic processes and their consequences.

Genetic structures of populations can be examined and characterised separately for populations. The observation of characteristic properties or significant deviations can lead to acceptance or rejection of working hypotheses about the genetic system. Much more detailed information can be achieved by comparisons of genetic structures. Spatial patterns or the dynamics of genetic structures within populations help to trace selection processes or assess adaptedness to particular environmental conditions.

1. Selection and genetic structures at marker gene loci

Unfortunately, the physiological function of the different alleles and their adaptive value has rarely been studied. Moreover, genetic variation at marker gene loci is often considered to be neutral. This would mean that presence or absence of particular genetic variants in an individual does not play any role for its fitness including survival abilities and gametic contributions to the subsequent generation. However, the dogma of general neutrality of enzyme gene loci has been impressively refuted. Enzyme gene loci obviously represent prime markers for studying selection and adaptation.

Genetic variation at enzyme gene loci can reflect relevant adaptation if, for example, the catalytic efficiency in biochemical pathways is differently affected by these variants. One example for this concerns the thermostability of the enzyme gene locus IDH-B enzyme variants (isocitrate dehydrogenase, NADP-dependent) in silver fir (*Abies alba*). Bergmann and Gregorius (1993) observed a latitudinal frequency cline for the allele frequencies at IDH-B. They proposed a correlation between allele frequencies and the north-south

C. Mátyás (ed.), Forest Genetics and Sustainability, Volume 63, 75–89.

temperature gradient. By investigating the thermostability of the allelic variants with extracts from the macrogametophyte, it turned out that B_1 (formerly designated as B_2) which was observed with larger frequencies in southern regions, clearly exceeded B_2 in thermostability and obviously represents an adaptational advantage under environmental conditions with higher temperatures.

A second example refers to Norway spruce (*Picea abies*) and the enzyme gene locus PEPC-A (coding for phosphoenolpyruvate carboxylase). Among the natural regeneration under old Norway spruce trees at contaminated sites in south-eastern Germany, the heterozygote A_1A_2 has been found more frequently among seedlings than among the old trees. The heterozygote therefore appears to be better adapted to the corresponding environmental conditions (Bergmann and Scholz 1989). Since PEPC plays a role in the carbohydrate turnover performed in the guard cells and mediates stomatal movements of needles, Rothe and Bergmann (1995) tried to determine kinetic properties of extracted PEPC variants from needles. Herewith, the heterozygous phenotype proved a superiority over the homozygotes concerning binding properties of phosphoenolpyruvate. Rothe and Bergmann (1995) proposed a relation between this superiority of the heterozygote and the heteromeric condition of the enzyme.

In a study on genetic response of European beech (*Fagus sylvatica* L.) seedling populations to stress dominated by acidification, superior viabilities were proven for a specific heterozygote at the enzyme coding gene locus MDH-C, and for a homozygote at the locus PER-B (Müller-Starck 1993).

Enzyme variants controlled by different enzyme gene loci can also interact in biochemical pathways or cycles which prevent independent behaviour of the selection regimes and can lead to complex interrelations at the multiple locus level. Such dependencies have been demonstrated by Gregorius and Ziehe (1995) for a special case of selection in a juvenile beech population, where the combination of the enzyme gene loci IDH-A and MDH-C showed complex adaptive behaviour.

On the other hand, evolutionary factors such as inbreeding (mixed mating) or drift (due to a restricted number of parents) may produce stochastic associations between different (not necessarily linked) loci of the genome so that also genotypic structures do not behave independently. Hence in such a case, structures at actually neutral (enzyme) gene loci also depend on the processes and changes at those gene loci where selection in fact operates. In particular under inbreeding, changes at these actually neutral gene loci may provide information about the directions of selection and the amount of adaptation that the population currently experiences (Ziehe 1999). In this sense, genetic structures and their dynamics at enzyme gene loci may even serve to monitor adaptive processes in populations at those gene loci which cannot be the subject of observations.

It is certainly not a good choice to generally shift to molecular markers just because of their larger variation. It is not the variation of markers alone determining the amount of information to be gained from genetic inventories. Nuclear markers which can be considered as codominant, polymorphic and adaptive at the same time are the exception rather than the rule. Some isoenzyme gene markers ("allozymes"), however, fulfil these requirements.

The present contribution focuses specifically on enzyme genetic structures within European beech populations. Comparisons will be made and analysed between structures of different ontogenetic stages (pre-germinated beech nuts and structures in two years old juvenile beech trees) as well as between stress tolerant and stress sensitive beech trees.

2. Methods to infer on adaptational processes and adaptational capacities

Selection generally leads to an adaptational process except in the case of selection equilibrium. Convergence against such an equilibrium has often been inferred from the observed heterozygote advantage of viability (Ziehe and Hattemer 1998).

Evidence for ongoing selection in beech has been presented in a growing number of instances. Selection can operate during reproduction; conclusions can then be drawn by comparisons between stands and their progeny (e.g. Müller-Starck 1996, Starke 1996, Ziehe *et al.* 1998). The study of selection during the diploid vegetative phase generally requires the comparison of collectives of trees in different ontogenetic phases (seeds, seedlings, adult trees) (see Starke *et al.* 1996, or Müller-Starck 1993). However, genetic differences between phenotypic classes can also provide information about the genetic background of adaptedness.

2.1. Adaptation of populations

The observation of changes of genetic structures at a particular gene locus does not necessarily allow for the conclusion that this gene locus possess an adaptive function. This may be different if, for example, in addition to clinal variation, a biochemical mechanism for an advantage or disadvantage of variants can be established. In addition, gene loci can also be subject to associated effects, i.e. they reflect different fitness parameters and changes of structures solely due to the fact that there are stochastic (genotypical) associations between these loci and those where selection operates. Criteria of the possible amount of changes of genetic structures due to such associations may help to test or evaluate the adaptivity of gene loci (Ziehe and Müller-Starck 1991, or Ziehe 1999).

Similar difficulties arise if criteria for ongoing adaptation are to be established. However, some evidence can be derived from observations of phenotypic traits (Hattemer and Ziehe 1997).

2.2. Adaptational capacity and adaptive potential of populations

Among all individuals which appear to be adapted to particular environmental conditions, those individuals adapted due to their genetic disposition belong to the **adaptational capacity** of the population. The consideration of adaptational capacity incorporates the possibility that some individuals are not sufficiently exposed to stress in order to show stress symptoms. Consequently, this implies heterogeneity of environmental conditions at the site under observation.

The estimation procedure of minimum adaptational capacities follows observations or experiments such as demonstrated subsequently. The comparison of the adaptational capacity of the same population under different environmental conditions (see example below) can allow for conclusions on at least parts of the **adaptive potential** of this population, i.e. those environmental conditions for which sufficient adaptational capacity may be present. This also offers a method to quantify or extrapolate the adaptive potential with respect to relevant environmental conditions.

3. Adult beech stands

3.1. Experimental design for adult stands

The observation of adaptedness or selection advantage is generally based on a phenotypic trait. If the distribution of genotypes over environmental situations is not sufficiently random, it becomes difficult to exclude that the variation of phenotypic trait expressions is controlled solely by environmental conditions. A precondition for reliable conclusions on correlations between genetic information and phenotypic trait expressions is stochastic independence between genotypes and environmental situations genotypes are exposed to. This is fulfilled in planted stands but not necessarily in stands originated from natural regeneration (as is generally the case with beech populations).

Nevertheless, using the pair method, it may be tested, whether phenotypic differences are likely to have a genetic component. Herewith, pairs of neighbouring trees are chosen in the same stand depending on the expression of an adaptive phenotypic trait. First, a tree with phenotypic trait expression X is picked, then its nearest neighbour with contrasting phenotypic expression Y (Gregorius 1989). It is essential that both partners could have emerged from the same seed pool which requires that the maximum distance among the partners of a pair falls short of usual seed transport distances from a single potential seed parent.

The genotypic structures between the two groups of different phenotypic trait expressions are compared for a set of gene loci. If, for a particular gene locus, similar frequency deviations are observed for the phenotypic classes over different populations, the genetic differences can be considered *consistent*. In this case either identical stochastic associations in the populations must have led to this result (for example, a correlation of heterozygosity after a species-specific amount of partial selfing) or otherwise the observed gene locus is involved in controlling the trait. If, on the other hand, the deviations are substantial, but vary with environmental conditions, then the adaptive pressure on the genetic variants may be different between the sites.

Gregorius (1995) presented a method which is appropriate to estimate minimum adaptational capacities by taking the maximum genetic distance (d_o, Gregorius 1974) between tolerant and sensitive individuals over all gene loci observed. If, for instance, 30% of the individuals of the population are tolerant and 70% are sensitive, and if the observed maximum genetic distance between the tolerant and sensitive subsets is 20%, then at least 20% of the 30% tolerant individuals are expected to show tolerance due to their genetic disposition. Hence the minimum adaptational capacity would amount to .30 x .20, which is finally 6%. Ziehe and Gregorius (1996) illustrated the sensibility of this procedure and its application (see also Müller-Starck and Ziehe 1999) in adult beech stands observed earlier by Müller-Starck (1989, 1993).

3.2. Results

The phenotypic trait expressions in these observations refer to tolerance or sensitivity to stress conditions in severely damaged natural populations on sites mostly at higher elevations and exposed to strong air pollution (Müller-Starck 1993). Phenotypic classification followed Roloff (1985) with class 0 for tolerant and class 3 for sensitive trees. Five populations located under environmental conditions with partly different stress characteristics (e.g. local emissions, ozone concentration etc.) were chosen in the western part of Germany (Harz Mountains, Black Forest etc.) and investigated by the pair method. Müller-Starck (1989, 1993) determined genetic structures at 15 enzyme gene loci in samples of between 43 and 53 tree pairs per population.

Figure 1 illustrates for two out of five locations the genetic differences (d_o, Gregorius 1974) at observed enzyme gene loci. The diagrams contain genotypic and allelic distances between the tolerant and sensitive subsets of the individual samples as well as of the bulked material of all 5 samples.

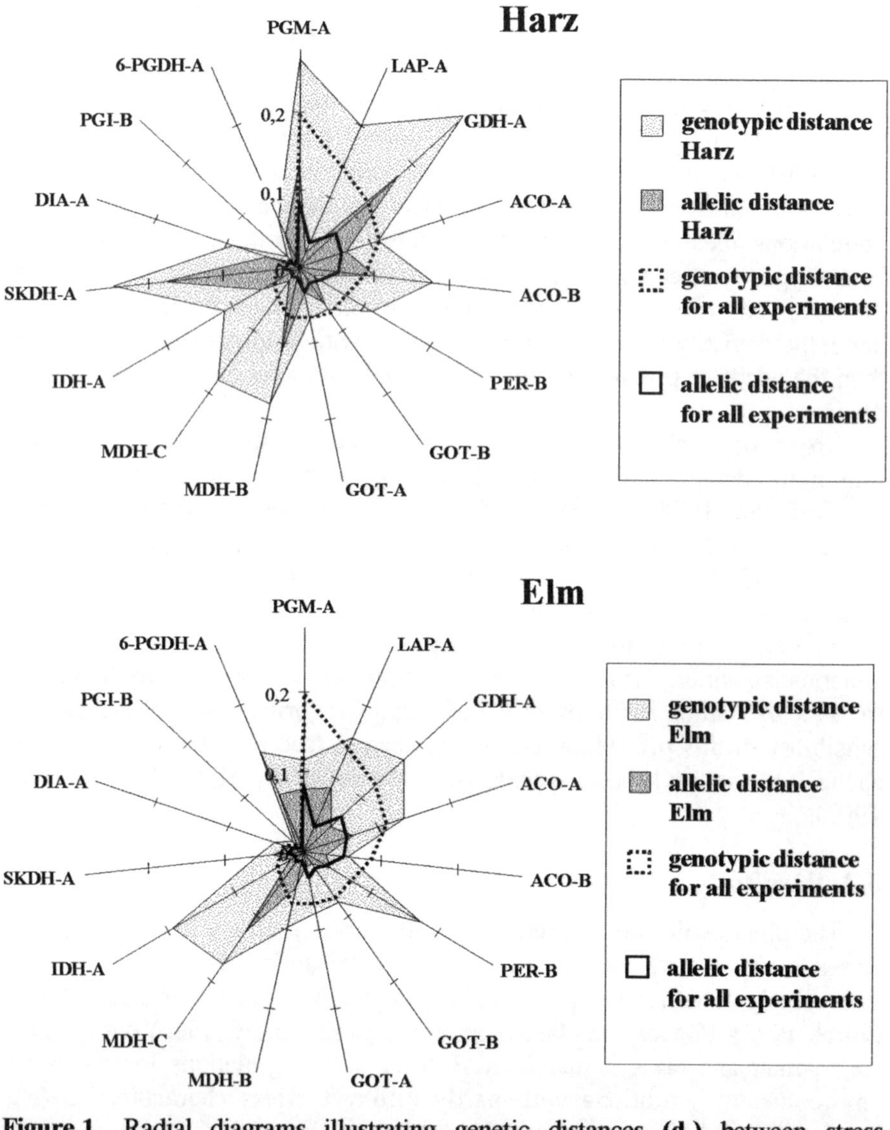

Figure 1. Radial diagrams illustrating genetic distances **(d_0)** between stress tolerant and sensitive adult beech trees for different enzyme gene loci (Müller-Starck and Ziehe 1999). Gene loci are ordered according to distance for genotypes of all samples.

In the Harz sample, gene loci PGM-A, LAP-A, GDH-A, and SKDH-A show genotypic distances of more than 20% between tolerant and sensitive trees. This indicates an adaptational capacity (with respect to the observed gene loci) of more than 20% among all trees classified as tolerant. The determination of the minimum adaptational capacity of the population requires additional knowledge about the percentage of tolerant beech trees in the population.

For PGM-A, LAP-A, and GDH-A, the genotypic distances over all samples are also substantially larger indicating that for these gene loci, the same genotypes appear to be more frequent among the tolerant beech trees as compared to the sensitive trees independently of special environmental site conditions. In contrast to this, SKDH-A shows a more specific response especially at this site which is hardly explained by special stochastic associations.

The Elm stand is located at lower altitude with different stress characteristics. In this stand, the adaptational capacity among tolerant beech trees appears to be smaller: distances do not exceed 20%. For PGM-A, the gene locus with the largest distances for the bulked material, the genotypic distance is smaller than in all other samples. There are either different or differently strong associations between PGM-A and the adaptive gene loci or there is less selection pressure at PGM-A under these particular conditions.

The Elm stand on the other hand shows a more specific response at the enzyme gene loci MDH-C and IDH-A. It becomes obvious that the shape of the radial diagrams and characteristic features may be useful to classify different environmental (stress) situations and factors controlling the observed phenotypic trait expressions by the response at enzyme gene loci.

The large deviations between genotypic and allelic distances especially for PGM-A is mainly explained by heterozygosity differences between tolerant and sensitive subgroups. Figure 2 presents these differences. It becomes immediately obvious that for the majority of enzyme gene loci the tolerant subpopulations show larger frequencies of heterozygotes than the sensitive group. With four of these gene loci (PGM-A, LAP-A, ACO-A, and GDH-A), the difference for the bulked material proves significant. Increased homozygosity among tolerant beech trees did not appear to be significant for the complete material. It is remarkable that a larger degree of heterozygosity as observed earlier (Ziehe and Müller-Starck 1991) is split into smaller effects at a larger number of gene loci. This may be explained by associated effects of heterozygote advantage at some non-observed gene loci and a correlation of heterozygosity in mixed mating populations (Ziehe 1999).

4. Selection among beech seedlings

Kim (1979, 1980, 1985) compared the selection response of beech nuts sown and germinated under field conditions and under temporally more homogeneous conditions in the greenhouse. In samples from Romania

(R) and Germany (D), he observed an increase of heterozygosity at the gene locus LAP-A until the age of 2 years under field conditions as well as a decrease under greenhouse conditions. In the field, mortality reduced the population size to a few percent.

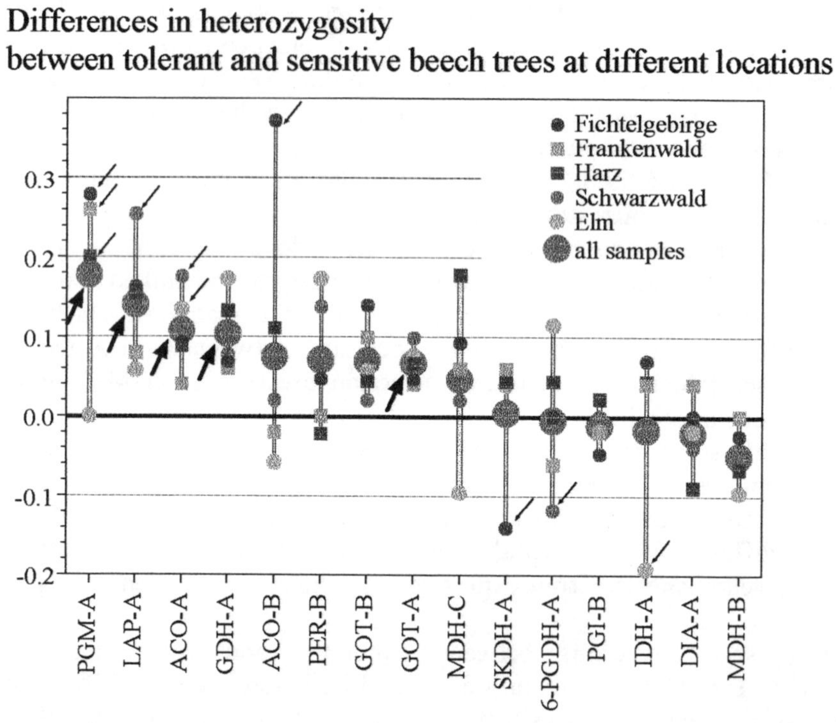

Figure 2. Heterozygosity differences between stress tolerant and sensitive adult beech trees for different enzyme gene loci (Müller-Starck 1993, Ziehe and Hattemer 1998, Ziehe 1999). Positive values stand for a surplus of heterozygotes among tolerant as compared to sensitive trees. Arrows indicate statistically significant differences ($P<0.05$).

In different experiments with a similar design and pre-germinated beech material from regions in western Germany, Müller-Starck and Ziehe (1991) found a similar trend at the gene locus LAP-A (compare Figure 3). Under several field conditions, Müller-Starck (1993) determined an increased frequency of the allele A_4 and of the heterozygotes carrying A_4 and he estimated relatively high viabilities for these types. Under greenhouse conditions, especially the heterozygote A_2A_4 shows inferiority, whereas the homozygote A_2A_2 displays exactly the opposite behaviour under field and greenhouse conditions.

In another series of experiments designed in order to investigate the effect of soil amelioration, plots were established in the Solling Hills north-west of the city of Göttingen, Germany, on a former Norway spruce site (Starke *et al.* 1996). Beech nut material from the same region (Uslar) and a mixed sample from different regions was exposed to different environmental conditions representing different levels of soil amelioration.

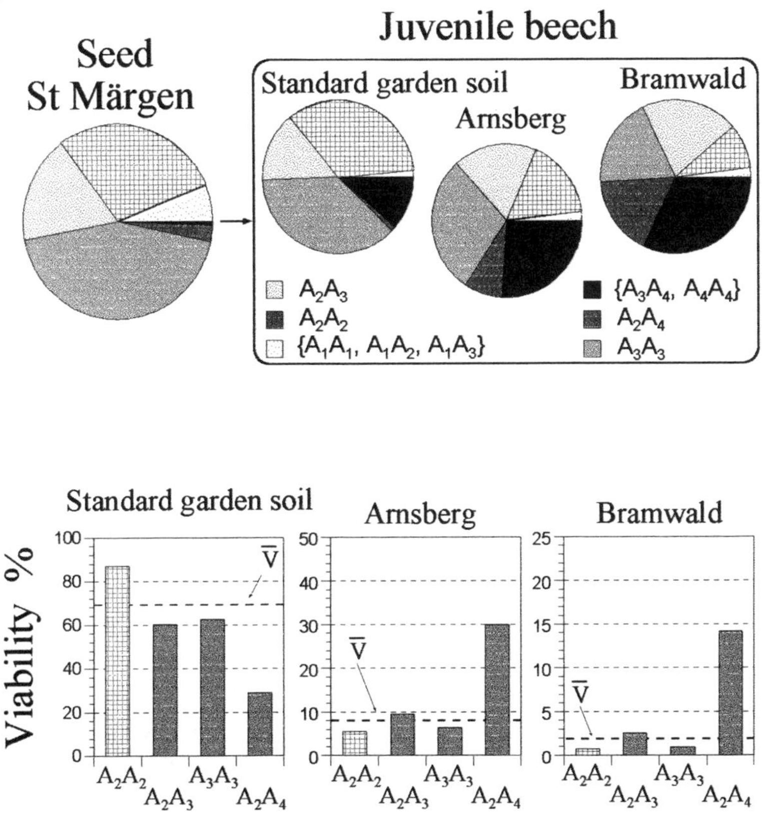

Figure 3. Relative frequencies and viabilities of European beech genotypes at the gene locus LAP-A (according to Müller-Starck and Ziehe 1991, Müller-Starck 1993, Hattemer *et al.* 1993). Seed material was collected near St. Märgen (Black Forest). Viabilities refer to the ontogenetic phase from pre-germinated seeds to the juvenile stage at the age of 2 years and to environmental conditions on standard garden soil in the greenhouse or under field conditions at two sites (Arnsberg, Bramwald).

The initial sizes of pre-germinated seed samples were 3,040 (Uslar sample) and 3,160 (mixed sample) for each treatment (non-manipulated ground, surface fertilisation and deep ploughing with fertilisation). The pH range of the topsoil of the three treatments were 2.80 - 2.86, 3.68 - 4.51 and 4.15 - 4.63, respectively (for details see Starke *et al.* 1996)

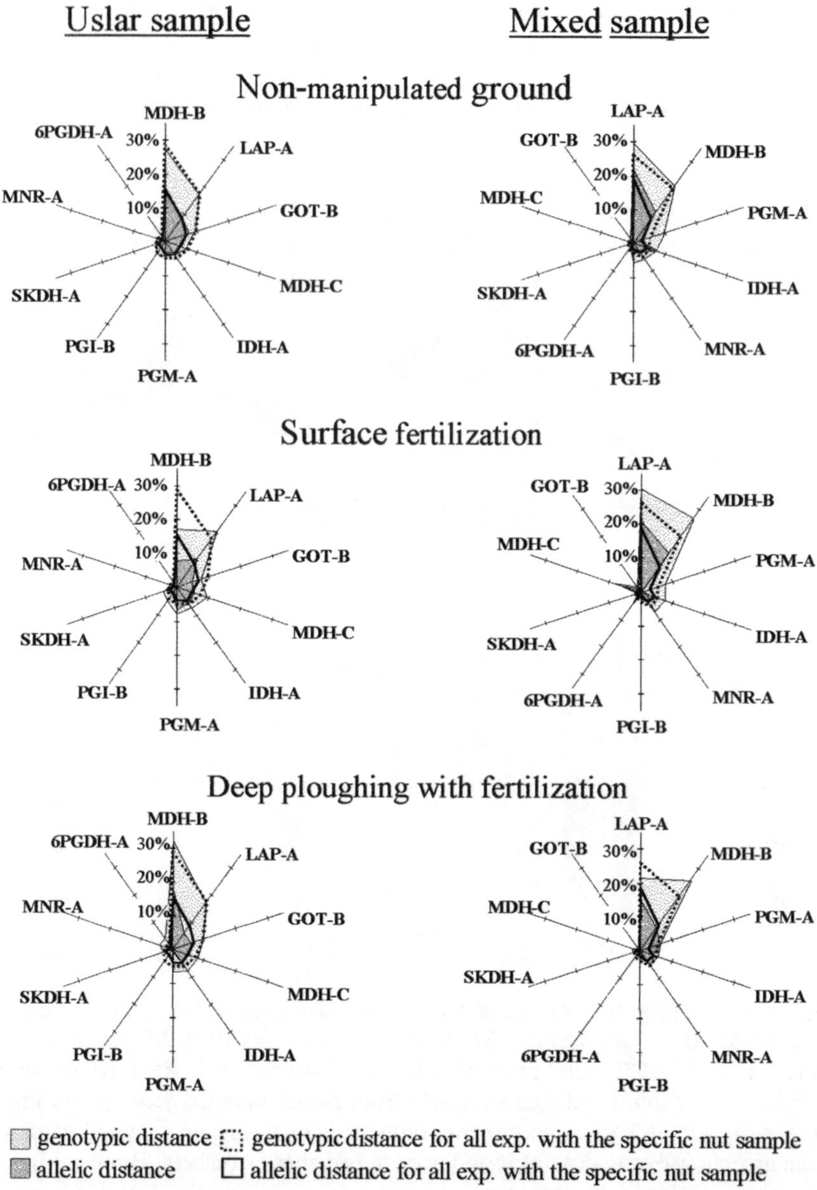

Figure 4. Radial diagrams illustrating genetic distances (d_0) between viable and non-viable beech seedlings for different enzyme gene loci (Müller-Starck 1993) under different amelioration measures. Gene loci are ordered according to genotypic distances observed with the complete material of the specific beech nut sample.

Figure 4 illustrates the results on viable and non-viable seedlings by radial diagrams. Two enzyme gene loci (LAP-A and MDH-B) obviously play a substantial role and show sufficient response to these amelioration measures.

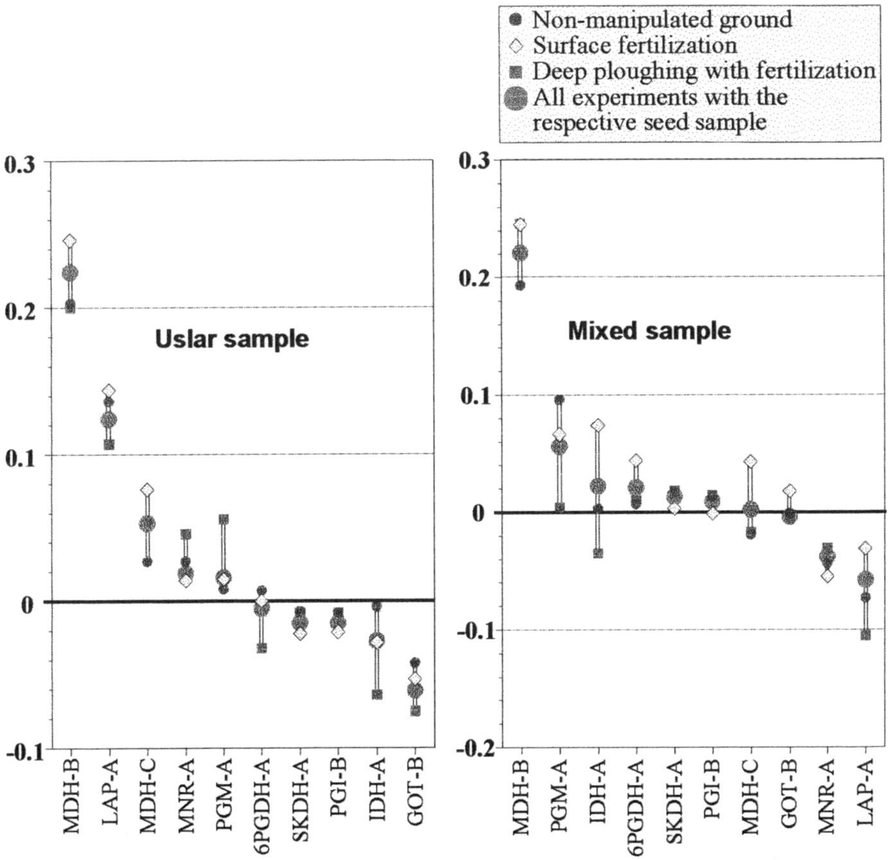

Figure 5. Heterozygosity differences between viable and non-viable pre-germinated seed material for different enzyme gene loci and two different seed samples (Starke *et al.* 1996).

For the non-manipulated ground as well as surface fertilisation, the mixed sample shows larger maximum observed distances. In the case of deep ploughing with fertilisation, the radial diagram for the Uslar sample indicates a maximum distance of 30%, whereas the mixed sample is inferior with 25%. However, under these conditions the percentage of surviving seedlings of the Uslar material was much smaller resulting eventually in a smaller minimum adaptational capacity as compared to the mixed sample. Hence, the mixed sample possesses the larger (observed) minimum adaptational capacity under all conditions illustrated in Figure 4 and, as a consequence, appears to show a larger adaptive potential.

Moreover, the results indicate that the ranking of survival abilities of genotypes at specific gene loci may change with the environmental and amelioration conditions. The most drastic changes have been observed for the genotype MDH-B_3B_4. A large proportion of the individuals carrying B_3B_4 survived under field conditions. This contributes to the high average heterozygosity of the surviving plants.

Figure 5 illustrates the differences in heterozygosity between viable and non-viable material. Here, MDH-B shows a consistent response, but for LAP-A heterozygosity surprisingly turns out to be disadvantageous in the mixed sample. Particular adaptedness properties present in the mixed sample may be responsible for the specific response.

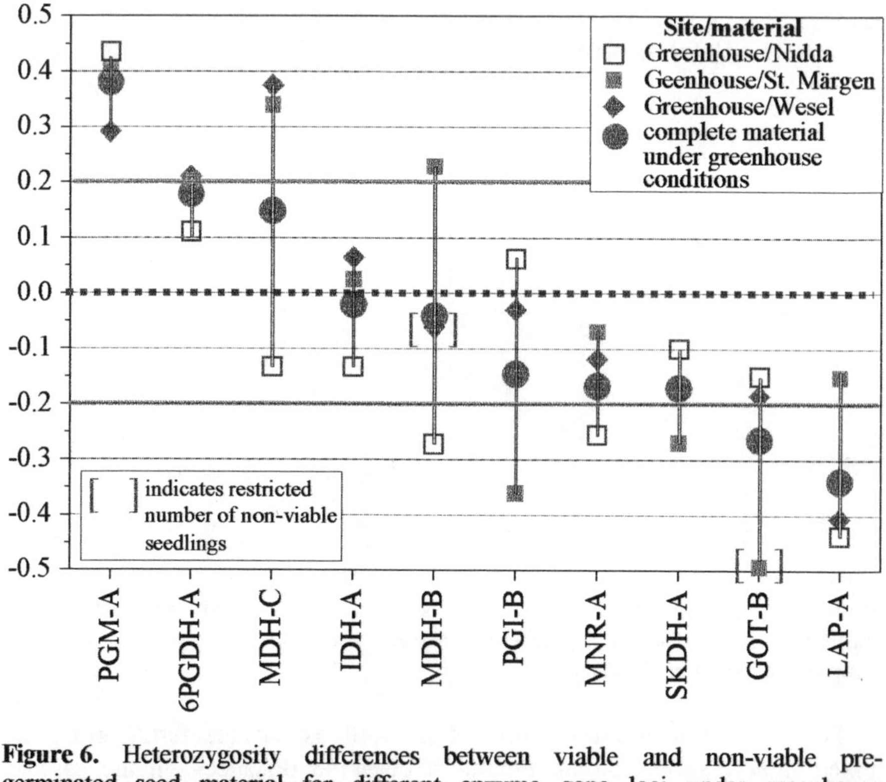

Figure 6. Heterozygosity differences between viable and non-viable pre-germinated seed material for different enzyme gene loci under greenhouse conditions (Müller-Starck 1993).

Results also indicate that methods to improve environmental conditions do not necessarily cause an increase of survival rates, less pronounced viability selection, or loss of genetic variation. The genetic response to environmental stress and to different amelioration regimes depends significantly on the genetic structures of the seed samples.

Figures 6 and 7 show heterozygosity differences between surviving and non-viable seedlings under greenhouse and field conditions from an extensive study conducted by Müller-Starck (1993). Again, under greenhouse conditions homozygosity was more advantageous but heterozygosity was an advantage under field conditions.

Once more, LAP-A shows completely different behaviour depending on environmental conditions. PGM-A, MDH-C, and 6PGDH-A also show substantial differences.

5. Concluding remarks

Different enzyme gene loci possess different metabolic functions and, therefore, **different suitability** as indicators for adaptedness or adaptation. In European beech, some loci such as PGM-A or LAP-A are useful for indicating general selection or stress pressure. Other enzyme gene loci indicate more specific effects depending on the particular environmental conditions.

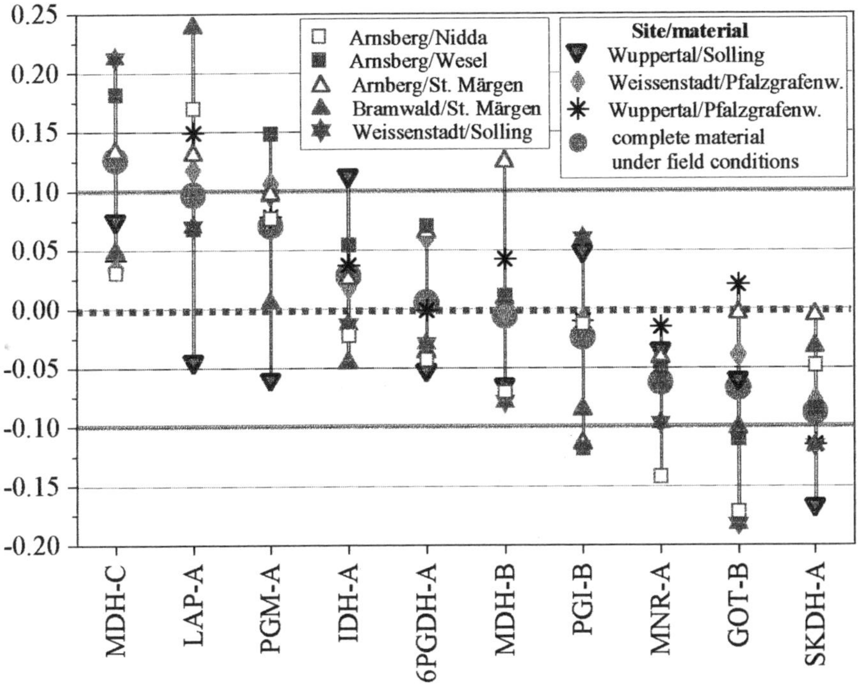

Figure 7. Heterozygosity differences between viable and non-viable pre-germinated seed material for different enzyme gene loci under field conditions (Müller-Starck 1993).

The **applicability** of enzyme gene loci as indicators for adaptation for a given species **depends** on a set of variables such as **initial genetic structures, genomic associations** and **environmental conditions**.

Generally, LAP-A, PGM-A, MDH-C and IDH-A play a substantial role in adaptational processes in European beech populations.

Heterozygosity is important under complex environmental stress conditions. However, under homogeneous conditions, homozygosity may be an advantage.

At present, it seems to be a widespread view that alloenzymes are without any adaptive relevance, and that they should be considered as neutral markers in population genetic studies. However, the experimental findings discussed in this paper do not support this view. Consequently, there is **a need for experiments** which are specifically designed in order to study the effects of certain enzyme gene loci on viability in other species and under different environmental conditions.

Literature cited

Bergmann, F. and H.-R. Gregorius 1993. Ecogeographical distribution and thermostability of isocitrate dehydrogenase (IDH-A) alloenzymes in European silver fir (*Abies alba*). *Biochem. Systematics and Ecology* 21, 597-605

Bergmann, F. and F. Scholz 1989. Selection effects of air pollution in Norway spruce (*Picea abies*) populations. *In* Scholz, F., Gregorius, H.-R., and Rudin, D. (eds.): *Genetic Effects of Air Pollutants in Forest Tree Populations.* Springer Verlag. 143-162

Gregorius, H.-R. 1974. Genetischer Abstand zwischen Populationen. I. Zur Konzeption der genetischen Abstandsmessung. *Silvae Gen.* 23, 22-27.

Gregorius, H.-R. 1989. The attribution of phenotypic variation to genetic or environmental variation in ecological studies. *In* Scholz, F., H.-R. Gregorius, and D. Rudin (eds.): *Genetic Effects of Air Pollutants in Forest Tree Populations.* Springer Verlag. 3-15

Gregorius, H.-R. 1995. Measurement of genetic diversity with special reference to the adaptive potential of populations. *In* Boyle, T.J.B. and Boontawee, B. (eds.): 1995. *Measuring and Monitoring Biodiversity in Tropical and Temperate Forests.* Bogor, Indonesia, Center for International Forestry Research. 145-175

Gregorius, H.-R. and M. Ziehe 1995. Detecting independence in viability selection on two traits. *Heredity* 74, 70-79

Hattemer, H.H., F. Bergmann und M. Ziehe. 1993. Einführung in die Genetik für Studierende der Forstwisssenschaft. Frankfurt am Main, J.D. Sauerländer's Verlag, 2nd ed.

Hattemer, H.H. and M. Ziehe. 1997. Genetic control of phenotypic traits with relevance to gene conservation in trees - a survey of methods. *In* Mátyás, Cs. (ed.): *Perspectives of Forest Genetics and Tree Breeding in a Changing World.* IUFRO World Series Vol. 6, Sopron, 135-148

Karp, A., Isaac, P.G., Ingram, D.S.(eds.): 1998. *Molecular Tools for Screening Biodiversity: Plants and Animals.* Chapman and Hall, London

Kim, Z.S. 1979. Inheritance of leucine aminopeptidase and acid phosphatase isozymes in beech (*Fagus sylvatica* L.). *Silvae Gen.* 28, 68-71

Kim, Z.S. 1980. Veränderung der genetischen Struktur von Buchenpopulationen durch Viabilitätsselektion im Keimlingsstadium. *Göttingen Res. Notes in Forest Genetics* no. 3, 88 pp.

Kim, Z.S. 1985. Viability selection at an allozyme locus during development in European beech (*Fagus sylvatica* L.). *Silvae Gen.* 34, 181-186.

Müller-Starck, G. 1989. Genetic implications of environmental stress in adult forest stands of *Fagus sylvatica* L.. *In* Scholz, F., H.-R. Gregorius, and D. Rudin (eds.): *Genetic Effects of Air Pollutants in Forest Tree Populations*. Springer Verlag. 127-142

Müller-Starck, G. 1993. Auswirkungen von Umweltbelastungen auf genetische Strukturen von Waldbeständen am Beispiel der Buche (*Fagus sylvatica* L.*). Schriften aus der Forstl. Fakultät der Univ. Göttingen und der Niedersächs. Forstl. Versuchsanstalt*. Bd. 112. German w. Engl.sum., 163 pages.

Müller-Starck, G. and R. Starke 1993. Inheritance of isoenzymes in European beech (*Fagus sylvatica* L.). *J. Hered.* 84: 291-296.

Müller-Starck, G. and M. Ziehe. 1991. Genetic variation in populations of *Fagus sylvatica* L., *Quercus robur* L. and *Q. petraea* Liebl. in Germany. *In* Müller-Starck, G. and M. Ziehe (eds.): *Genetic Variation in European Populations of Forest Trees*. Frankfurt am Main., J.D. Sauerländer's Verlag, 125-140

Müller-Starck, G. and M. Ziehe. 1999. Genetic response of populations of European beech (*Fagus sylvatica* L.) to environmental stress - a case study. *In* Yeh, F. and F. Scholz (eds.): *Diversity and Adaptation in Forest Ecosystems in a Changing World*. Kluwer Academic Publishers, Dordrecht, submitted.

Müller-Starck, R. 1996. Genetische Aspekte der Reproduktion der Buche (*Fagus sylvatica* L.) unter Berücksichtigung waldbaulicher Gegebenheiten. *Ber. Forschungszentrum Waldökosysteme*, Reihe A, Bd. 135. Göttingen, 103+XII

Roloff, A. 1985. Schadstufen bei der Buche - Vorschlag für eine bundeseinheitliche Einordnung der Buche in 4 Schadstufen bei terrestrischen Aufnahmen. *Forst- und Holzwirt* 40: 131-134.

Rothe, G. and F. Bergmann 1995. Increased efficiency of Norway spruce heterozygous phosphoenolpyruvate carboxylase phenotype in response to heavy air pollution. *Angew. Bot.* 69, 27-30

Starke, R. 1996. Die Reproduktion der Buche (*Fagus sylvatica* L.) unter verschiedenen waldbaulichen Gegebenheiten. S. 135-159 *In* Müller-Starck, G. (Hsg.) Biodiversität und nachhaltige Forstwirtschaft. Ecomed Verlag, Landsberg

Starke, R., M. Ziehe, and G. Müller-Starck 1996. Viability selection in juvenile populations of European beech (*Fagus sylvatica* L.). *Forest Gen.* 3, 217-225

Ziehe, M. 1999. Genomische Assoziationen durch Selbst- und Fremdbefruchtung und ihre Bedeutung für die Interpretation genetischer Strukturen am Beispiel der Buche (*Fagus sylvatica* L.). Habilitationsschrift Universität Göttingen, submitted

Ziehe, M. and H.-R. Gregorius 1996. Beurteilung der Gefährdung genetischer Ressourcen anhand von Streßmerkmalen. *In* Müller-Starck, G. (Hsg.): *Biodiversität und nachhaltige Forstwirtschaft*. Ecomed Verlag, Landsberg, 300-317

Ziehe, M. and H.H. Hattemer 1998. The significance of heterozygosity in tree breeding and gene conservation. *Forest Tree Improvement* no. 26, 3-25

Ziehe, M. and G. Müller-Starck 1991. Changes of genetic variation due to associated selection. *In* Müller-Starck, G. and Ziehe, M. (eds.): *Genetic Variation in European Populations of Forest Trees*. Frankfurt am Main, J.D. Sauerländer's Verlag, 259-271

Ziehe, M., Starke, R., Hattemer, H.H. und Turok, J. 1998. Genotypische Strukturen in Buchen-Altbeständen und ihren Samen. *Allgem. Forst- u. Jagdztg*. 169, 91-99

Assessment of population history and adaptive potential by means of gene markers

Reiner Finkeldey and Gábor Mátyás

Introduction

The adaptedness of a biological entity to a specific environment implies its capacity of survival. Changes of environmental conditions require reactions which ensure the continued existence of the entity without loosing its identity. Adaptation processes are reactions to modified environmental conditions ultimately leading to a state of adaptedness and adaptability refers to the ability to react in a way which ensures survival (Finkeldey and Gregorius 1994).

Environmental variability is experienced by all organisms. Thus, adaptation is a ubiquitous phenomenon in biology. It is important to distinguish between two types of adaptations at different levels of biological organisation in this context: physiological adaptations at the level of individual organisms and evolutionary adaptations at the level of populations.

Physiological adaptations are reactions of an organism ensuring its survival and potential for reproduction in response to environmental fluctuations or directional environmental change. The adaptive potential of an organism is often described as its phenotypic plasticity (Schlichting 1986). The outcome of a successful adaptation process is the survival of the organism and the maintenance of its reproductive capacity. Phenotypic plasticity is the main force ensuring the survival of organisms in changing environments. Different genotypes are expected to differ with regard to their phenotypic plasticity, i.e. physiological adaptability has a genetic basis. Studies on phenotypic plasticity are of great interest both for ecologists and breeders. However, physiological adaptations do not result in any changes of heritable information and should accordingly be clearly distinguished from evolutionary processes.

Evolutionary adaptations are responses of populations to environmental change. The fitness of individuals is influenced by the environment and changes of the relative fitness of genotypes are likely to occur in case of environmental change. Evolutionary adaptation is often accompanied by viability selection and thus the death of some, but not all individuals of a population. It can only be studied at the population level and results in

C. Mátyás (ed.), Forest Genetics and Sustainability, Volume 63, 91–104.

changes of the genetic structures of the population at least at a single gene locus.

Genetic variation is a prerequisite for evolutionary adaptability, but not for physiological adaptability. Both genetic structures and environmental conditions determine the adaptedness of a population. This implies that the status of adaptedness of a population can be assessed only in a given environment. However, the evolutionary adaptability of a population exclusively depends on its genetic information and can principally be assessed by an analysis of its genetic structures disregarding specific environments (Finkeldey 1993).

1. Traits to measure adaptive potentials

Variation pattern at certain traits must be observed in order to assess and compare the adaptive potential of populations. Two types of traits are frequently observed for this purpose: phenotypic, usually quantitative traits closely correlated to adaptedness or even directly measuring adaptedness to a given environment, and genetic markers, which are environmentally stable but often not directly related to adaptedness. It is our main purpose to examine the role of genetic markers for an assessment of adaptive potentials. However, we will first briefly discuss the suitability of phenotypic, quantitative traits as measured in field trials for an evaluation of the adaptedness and adaptive potentials of populations.

The observation of phenotypic traits characterizing for example the health status, growth, and reproduction of a population allows to conclude on the status of adaptedness of the population to a specific environment. Populations are exposed to a given environment in a field trial in order to compare their relative states of adaptedness.

It is usually impossible to conclude on the adaptive potential of populations by an assessment of phenotypic traits at a single trial site. For this purpose replicated field trials are established in a variety of environmental conditions. Populations performing well in the majority of field trials are regarded as having a high adaptability (Zobel and Talbert 1984). However, it is usually not the evolutionary adaptive potential of populations which is assessed in replicated field trials but rather the individual physiological adaptabilities of the plants. "Adaptable" populations are those, which contain a large proportion of plants with a high physiological adaptability, while populations with a high evolutionary adaptability may carry a considerable genetic load (Gregorius 1991) in all environments and may perform poor on average.

Trials containing identical clones or families from several populations at multiple sites can be used to separate physiological from evolutionary adaptation processes. The distinction between evolutionary and physiological adaptability is far from being purely academic since the inheritance and transmission of adaptations to consecutive generations critically depends on the type of adaptation process.

Variation patterns at quantitative traits may be useful indicators also of the evolutionary adaptive potential of populations if trials are specifically designed for the purpose to separate physiological and evolutionary adaptation processes. For example, Larsen (1986) formulated a hypothesis on a lack of evolutionary adaptability in Central European populations of silver fir (Abies alba) mainly based on the observation of polygenic traits of seedlings (e.g. Larsen and Mekic 1991).

2. Genetic markers

The application of a rapidly increasing number of types of genetic markers to forest tree species confirms the insight that most long-living woody plant species are genetically variable and that the spatial and temporal dynamics of genetic information is complex (Namkoong 1989). Evolutionary forces acting on different gene loci of the same population are variable. Spatial patterns of genetic variation at a particular marker locus are primarily influenced by its mode of inheritance, gene flow through seed and pollen, selective constraints, mutation rates, and population sizes. The resolving power of the method used to observe a polymorphism has an additional influence on the observed patterns of genetic variation.

The spatial organization of genetic variation is, for example, usually very different for a uniparentally inherited cp- or mt-DNA marker as compared to a nuclear microsatellite or an isozyme gene locus (e.g. Dumolin-Lapègue 1997, Streiff *et al.* 1998). Even different loci of the same marker type often show contrasting patterns of genetic variation. For example, populations which are highly polymorphic at certain isozyme gene loci are often fixed on a single allele at other enzyme loci.

A widely applied approach to conclude on the total magnitude of genetic variation present in a population is to observe variation patterns at several gene loci and to compute average or multi-locus variation measures. This procedure allows to compare levels of genetic variation of different species or populations (Hamrick *et al.* 1992).

Genetic variability is a necessary requirement for evolutionary adaptability. However, genetic variation is measured at a few marker loci, which are unlikely to be directly involved in future adaptation processes and represent only a very small fraction of the overall genetic variation present in populations. Thus, the assumption of a positive correlation between the amount of genetic variation at certain marker loci and the adaptive potential of populations needs justification.

3. Selectivity and adaptability

Controversial discussions have centred on the evolutionary forces acting on biochemical and molecular gene marker loci. Three issues deserve particular attention within the context of measuring adaptive potentials:

- The potential for selection acting directly on biochemical and molecular marker loci;
- Associations between marker loci and adaptive loci;
- Selective control and its significance for evolutionary adaptability.

3.1. Direct selection acting on marker loci

The neutralist-selectionist controversy on the relative importance of evolutionary factors for shaping genetic structures at the molecular level will not be reviewed here (see e.g. Lewontin 1991). Strong evidence has been gained concerning selection acting on particular biochemical marker gene loci for certain species. In many other cases it proved to be impossible to test the hypothesis of the selective control of genetic structures.

Many types of molecular gene markers are very unlikely to be selectively controlled. A polymorphism caused by a single point mutation in an intron of cp-DNA is, for example, highly likely to be selectively neutral just as the length variation of short DNA-fragments due to a different number of sequence-repeats (microsatellites).

Biochemical gene markers like isoenzymes are at least potentially selective. Variation of isozymes as observed by standard procedures like gel-electrophoresis is due to differences in the charge and/or spatial structure of the enzymes caused by the substitution of at least one amino acid. The exact differences of amino acid sequences are rarely known but different physiological and metabolic features of allozymes have been shown in a number of cases. Most of the enzymes routinely observed in genetic inventories based on isozyme gene markers catalyze key reactions in the metabolism of plants (Bergmann and Hattemer 1998). Thus, it cannot be ruled out that the occurrence of different allozymes has a direct impact on the fitness of the respective carriers of the controlling genetic information at least under particular environmental conditions.

Low levels of genetic differentiation among populations are frequently observed at isozyme gene loci in spite of considerable differentiation at other traits including those of adaptive significance. Qualitative differentiation is confined to rare alleles. Similar patterns of allelic profile bundles, i.e. allelic frequency distributions at isozyme gene loci in several populations, were found in *Picea abies* and *Drosophila melanogaster* (Table 1). It seems that these figures, i.e. a dominance of profile bundles with one frequent allele (monomorphism and minor polymorphism) or two (rarely three) frequent alleles (major polymorphism) in all populations are the rule for species composed of populations which did not experience very small effective population sizes in their recent evolutionary past.

The observation of low differentiation is frequently explained by the presumed selective neutrality of the markers and extensive gene flow among populations (Muona 1990). However, an alternative explanation for the absence of quantitative differentiation and low levels of qualitative differentiation at isozyme gene loci is a selective regime favouring the same few allele(s) in all populations, i.e. in different environmental conditions.

Table 1. Qualitative description of allelic frequency distributions at isozyme gene loci in populations of *Picea abies* (Finkeldey 1993) and *Drosophila melanogaster* (Singh and Rhomberg 1987). Relative frequency of types of allelic profile bundles are given in percents; very rare alleles ($p_i \leq 0.01$) are disregarded.

Description	*Picea abies*	*Drosophila melanogaster*
Number of investigated loci	13	117
Monomorphism, no differentiation	46%	48%
Two (rarely three) frequent alleles in all populations (major polymorphism) without qualitative differentiation	31%	20%
One frequent and a series of rare alleles (minor polymorphism); no qualitative differentiation of the frequent type	15%	25%
Other types including qualitative differentiation for frequent alleles	8%	7%

Considerable environmental heterogeneity among populations may lead to adaptive differentiation. However, trees are like all other organisms capable of keeping the physical and chemical milieu within cells remarkably constant. In consequence, the direct environmental conditions experienced by enzymes within cells are much less variable than usually thought. Selection may favour the same few alleles in all populations growing under various environmental conditions. These alleles code for enzymes evolutionarily optimized for their highly specific function in a stable environment (Finkeldey 1993). This perception is consistent with the occurrence of a small number of frequent alleles (one or two, rarely three) at isozyme gene loci and the absence of qualitative differentiation among populations with regard to these frequent alleles. Strong differentiation due to disruptive selection can be expected at other loci including those regulating gene activity.

3.2. Associations between loci and hitch-hiking

Selection does not only act directly on particular gene loci, but may also affect associated gene loci. Associations may arise due to linkage and linkage disequilibria among primarily selected and "hitch-hiking" gene loci, but also due to evolutionary factors affecting the complete genome, e.g. genetic drift.

Both "neutral" and (potentially) selective gene loci may be affected by stochastic associations. Studies aimed at the identification of quantitative trait loci (QTLs) proved that associations between marker loci and loci influencing economically important, often adaptive traits exist within families of forest tree species and are strong enough to be detected (e.g. Grattapaglia *et al.* 1996). These associations are due to linkage between loci. The significance of these findings for associations at the level of populations remains to be evaluated.

Changes of genetic structures at marker loci within forest tree populations have been observed for a number of gene loci particularly during reproduction and in early life stages (Kim 1985) as well as in populations under severe environmental stress (Müller-Starck 1985). This temporal dynamics is likely to be caused more often by stochastic associations and "hitch-hiking" rather than by direct selection at the marker loci. Thus, even "neutral" gene markers are useful tools to monitor selective processes in forest tree populations.

3.3. Selection and adaptive potentials

The possibility to monitor selective processes at marker loci suggests their suitability to assess and compare the evolutionary adaptive potential of forest tree populations (Hattemer and Ziehe 1997). However, the following considerations need to be taken into account in this context:

- The relation between selection and adaptive potential of populations is complex. Selection does not necessarily increase the adaptedness of populations to a specific environment. In special cases selection favouring particular types decreases population fitness and may even cause extinction of populations (Gregorius 1984). However, this effect occurs only rarely in nature. In most cases selection will increase the adaptedness of a population in particular after environmental change. Genetic variation is a necessary requirement both for selection and evolutionary adaptation as well as for the maintenance of adaptive potential.
- Monitoring of selection after environmental change, i.e. observation of changes of genetic structures at marker loci, proves the importance of evolutionary adaptation processes for the survival of populations. However, this does not necessarily imply the possibility to quantify and compare the adaptive potential of populations by means of gene marker studies.
- Patterns of genetic differentiation at marker loci do not necessarily reflect adaptive potentials and adaptive differentiation among populations due to variation at other loci. This holds irrespective of the potential or actual

selectivity of marker loci. However, a uniform selective regime is likely to favour the same alleles in populations experiencing different environmental conditions resulting in poor genetic differentiation among populations. This proposed type of selective control is peculiar for enzyme gene loci and does not reflect spatial patterns of genetic variation at other gene loci which are determined by other evolutionary factors. Thus, variation patterns at isozyme gene loci may be of limited use to directly quantify and compare adaptive potentials of populations (Finkeldey 1993).

In summary, it seems that levels of genetic variation at gene marker loci are poorly suited to directly measure the evolutionary adaptive potentials of forest tree populations. This conclusion is not due to the frequently presumed "neutrality" of gene markers, which are suitable tools to monitor selective processes in populations. However, the magnitude and spatial organisation of evolutionary adaptive potentials within species is too complex to be directly measured by variation at gene marker loci, which represent only a minute proportion of the extant genetic variation.

4. Gene markers and population history

Inheritance of a trait implies its suitability to trace back historic events. Thus, gene markers are useful tools to elucidate the evolutionary past of populations. The choice of markers with different modes of inheritance (maternal, paternal, biparental) and different time horizons for evolutionary change allows to investigate aspects of population history at different levels of biological organisation (e.g. speciation, population migration, separation, and hybridization) and on different time scales. The availability of an increasing number of different, mainly molecular marker types greatly promotes the possibilities to gain insights into the history of forest tree populations (Schaal *et al.* 1998, Sperisen *et al.* 1998).

Genetic markers are particularly suitable to identify events caused by evolutionary factors affecting all gene loci in principally the same way. Drift as a consequence of small effective population sizes affects all gene loci, i.e. gene markers and loci of adaptive significance. Inbreeding and other mating types affect genotypic structures of all gene loci in characteristic ways. Gene flow including hybridization among species can also be investigated by observation of genetic structures at marker loci, although its effect at single loci both depends on and determines the amount of genetic differentiation among populations. Mutation and in particular selection are the principal evolutionary factors shaping genetic structures of populations mainly at single loci. Consequently, their role is more difficult to assess by means of gene marker studies.

The recent evolutionary past of populations is of special interest for an assessment of their evolutionary adaptive potential. The following two examples illustrate the suitability of gene marker studies to draw conclusions on the evolutionary past of forest tree populations which are of relevance for an assessment of their adaptive potentials.

5. Case Studies

5.1. The evolutionary past of *Paulownia* spp. on Taiwan

Three species of the genus *Paulownia* (Scrophulariaceae) are native to Taiwan (Li 1978). The taxon *P. taiwaniana* is endemic, but known only in cultivation on Taiwan and was planted extensively during the 1970's. In the late 1970's most plantations were severely attacked by a witches-broom disease caused by MLOs. *P. taiwaniana* is like other *Paulownia* spp. propagated mainly vegetatively by means of root cuttings. It is assumed to be a natural hybrid between the other two species native to the island, which have a wide distribution on the Chinese mainland: *P. fortunei* and *P. kawakamii* (Lin and Wang 1991). Both species are extremely rare on Taiwan. Only 8 respectively 20 trees could be found in one year of intensive search.

Table 2. Genetic structures at 13 isozyme gene loci in Taiwanese *Paulownia* populations. No variation was observed within species (Finkeldey 1992a); N: sample size

Enzyme Systems	*P. kawakamii*	*P. fortunei*	*P. taiwaniana*
N:	20	8	203
SKDH-A	A_1A_1	A_2A_2	A_1A_2
MDH-C	C_1C_1	C_2C_2	C_1C_2
MDH-D	D_1D_1	D_0D_o	D_0D_1
GOT-A	A_2A_2	A_1A_1	A_1A_2
PGI-B	B_2B_2	B_1B_1	B_1B_2
IDH-B	B_2B_2	B_1B_1	B_1B_2
MDH-B	B_2B_2	B_1B_1	B_1B_2
IDH-A	A_1A_1	A_1A_1	A_1A_1
LAP-A	A_1A_1	A_1A_1	A_1A_1
MDH-A	A_1A_1	A_1A_1	A_1A_1
LAP-B	B_1B_1	B_1B_1	B_1B_1
GOT-B	B_1B_1	B_1B_1	B_1B_1
MDH-F	F_1F_1	F_1F_1	F_1F_1

Results of a genetic inventory at 13 isozyme gene loci are presented in Table 2 (Finkeldey 1992a). No variation was found within any of the species. 203 trees of *P. taiwaniana* proved to be uniformly heterozygous at seven gene loci. Segregation within sexually produced *P. taiwaniana* offspring at these loci confirmed the genetic interpretation of the isozyme phenotypes (Finkeldey 1992b). *P. fortunei* and *P. kawakamii* were fixed on different alleles at these loci. No variation within or among species was found at six further putative enzyme gene loci.

These results confirm the hybrid nature of the taxon *P. taiwaniana* and strongly suggest an evolutionary past of the genus on Taiwan as follows: The island was colonized by a few trees of the "parent" species *P. kawakamii* and *P. fortunei* from the Chinese mainland. During the recolonization or thereafter both species lost genetic variation due to drift. Natural hybridization occurred, and the genetically uniform first-generation hybrid was cultivated. Propagation was exclusively vegetative and resulted in the large-scale establishment of plantations without or with only negligible amounts of genetic variation.

There is no potential for evolutionary change of the genetically impoverished relics of *P. kawakamii* and *P. fortunei* on Taiwan. This may at least partially explain their status as almost extinct taxa on Taiwan. Their hybrid *P. taiwaniana* was cultivated in plantations because a heterosis effect contributed to its superior performance. However, the large-scale establishment of genetically uniform plantations resulted in a serious disease which eventually precluded further Paulownia plantation establishment on Taiwan.

The adaptive potential of *P. taiwaniana* is also very limited since no more than two alleles are expected at most, if not all gene loci for all plantations throughout Taiwan. A breakdown of co-adapted gene complexes in the sexual offspring of *P. taiwaniana* trees further reduces the potential for evolutionary adaptations.

In summary, the genetic inventory of the native *Paulownia* spp. on Taiwan gives detailed insights into the evolutionary past of the genus on Taiwan (drift, hybridization), poses questions regarding the taxonomic status of *P. taiwaniana*, contributes to the explanation of important silvicultural features of *P. taiwaniana* (heterosis, susceptibility to diseases), and strongly suggests a very limited evolutionary adaptive potential of the genus on Taiwan.

5.2. Postglacial migration of *Quercus* spp. to Switzerland

In Switzerland pedunculate oak (*Quercus robur*), sessile oak (*Q. petraea*) and pubescent oak (*Q. pubescens*) are the most abundant oak species. The three species are interfertile (Aas 1998) and survived the last glacial era in refugia in the South of Europe (Dumolin-Lapègue *et al.* 1997, Ferris *et al.* 1998) and migrated from these refugia to the North.

Cp-DNA of oaks is maternally inherited (Dumolin *et al.* 1995) and shows low mutation rates. Thus, polymorphism of cp-DNA is a suitable marker to trace postglacial migration routes of oaks. Ferris *et al.* (1993) used a single base substitution (point mutation) in an otherwise highly conserved intron of the trnL(UAA) gene to distinguish two haplotypes by means of a PCR-based restriction enzyme assay (PCR-RFLP). A large-scale inventory revealed a divide across Europe. The "white" haplotype showing the point mutation occurs mainly in eastern Europe, the non-mutated "black" haplotype occurs in western Europe. This distinct geographical pattern is explained by the recolonization of Europe from at least two refugia in the south-west and the south-east of Europe.

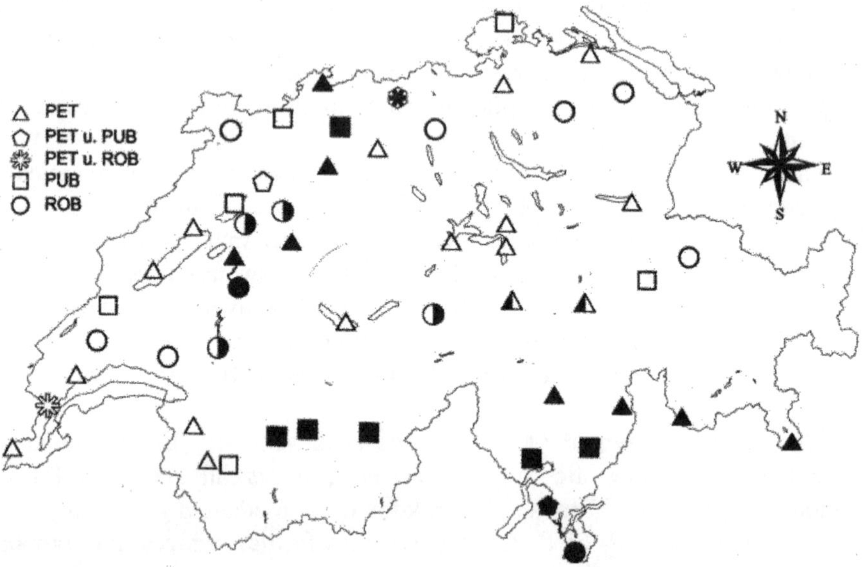

Figure 1. Spatial distribution of a cp-DNA gene marker of oaks (*Quercus* spp.) in Switzerland. Different shapes of the symbols refer to different species (PET: *Q. petraea*; PUB: *Q. pubescens*; ROB: *Q. robur*), different colours (black or white) denote the two haplotypes observed in Switzerland. Populations containing both haplotypes are illustrated by black/white symbols.

We used the same marker to observe spatial patterns of genetic variation of oaks in Switzerland. We investigated material from 53 sites all over Switzerland. Five trees were sampled at each collection site. The spatial pattern of genetic variation in Switzerland is depicted in Figure 1. Results can be summarized as follows:

- Both haplotypes are found in Switzerland. This indicates an origin of *Quercus* spp. in Switzerland through at least two different recolonization routes.
- The geographical structure of the two haplotypes is not at random. The haplotypes are mainly found in separate regions. The non-random distribution of the haplotypes indicates only limited transfer of seeds by humans among regions with different natural history.

– The spatial distribution of the haplotypes is similar for the three species. Each main region is characterized by a single haplotype which is found in most trees irrespective of their species status. This result indicates hybridization and introgression in the evolutionary past. Similar results were obtained in French populations of *Q. robur* and *Q. petraea* (Petit *et al.* 1997).

The maternally inherited cp-DNA marker is not suitable to assess the consequences of gene flow through pollen which seems to be much more efficient than seed migration in oaks (Ennos 1994; Dow and Ashley 1996). This severely limits the applicability of the results to assess patterns of variation at nuclear genes and thus of adaptive potentials. However, extensive gene flow through pollen among regions as widely separated as, for example, Ticino ("black" haplotype) and the region west of the Lake Geneva ("white" haplotype) is unlikely. Thus, we presume a different evolutionary adaptive potential of widely separated populations with differing haplotypes due to the long separation of these populations during the last glacial era. It is, however, impossible to directly compare the magnitude of the respective adaptive potentials based on the observed cp-DNA variation.

6. Conclusions and outlook

The application of genetic marker studies for a direct measurement of the adaptive potential is impeded by the complex organization of genetic variation in forest tree species. Genetic variation of adaptive significance may not be appropriately reflected at marker loci. On the other hand, gene markers are powerful tools to elucidate the evolutionary past of populations, and the evolutionary adaptive potential of populations is determined by genetic processes in their distant or recent history. Thus, the main value of gene markers within the context of a measurement of evolutionary potentials is their importance to gain insights into population history and specifically to identify the main evolutionary factors which shaped the genetic structures of populations.

Detailed conclusions concerning the evolutionary adaptive potential of Taiwanese *Paulownia* species could be made based on patterns of variation at comparatively few isozyme gene markers. Drift and hybridization, the main evolutionary factors responsible for the observed structures, affect all loci of the genome in principally the same way. The insights gained into the evolutionary past of Taiwanese *Paulownia* spp. suggest an extraordinarily simple distribution of genetic variation not only at the investigated marker loci, but at all nuclear gene loci and a very low potential for adaptive evolutionary change.

The situation of the *Quercus* spp. in Switzerland is more complex since the extant spatial structure of nuclear genetic information is influenced by many evolutionary factors including selection, hybridization, and migration through seed and pollen. First insights into the evolutionary past of *Quercus* spp. in Switzerland suggest the recolonization through at least two different recolonization routes. Differentiation of the adaptive potential between widely separated populations with different evolutionary history is likely.

Both examples illustrate the applicability of gene marker studies to draw conclusions on the adaptive potential of populations by gaining insights into their evolutionary history. The magnitude of necessary experimental work as well as the precision and meaningfulness of results mainly depend on the principal evolutionary factors responsible for genetic structures. Evolutionary factors influencing all loci of the genome (drift, mating system) are more easy to detect and have more straightforward implications for adaptive potentials than factors shaping genetic structures primarily at single loci (in particular selection and mutation).

Complex spatial patterns of genetic variation influenced by several evolutionary factors including selection are the rule for forest tree species. Recent selective processes confined to few gene loci have been monitored at marker loci. However, many changes of genetic structures due to selection may go unnoticed by gene marker studies but have considerable influence both on the status of adaptedness to a particular environment and the evolutionary adaptability of a population.

Our understanding of the evolutionary adaptive potential of forest tree populations is still poor. New insights are to be expected from the development and application of new types of molecular markers. Polymorphic molecular gene markers within coding regions of the DNA with known functions and a potential direct impact on adaptive processes are likely to be of particular relevance in this context although the identification of all polymorphic genes with a potential importance for adaptive processes is illusory even for a single population. Our knowledge on evolutionary adaptive potentials of forest tree populations will be further improved by the combination of gene markers with different properties, e.g. with a different mode of inheritance or a different potential to detect selective processes. Another rewarding field of forest genetic research is the combination of gene markers and studies measuring variation at polygenic, adaptive traits in controlled environments at the level of populations.

Acknowledgement

The authors are grateful to Tore Skrøppa for his comments on the manuscript.

Literature cited

Aas, G. 1998. Morphologische und ökologische Variation mitteleuropäischer *Quercus*-Arten: Ein Beitrag zum Verständnis der Biodiversität. Libri Botanici, Band 19. IHW-Verlag, Eching, Germany, 221p.

Bergmann, F. and H.H. Hattemer 1998. Isozymes in forest genetics research. In Mandal, A.K. and Gibson, G.L. (eds.): *Forest Genetics and Tree Breeding*. CBS Publ. and Distrib., New Delhi. pp. 227-238.

Dow, B.D. and M.V. Ashley 1996. Microsatellite analysis of seed dispersal and parentage of saplings in bur oak, *Quercus macrocarpa*. *Molecular Ecology 5*: 615-627

Dumolin, S., B. Demesure and R.J. Petit 1995. Inheritance of chloroplast and mitochondrial genomes in pedunculate oak investigated with an efficient PCR method. *Theoretical and Applied Genetics 91*: 1253-1256

Dumolin-Lapègue, S., B. Demesure, S. Fineschi, V. Le Corre and R.J. Petit 1997. Phylogeographic structure of white oaks throughout the European continent. *Genetics 146*: 1475-1487

Ennos, R.A. 1994. Estimating the relative rates of pollen and seed migration among plant populations. *Heredity 72*: 250-259

Ferris, C., R.P. Oliver, A.J. Davy and G.M. Hewitt 1993. Native oak chloroplasts reveal an ancient divide across Europe. *Molecular Ecology 2*: 337-344

Ferris, C., R.A. King, R. Väinölä and G.M. Hewitt 1998. Chloroplast DNA recognizes three refugial sources of European oaks and suggests independent eastern and western immigrations to Finland. *Heredity 80*: 584-593

Finkeldey, R. 1992a. The hybrid origin of *Paulownia taiwaniana* Hu and Chang - evidence from isozyme gene markers. *Silvae Genetica 41*: 278-282

Finkeldey, R. 1992b. Inheritance of isozyme phenotypes of the native *Paulownia* spp. in Taiwan. *The Journal of Heredity 83*: 140-143

Finkeldey, R. 1993. Die Bedeutung allelischer Profile für die Konservierung genetischer Ressourcen bei Waldbäumen. *Göttingen Research Notes in Forest Genetics*, vol. 14, Göttingen. 176p.

Finkeldey, R. and H.-R. Gregorius 1994. Genetic resources: selection criteria and design. In Kim, Z.-S. and H.H. Hattemer (eds.): *Conservation and Manipulation of Genetic Resources in Forestry*. Kwang Moon Kag Publ. Co., Seoul. pp. 322-347.

Grattapaglia, D., F.L.G. Bertolucci, R. Penchel and R.R. Sederoff 1996. Genetic mapping of quantitative trait loci controlling growth and wood quality traits in *Eucalyptus grandis* using a maternal half-sib family and RAPD markers. *Genetics 144*: 1205-1214

Gregorius, H.-R. 1984. Selection with two alleles and complete dominance. *Biological Journal of the Linnean Society 23*: 157-165

Gregorius, H.-R. 1991. Gene conservation and the preservation of adaptability. In Seitz, A. and V. Loeschke (eds.): *Species Conservation: a Population Biological Approach*. Birkhäuser Verlag, Basel. pp. 31-47.

Hamrick, J.L., M.J.W. Godt and S.L. Sherman-Broyles 1992. Factors influencing levels of genetic diversity in woody plant species. *New Forests 6*: 95-124

Hattemer, H.H. and M. Ziehe 1997. Genetic control of phenotypic traits with relevance to gene conservation - a survey of methods. In Mátyás, Cs. (ed.): *Perspectives of Forest Genetics and Tree Breeding in a Changing World*. IUFRO World Series Vol. 6, Vienna. pp. 135-148.

Kim, Z.-S. 1985. Viability selection at an allozyme locus during development in European beech (*Fagus sylvatica* L.). *Silvae Genetica 34*: 181-186

Larsen, J.B. 1986. Das Tannensterben: Eine neue Hypothese zur Klärung des Hintergrundes dieser rätselhaften Komplexkrankheit der Weißtanne (*Abies alba* Mill.). *Forstwissenschaftliches Centralblatt 105*: 381-396

Larsen, J.B. and F. Mekic 1991. The geographic variation in European silver fir (*Abies alba* Mill.). Gas exchange and needle cast in relation to needle age, growth rate, dry matter partitioning and wood density by 15 different provenances at age 6. *Silvae Genetica 40*: 188-198.

Lewontin, R.C. 1991. Twenty-five years ago in *Genetics*: electrophoresis in the development of evolutionary genetics: milestone or millstone? *Genetics 128*: 657-662

Li, H.-L. 1978. Scrophulariaceae. In Anonymous (ed.): *Flora of Taiwan*, vol. 4. Epoch Publ. Co., Taipei, Taiwan. pp. 581-584.

Lin, T.-P. and Y.-S. Wang 1991. *Paulownia taiwaniana*, a hybrid between *P. fortunei* and *P. kawakamii* (Scrophulariaceae). *Plant Systematics and Evolution 178*: 259-269

Müller-Starck, G. 1985. Genetic differences between "tolerant" and "sensitive" beeches (*Fagus sylvatica* L.) in an environmentally stressed adult forest stand. *Silvae Genetica 34*: 241-247

Muona, O. 1990. Population genetics in forest tree improvement. In Brown, A.H.D., M.T. Clegg, A.L. Kahler and B.S. Weir (eds.): *Plant Population Genetics, Breeding, and Genetic Resources.* Sinauer Ass., Sunderland, Mass. pp. 282-298.

Namkoong, G. 1989. Systems of gene management. In Gibson, G.I., A.R. Griffin and A.C. Matheson (eds.): *Breeding Tropical Trees: Population Structure and Genetic Improvement Strategies in Clonal and Seedling Forestry.* Oxford Forestry Inst. and Winrock Int., Oxford, UK, and Arlington, Virginia. pp. 1-8.

Petit, R.J., E. Pineau, B. Demesure, R. Bacilieri, A. Ducousso and A. Kremer 1997. Chloroplast DNA footprints of postglacial recolonization by oaks. *Proceedings of the National Academy of Sciences USA 94*: 9996-10001

Schaal, B.A., D.A. Hayworth, K.M. Olsen, J.T. Rauscher and W.A. Smith 1998. Phylogeographic studies in plants: problems and prospects. *Molecular Ecology 7*: 465-474

Schlichting, C.D. 1986. The evolution of phenotypic plasticity in plants. *Annual Review of Ecology and Systematics 17*: 667-693

Singh, R.S. and L.R. Rhomberg 1987. A comprehensive study of genetic variation in natural populations of *Drosophila melanogaster*. II. Estimates of heterozygosity and patterns of geographic differentiation. *Genetics 117*: 255-271

Sperisen, C., U. Büchler and G. Mátyás 1998. Genetic variation of mitochondrial DNA reveals subdivision of Norway spruce (*Picea abies* (L.) Karst.). In Karp, A., P.G. Isaac and D.S. Ingram (eds.): *Molecular Tools for Screening Biodiversity.* Chapman and Hall, London. pp. 413-417.

Streiff, R., T. Labbe, R. Bacilieri, H. Steinkellner, J. Glössl and A. Kremer 1998. Within population genetic structure in *Quercus robur* L. and *Quercus petraea* (Matt.) Liebl. assessed with isozymes and microsatellites. *Molecular Ecology 7*: 317-328

Zobel, B. and J. Talbert 1984. *Applied Forest Tree Improvement.* J. Wiley and Sons, New York, Chichester, Brisbane, Toronto, Singapore. 505p.

Adaptability of relic populations: a case study in *Pinus leucodermis* Ant.

Gabriele Bucci and Giovanni G. Vendramin

Introduction

Pinus leucodermis Ant. is a wind-pollinated coniferous species with a discontinous and restricted distribution. Its range consists of small and disjunct populations located both in the Balkan peninsula (main range) and in southern Italy. *Pinus leucodermis* in Italy occurs in only four small groups of populations located in Basilicata and Calabria regions. A drastic reduction of its distribution occurred during the last centuries because of the strong human impacts on this species. Due to the restrictedness of its actual range and to the frequent occurrence of fires in these areas, it can be considered as a highly endangered species. This species shows a great adaptability to extreme environmental conditions and a great colonising potential (Avolio 1986), thus making it suitable for forestation of extensive dry areas in southern Italy.

Several studies were recently performed with the aim of determining the level and distribution of the genetic resources as well as analysing the mating pattern of this species using isozymes and molecular markers. The main results are summarized following.

1. Genetic divergence of peripheral, endangered populations

Peripheral, isolated populations are expected to genetically differ from main range stands as a consequence of both genetic drift and/or local adaptation. Divergence among Italian stands (peripheral populations) was as large as that between Italian and Greek stands nuclei, thought to have been separated more than 10,000 years ago (Bucci *et al.* 1997). The comparison of pairwise divergence between populations using allozymes and RAPD markers showed a good agreement. If genetic drift is responsible of the above divergence, lowered levels of genetic diversity would be expected for small marginal populations (Hamrick *et al.* 1989; Millar and Marshall 1992). Different results were obtained depending on the markers used. Nuclear RAPD and isozymes markers revealed similar values of within-population diversity for the Greek and Italian stands (Boscherini *et al.* 1994a; Bucci *et al.* 1997), while chloroplast microsatellite markers revealed a larger amount of diversity for the Greek stands (Powell *et al.* 1995). The different

C. Mátyás (ed.), Forest Genetics and Sustainability, Volume 63, 105–110.

pattern of genetic diversity observed by the analysis of the two genomes may be due to a stronger effect of genetic drift on the chloroplast genome (uniparentally inherited in conifers). The large, significant randomly-generated disequilibria detected between unlinked RAPD markers in the Italian populations (Bucci *et al.* 1997), seems to support a possible role of genetic drift in determining their high level of divergence. In fact, Italian stands are partly characterized by low density and fire is one of the main ecological factors of this area. Both factors may determine a high level of drift by population bottlenecks, generating random linkage disequilibrium between loci. Estimates of effective population size obtained by RAPD markers (Bucci *et al.* 1997) were quite small for three of the five Italian populations considered. Loss of genetic variability may be triggered a reduction of population size (Prober and Brown 1994) and small effective population size may generate an increase of inbreeding through generations (Falconer 1989). As a consequence, at least two of the five Italian *Pinus leucodermis* populations analyzed are prone to loose genetic variation by drift in the next generations (Bucci *et al.* 1997).

2. Mating system of natural and artificial stands

Mating system parameters were analyzed in two natural and two artificial stands of the species (Morgante *et al.* 1991 and 1994) using 4 and 5 isozyme markers, respectively. The estimated multilocus outcrossing rates were 0.76 and 0.84 in the two natural populations. Both the regression of individual pollen allele frequencies and the analysis using the method of unique allele revealed that the low level of outcrossing rate is due to both mating among relatives and selfing, the latter being significantly larger than what is usually found in predominantly outcrossing coniferous species. A lowered multilocus outcrossing rate was also detected in both artificial stands ($t < 86\%$), where the density is higher and neighbourhood structure should be less pronounced (Morgante *et al.* 1994). The high level of self-fertilisation found in *Pinus leucodermis* could be explained by assuming a reduction of the recessive embryonic lethals as a results of genetic drift. *P. leucodermis* is a pioneer species for which a relatively high level of self-fertilisation may represent a selective advantage, particularly when pollen is in short supply, such as during the first phases of a recolonization process when the population is small and plant density low. This strategy could assume particular importance after fire, which often occurs in the Italian natural stands of this species.

3. Inbreeding and selection against inbreds

Positive fixation index was observed in the seed stage due to partial selfing but not in the adult stage, and selection against inbreds has been invoked to explain this evidence. Two populations with different levels of selfing were studied (Morgante *et al.* 1993). In both populations, a deficiency of heterozygotes as a result of selfing was observed in dormant and germinating embryos, while similar values of heterozygote excess was found in 5-year-old plantlets and in adult trees. Analysis of inbreeding coefficients revealed that selection against inbreds did not act during the germination of seeds in experimental conditions. Inbreeding depression acted between germination and the 5-year-old plantlets stage. Similar trend has been found in both natural and artificial conditions. The above results indicate that inbreds are efficiently eliminated before the plantlets are 5 years old and that heterozygotes are favoured in both the nursery and in natural conditions, although very likely different selective pressures occurred.

4. Linkage disequilibrium analysis of peripheral populations

In a previous investigation on the populations described above (Bucci *et al.* 1997), a large amount of randomly-generated linkage disequilibrium (probably due to drift - Hill 1981) was detected and used to estimate the effective population size (Waples 1991). On the other hand, a relevant percentage of marker pairs discarded from the above analysis (19 out of 253 - 7.51%), showed a significant, homogeneous genotypic disequilibrium over all populations. The above evidence may be interpreted as a hint of epistatic selection. Nonetheless, linkage disequilibrium between unliked loci has been reported to occur in natural populations due to several factors: i) physical association between markers (Hartl 1989); ii) non-random mating within populations (Allard 1975); iii) subdivision of the populations analyzed (Prout 1973); iv) "hitch-hiking" effect of selection on loci linked to markers (Hedrick 1980); v) epistatic selection (Hastings 1990). Linkage disequilibrium can be generated by a combination of the above factors and their balance in natural population is an interesting topic for genetic studies.

In order to throw light on the relative importance of the processes described above, linkage disequilibrium analysis was carried out on 23 RAPD loci previously described (Boscherini *et al.* 1994b). Linkage analysis carried out on haploid, single-tree megagametophytes revealed no significant association between any pair of markers.

5. Dilocus disequilibria at haplotypic and genotypic level

Megagametophyte of seeds from seven natural populations from the whole range were characterized for the above RAPD loci (Bucci *et al.* 1997). For five Italian populations, single-tree seedlots were available and maternal genotypes were inferred from progeny-array data. For the two Greek populations, only bulk seedlots were available and typization of megagametophytes from single-seeds based on the above 23 markers was carried out.

Gametic linkage disequilibrium for pairs of loci was estimated at the haplotypic level on the Greek populations following Nei (1987). Dilocus composite genotypic disequilibria were estimated at the genotypic level for the Italian populations according to Cockerham and Weir (1977). Pooling data from Italian stands, the proportion of significant linkage disequilibria between pair of loci was larger than what expected by chance alone (20.9 %). On the other hand, significant dilocus gametic disequilibria on the pooled dataset from Greek populations was smaller (3.2 %). Gametic disequilibria between pair of loci does not seem to explain completely the extensive composite genotypic disequilibrium found for the Italian populations (Mantel Test: $Z = 4.936$; $r = 0.167$; $p < 0.001$). Whatever the cause of the large proportion of significant cases detected at the genotypic level, this cannot be due to linkage or any other factor acting at the haplotypic level. Moreover, Ohta's (1982) components of dilocus disequilibria were estimated, revealing that most of the dilocus allelic correlation resides within populations (data not shown).

6. Multilocus analysis and test for selective neutrality

Multilocus disequilibrium analysis was carried out on all possible combinations of 3 and 4 of the 23 RAPD loci analyzed. Permutation tests ($n = 1000$) were carried out by shuffling alleles among genotypes and comparing observed mutilocus disequilibrium to null distributions obtained (GDA© statistical package, Lewis and Zaykin 1997). The above analysis revealed a large proportion of significant cases of 3-locus and 4-locus disequilibrium (25.1 % and 31.7 %, respectively) at the genotypic level not observed at the haplotypic level (5.4 % and 7.9 %, respectively). In particular, a subset of 4 RAPD loci (*OPG02B*, *OPG06A*, *OPG06D*, *OPG06G*) showed strong, significant dilocus, trilocus and tetralocus interactions in all the Italian populations analysed, while no evidence of association was found at the haplotypic level for the same subset of loci.

Maximum-likelihood (ML) frequencies of haplotypes at the above four loci in the Italian populations were estimated using the Arlequin© package (Schneider *et al.* 1997). Large differences in multilocus make-up of the Italian populations were detected. The haplotype frequency distribution of ML-estimates at the mentioned four loci in each population was then compared with that obtained by simulating random neutral samples having the same number of genes and haplotypes, using the algorithm of Steward (1977). Significant deviations from selectively-neutral random distribution of haplotype frequencies were detected only for the La Spina population, while no deviations were observed in the Greek populations (Table 1). The above evidence provides a hint of non-neutral association of multilocus genotypes in this population. Further analysis are needed to throw light on the importance of epistatic interactions among loci in natural populations of *Pinus leucodermis*.

Table 1. Simulation test for random distribution of selectively-neutral haplotype frequencies in the seven *P. leucodermis* populations analyzed. Only the four unlinked loci (*OPG02B, OPG06A, OPG06D, OPG06G*) showing strong significant dilocus, trilocus and tetralocus linkage disequilibrium were considered. Maximum-likelihood haplotype frequencies in each population were estimated using Arlequin package version 1.1 (Schneider *et al.* 1997). Deviation from distribution of allele frequencies expected under neutrality was significant for the La Spina population only.

Population	No. of Observed Haplotypes	No. of Simulated Samples	Observed F Value	Expected F Value	Slatkin's Exact P Value
Genotypic Level (Italian Populations)					
La Spina	11	10,000	0.1146	0.1488	0.0019
Cima Pollino	12	10,000	0.2961	0.1441	0.9849
Valle Torta	9	10,000	0.1482	0.1701	0.2778
Valle Gaudolino	11	10,000	0.1080	0.1313	0.1095
Pollinello	12	10,000	0.1338	0.1217	0.6806
Haplotypic Level (Greek Populations)					
Olimpos	9	10,000	0.2377	0.2576	0.515
Pindos	9	10,000	0.2456	0.2534	0.304

7. Conclusions

1. The large divergence detected for Italian populations makes them interesting as target for conservation of adaptive potential in *Pinus leucodermis* Ant.
2. Previously reported evidence based on isozymes supports the evidence of selection against inbreds in early life-cycle stages in natural populations of *Pinus leucodermis.*
3. Linkage disequilibrium analysis revealed the existence of large non-random association of multilocus genotypes at genotypic level (Italian populations) not found at the haplotypic level (Greek populations).
4. Large differences in multilocus make-up of the populations for a subset of 4 loci showing extensive dilocus, 3-locus, and 4-locus disequilibria were detected. Simulation test based on Ewens (1972) sampling theory of neutral alleles provide a hint of non-neutral association of specific multilocus genotypes in the populations analyzed.
5. Studies carried out in this investigation can provide useful information in order to throw light on the molecular basis of adaptation in natural populations of forest trees.

Literature cited

Allard, R.W. 1975. The mating system and microevolution. *Genetics* 79: 115-126.

Avolio, S. 1984. Il pino loricato (*Pinus leucodermis* Ant.). *Ann. Ist. Sper. Selv.* 17: 79-153.

Boscherini, G., Morgante, M., Rossi, P. and G.G. Vendramin 1994a. Allozyme and chloroplast DNA variation in Italian and Greek populations of *Pinus leucodermis*. *Heredity* 73: 284-290.

Boscherini, G., Morgante, M., Rossi, P., Vendramin, G.G. and F. Vicario 1994b. Detection of DNA polymorphism in *Pinus leucodermis* Ant. using random amplification. *Forest Genetics* 1: 131-137.

Bucci, G. , Vendramin, G.G., Lelli, L. and F. Vicario 1997. Assessing the genetic divergence of *Pinus leucodermis* Ant. endangered species: use of molecular markers for conservation purposes. *Theor. Appl. Genet.* 95: 1138-1146.

Cockerham, C.C. and B.S. Weir 1977. Digenic descent measures in a finite population. *Genet. Res.* 30: 121-147.

Ewens, W.J. 1972. The sampling theory of selectively neutral alleles. *Theor. Popul. Biol.* 3: 87-112.

Falconer, D.S. 1989. Introduction to quantitative genetics. 3rd edition, Longman, London.

Hamrick, J.L., Blanton, H.M. and K.J. Hamrick 1989. Genetic structure of geographical marginal populations of ponderosa pine. *Am J. Bot.* 76: 1559-1568.

Hartl, D.L. 1989. *Principle of Population Genetics. 2nd Edition*, Sinauer Associates, Inc., NY.

Hastings, A. 1990. The interaction between selection and linkage in plant populations. *In* Brown A.H.D., Clegg M.T., Kahler A.L., and Weir B.S. (eds.): *Plant population genetics, breeding and genetic resources*. Sinauer Ass. Inc., Sunderland, MA. 163-180.

Hedrick, P.W. 1980. Hitchhicking: a comparison of linkage and partial selfing. *Genetics* 94: 791-808.

Hill, W.G. 1981. Estimation of effective population size from data on linkage disequilibrium. *Genet. Res.* 38: 208-216.

Lewis, P.O and D. Zyking 1997. *Genetic data analysis: computer program for the analysis of allelic data*. Version 1.0. http://chee.unm.edu/gda.

Millar, C.I. and K.A. Marshall 1992. Allozyme variation in port-Oxford cedar (*Chamaecyparis lawsonii*): implication for genetic conservation. *For. Sci.* 37: 1060-1077.

Morgante, M., Vendramin, G.G. and A.M. Olivieri 1991. Mating system analysis in *Pinus leucodermis* Ant.: detection of self-fertilisation in natural populations. *Heredity* 67: 197-203.

Morgante, M., Vendramin, G.G., Rossi, P. and A.M. Olivieri 1993. Selection against inbreds in early-cycle phases in *Pinus leucodermis* Ant. *Heredity* 70: 622-627.

Morgante, M., Rossi, P., Vendramin, G.G. and G. Boscherini 1994. Low levels of outcrossing in *Pinus leucodermis* Ant.: further evidences in artificial stands. *Can. J. Bot.* 72: 1289-1293.

Nei, M. 1987. Molecular evolutionary genetics. *Columbia University Press, New York.*

Ohta, T. 1982. Linkage disequilibrium due to random genetic drift infinite subdivided populations. *Proc. Natl. Acad. Sci. USA* 79: 1940-1944.

Powell, W., Morgante, M., McDewitt, R., Vendramin, G.G. and A.J. Rafalski 1996. Polymorphic simple sequence repeat regions in chloroplast genome: application to the population genetics of pines. *Proc. Natl. Acad. Sci. USA* 92: 7759-7763.

Prober, S.M. and A.H.D. Brown 1994. Conservation of the grassy white box woodlands: population genetics and fragmentation of *Eucalyptus albens*. *Conser. Biol.* 8: 1003-1013.

Prout, T.J., Bundgaard, J. and S. Bryant 1973. Population genetics of modifiers of meiotic drive. I. The solution of a special case and some general implications. *Theor. Pop. Biol.* 4: 446-465.

Schneider, S., Kueffer, J.M., Roessli, D. and L. Excoffier 1997. ARLEQUIN ver 1.1. A software for population genetic data. http://anthropologie.unige.ch/arlequin.

Steward, F.M. 1977. Computer algorithm for obtaining a random set of allele frequencies for a locus in an equilibrium population. *Genetics* 86: 482-483.

Waples, R.S. 1991. Genetic methods for estimating the effective population size of cetacean populations. *In* Hoelzel A.R. (ed.): *Genetic ecology of whales and dolphins*. International Whaling Commision (Special issue no. 13).

4. Genetic Resources and Changing Environment

Genetic implications for forest trees of increasing levels of greenhouse gases and UV-B radiation

David F.Karnosky, Kevin E. Percy, Blanka Mankovska, Richard E. Dickson, Judson G. Isebrands and Gopi K. Podila

Introduction

Globally, the environment is changing and deteriorating as greenhouse gases such as carbon dioxide (CO_2) and tropospheric ozone (O_3) continue to increase at a rate of about 1% per year (Keeling *et al.* 1995, Chameides *et al.* 1995). The increase in these gases is directly related to anthropogenic activities (Chameides *et al.* 1995, Mooney *et al.* 1991) and is likely inducing subtle but substantial changes in the earth's surface temperatures and weather (Cha 1997; Martin 1996). In addition, anthropogenic activities have been linked to decreasing levels of stratospheric O_3 and concomitant increases in ultraviolet-B radiation (UV-B) passing through to the earth's surface. Thus, man is creating a constantly changing global environment for which tree breeders must attempt to develop genotypes and races suitable for future forests. The purpose of this paper is to examine the genetic implications for forest trees of increasing greenhouse gases and UV-B and to suggest where tree breeders need to be concerned about the changing environment. Since very little is known about the impact of greenhouse gases and/or UV-B on genetic population structure, we will not discuss population structure or the effects of selection on population structure.

1. Co-occurrence of stresses

Increasing CO_2 is a global phenomenon with similar increases being observed around the world. In contrast, increases in tropospheric O_3 and decreases in stratospheric O_3 are regional in nature and more patchy (Miller 1993). Thus, future forests are not likely to be exposed to all three of these stresses simultaneously (Kickert and Krupa 1990). Complicating the matter further is that some areas will have increased or decreased temperatures, increased or decreased drought conditions, or increasingly extreme weather

C. Mátyás (ed.), Forest Genetics and Sustainability, Volume 63, 111–124.

conditions. Thus, it is difficult to predict the suite of environmental conditions under which populations will function, and it is equally difficult to predict the impacts of changing global conditions on the genetics of forest trees.

2. Genetic consequences of individual stresses

2.1. CO_2

The pre-industrial concentration of atmospheric CO_2 is expected to double in the next thirty years (Barnola *et al.* 1995). This is may trigger a global mean temperature increase of about 2 to 3°C above the pre-industrial baseline (Kelly 1990). Thus, there are two major concerns about the impacts of elevated CO_2 on the genetics of forest trees: (1) natural selection for CO_2 responsiveness, and (2) adaptability to changing climate.

Firstly, trees respond in varying degrees to elevated CO_2. While there is a general stimulation of growth and productivity with increasing CO_2 (Eamus and Jarvis 1989; Körner 1995), some species have minimal responses to elevated CO_2 while others are stimulated dramatically. Similarly, there are markedly differing responses within species in response to elevated CO_2 with growth increases ranging from 38 to 68% (Ceulemans and Mousseau 1994). Similarly, water use efficiency can be changed dramatically with increasing for some species and some genotypes as stomatal conductivity is generally decreased in the presence of elevated CO_2. With this strong genetic variability in CO_2 responsiveness, it seems likely that natural selection will occur (Miao *et al.* 1992, Woodward *et al.* 1991). To our knowledge, however, no one has characterized heritability of CO_2 responsiveness in trees and no one has conducted long-term, community-level CO_2 studies to examine genetic change. Reproductive fitness under elevated CO_2 is also largely unknown (Bazzaz and McConnaughay 1992). Certainly, these are important gaps in the current knowledge base.

Secondly, as temperature and soil moisture levels change with increasing atmospheric CO_2, it is likely that subtle population changes will occur. Present techniques available for explaining the cause and possible consequences of global warming permit only broad generalizations concerning both the temporal and spatial scales over which changes are expected to take place (Miller 1993), as these projections are based primarily on large-scale global circulation models (GCM's). While some migration of forest trees is expected with the changing climate, the magnitude of these migrations is largely uncertain (Cha 1997). Conservation of forest genetic resources via preserves is certainly complicated by climatic change (Martin 1996), as reserves with suitable climate today may not have a suitable climate in 50 to 100 years.

2.2. Ozone

Background global tropospheric O_3 concentrations are increasing at a rate of about 1 to 2% per year (Chameides *et al.* 1995; Finlayson-Pitts and Pitts 1997). A highly phytotoxic compound, O_3 negatively affects trees in many ways including decreasing photosynthesis, accelerating foliar and branch senescence, and inhibiting growth (Pye 1988; Hogsett *et al.* 1997; Chappelka and Samuelson 1998). In eastern North America, the annual mean aboveground growth loss due to elevated O_3 is currently estimated to be in the range of 0 to 10% (Chappelka and Samuelson 1998), but it appears to be more in some sensitive species such as trembling aspen (*Populus tremuloides*) (Wang *et al.* 1986, Hogsett *et al.* 1997).

The impacts of O_3 on forest tree populations has been studied in considerable detail. One of the first such problems to be diagnosed was the oxidant damage to ponderosa pine (*Pinus ponderosa*) over large areas of the San Bernardino Mountains in southern California (Miller *et al.* 1963). Community changes related to natural succession caused by interspecific variability in response to oxidants were initially described in this region by Miller (1973). He noted that mixed forests of ponderosa pine, sugar pine (*Pinus lambertiana*) and white fir (*Abies concolor*) were changing to predominantly fir because of the greater sensitivity of the pines to oxidants. Similar results have been described for Jeffrey pine (*Pinus jeffreyi*) and ponderosa pine at several locations along the western slope of the Sierra Nevadas (Peterson *et al.* 1989, Miller *et al.* 1996).

More recently, similar O_3 -induced population impacts have been noted in the mountain pine forests surrounding Mexico City where O_3 remains at exceedingly high levels (100-200 ppb peaks or more) throughout the year (Miller 1993). In this area, *Pinus hartwegii* appears to be the most highly sensitive to O_3 and has been severely impacted by O_3 since the 1970's (Hall *et al.* 1996) with widespread dieback and decline resulting in its replacement in an extensive forest area surrounding Mexico City.

In the eastern United States, O_3 has been linked to visible foliar injury and growth decrease (Dochinger and Seliskar 1970), decreased reproduction (Benoit *et al.* 1983), and increased mortality rates (Karnosky 1981) for eastern white pine (*Pinus strobus*). Since the responses of eastern white pine appear to be highly heritable, the components are in place for Phase I of natural selection, that is the elimination of sensitive genotypes. Since O_3 sensitive genotypes make up a relatively small portion of natural eastern white pine stands and the selection pressure conveyed by O_3 is rather low (Taylor and Pitelka 1991), the question has been raised does it really matter if we lose these sensitive genotypes. Surely, this remains an openly debated and important research topic.

Similar responses to O_3 for sensitive genotypes of trembling aspen are expected. Evidence for population changes induced by O_3 in trembling aspen in the eastern United States are the studies by Berrang *et al.* (1986, 1989, 1991) which have shown a strong positive correlation between O_3 concentration at the population origin and the mean O_3 tolerance of the population. Populations from more heavily polluted areas tended to be more tolerant of O_3 than did populations from relatively pristine areas. As in eastern white pine, O_3 responses in aspen are highly heritable (Karnosky 1977). Ozone has been shown to decrease aboveground biomass accumulation by 20 to 40% or more for sensitive genotypes (Wang *et al.* 1986, Karnosky *et al.* 1996a, b).

Ozone can also affect the relative abundance of understory vegetation in forests. Barbo *et al.* (1998) showed that O_3 exposures can cause shifts in the competitive interactions between plant species, thereby altering community structure. These understory plant interactions could also influence the ability of forest trees to naturally regenerate, grow and reproduce.

2.3. UV-B

It is well known that stratospheric ozone concentrations have declined during the past 25 years owing to emission of chlorine- and bromine-containing compounds into the atmosphere. This decline has been especially intense since 1984 and is already having important consequences for biological systems. Firstly, a decrease in lower stratospheric ozone concentrations can cool the climate system. Secondly, as ozone naturally absorbs solar ultraviolet (UV) radiation, decreasing concentrations are leading to increasing intensity of UV at the terrestrial surface (Pyle 1997). Documented trends in increasing UV have been strongest at wavelengths within the UV-B (280-320 nm) portion of the spectrum. Cellular processes in biological systems have been shown to be particularly sensitive at wavelengths less than 300 nm (SCOPE 1992).

Forests are long-lived and of critical importance to the maintenance of global cycles, including carbon (C). There are several recent reviews available which summarize known and predicted impacts of enhanced UV-B on various trophic levels within forests (Percy and Gordon 1998, Percy and Cameron 1997, Mcleod and Newsham 1997, Manning and Tidemann 1995, Bornmann and Teramura 1993) to which the reader is referred for in-depth discussion of effects.

It is of concern to long-lived systems like forests that, even under the most optimistic scenarios, stratospheric ozone concentrations are not predicted to recover until at least 2050 (Madronich *et al.* 1995). Deleterious effects have been reported on : 1) physicochemical changes to the first UV-B attenuating barrier, the leaf surface (Gordon *et al.* 1998a,b); 2) photosynthesis, manifested as reduced C assimilation (Sullivan and Teramura 1989), and more negative 13C isotope ratio (Naidu *et al.* 1993); 3) activation of UV-B-absorbing pigments in the epidermis (Schnitzler *et al.*

1996) and trichomes (Shaltsa *et al.* 1994); 4) changes in leaf expansion (Sullivan *et al.* 1996) and morphology (Sullivan *et al.* 1994); and 5) effects on phylloplane fungi (Newsham *et al.* 1997) and foliar saprophytes (Ayres *et al.* 1996).

2.3.1. Genetic implications of UV-B

Assessment of the full impacts of UV-B on genetic variation is not possible at this time. Few experimental response data are available and were largely gathered from short duration (<single season) exposure of seedlings (<2 yrs) to enhanced UV-B at high doses under artificial conditions. Most data, therefore, cannot be extrapolated directly to the field situation. Nevertheless, evidence (Sullivan and Teramura 1992) pointing to the importance of cumulative (several seasons) exposure in determining direction and magnitude of response at the plant level (biomass, growth, C allocation) is building.

The way in which UV-B influences gene expression is yet unclear (Mackerness *et al.* 1997). Based on three lines of evidence, UV-B induced down-regulation of gene expression is stated by the these authors to not be a simple consequence of non-specific damage to DNA.

SCOPE (1992) identified pollination as a potentially UV susceptible stage in plant reproduction. In other life forms such as monocotyledonous and dicotyledonous ephemerals , this ecologically critical developmental stage has been shown to be particularly sensitive to UV-B damage (Musil 1995). UV-B induced damage to the plant genome has also been reported to be inherited by successive generations (Musil 1996). Using fluctuating leaf asymmetry as a variable, the same group later confirmed that accumulated UV-B damage is indeed heritable (Midgley *et al.* 1998).

2.3.2. Cellular and organ levels

Taylor *et al.* (1997) concluded that DNA is the primary absorbing chromophore in the cell in the UV-B region of the spectrum. Although action spectra for DNA indicate maximum absorption at around 260 nm (UV-C), absorption occurs up to 320 nm. Dimeric photoproducts which make up the majority of lesions formed after UV irradiation include cyclobutane pyrimidine dimers (CPD's) and pyrimidine (6-4) pyrimidone photoproducts or (6-4) photoproducts which are toxic and mutagenic (Taylor *et al.* 1997). Major differences exist between monocots and dicots in photorepair of CPD's.

2.3.3. Provenance and species level

As is the case with many stresses, response differs widely between species and groups. Unfortunately, large screenings for within species variance has not occurred with UV-B. Early work by Kossuth and Biggs (1981) on pines showed deleterious effects of UV-B on growth in 5 species, stimulation in 1, and decreases in height for 6 species. Magnitude of effect was species dependent. Sullivan and Teramura (1988) exposed seedlings of 10 pine species and noted stunting or needle discoloration in three. Height was decreased in only 3 species. Biomass increased in 1 species, decreased in 6, while 3 species were unaffected. Gordon, Percy and Riding (1998a), using older seedling material, reported negative leaf surface effects on three of four Picea species tested.

Differences in response are especially large at the life form level. Leaf optical properties, anatomy, phenology and canopy architecture are critical factors influencing group response. Conifer needles attenuate UV-B much more efficiently than do woody and herbaceous dicot leaves (Day *et al.* 1992). Less than 0.5% of UV-B is transmitted by conifer needles, while woody and herbaceous dicots may transmit up to 12% and 41% respectively. UV-B penetration is also much deeper in dicots. Conifers, therefore, are much more efficient and uniform filters for UV-B than other groups (Day *et al.* 1993).

Of prime concern to breeding programs, of course, is origin of parent material. It has been implied for some time that trees growing at higher elevations were more resistant to enhanced UV-B due to naturally higher levels with elevation. Early evidence for this was produced by Sullivan and Teramura (1988) and later corroborated by Binder and L'Hirondelle (1996) using material collected from 5 elevations (850-1700 m). Differences can be partly attributed to needle anatomy (cuticle, epidermis) and pigmentation (Bornmann and Vogelmann 1988) which together influence degree of UV-B light attenuation and depth of penetration within the leaf. The window for damage to sensitive physiological processes in conifers is about 4 weeks during the early part of the needles' first growing season, prior to development of the cuticle/epidermal complex (DeLucia *et al.* 1992).

Latitude is also an important factor. Using loblolly pine (Pinus taeda) seed sources collected from 7 U.S. latitudes (31° to 39°N) selected to mimic a natural UV-B gradient, Sullivan and Teramura (1992) observed biomass reductions in 2 of 7 (28%) sources after one year of field exposure to enhanced UV-B dose. In year two, 50% of sources tested decreased biomass and after three years, 75% of sources had reduced biomass. It is important to note that negative effects were equal in both above- and below-ground portions.

2.3.4. Community level

There have been no stand- or ecosystem-level enhancement experiments conducted. Only one report exists of visible UV-B damage on mature, field-grown trees (Loucks 1994). However, given the large differences which exist in response between life forms and between tree species at the leaf and seedling levels, it is reasonable to conclude that current and predicted UV-B levels pose a real and long-term threat to forest genetic variability and consequently, forest sustainability on a global scale.

State of science suggests that subtle, cellular- and organ-level effects may be occurring now in many forested regions, particularly in forests growing at higher latitudes. These may be leading to impacts on leaf morphology, surface physicochemical characteristics and physiology. Perturbation of essential processes would be expected to result in differential effects determined by species composition, age, phenology and local edaphic conditions.

Alteration of canopy architecture through crown thinning with resultant increased transparency and transmission of damaging UV-B to ultra-sensitive ground flora may have long-term consequences for biodiversity and forest ecosystem structure. Certainly, there are indications that host-pathogen interactions will be affected (Ayres *et al.* 1996). Resilience to UV-B is also diminished in the presence of other co-occurring global stressors such as extreme climatic events (temperature extremes) and increasing atmospheric CO_2 (Yakimchuck and Hoddinott 1994) and tropospheric O_3 (Zeuthen *et al.* 1997).

3. Genetic consequences of interacting stresses

Although little research has been conducted with interacting global change stresses, there are reasons to suggest that genetic responses to multiple stresses will be difficult to predict as is demonstrated in the next few examples. The combination of UV-B and ozone reduces pollen tube growth more than either stress alone (Feder and Shrier 1990). Similar results have been shown for visible foliar injury in eastern white pine induced by SO_2 plus O_3 (Dochinger and Seliskar 1970). Kull *et al.* (1996) found an interesting response when exposing trembling aspen clones to a combination of elevated CO_2 and O_3. While the predicted outcome was for the elevated CO_2 to alleviate the O_3 stress, this study actually found more injury by O_3 when CO_2 was present (Kull *et al.* 1996). The authors speculated that antioxidants may be reduced in the presence of elevated CO_2, as reported by Badiani *et al.* (1997), thereby increasing the O_3 sensitivity of otherwise tolerant genotypes. Antioxidants appear to play a role in determining O_3 tolerance in aspen (Sheng *et al.* 1997). Subsequent growth studies have also suggested that CO_2 did not alleviate O_3 effects in aspen (Karnosky *et al.* 1998).

It is well known that plant responses to elevated CO_2 are very sensitive to nutrition, particularly to nitrogen levels in the soil (Lloyd and Farquhar 1996). Thus, it is likely that regions of the world experiencing high levels of anthropogenic nitrogen input would also be likely places to see the maximum CO_2 effects.

4. Novel approaches to study the impacts of atmospheric pollutants on forest tree populations

4.1. Pollution gradients

Inferences of genetic change can be made from examination of natural pollutant gradients. Berrang *et al.* (1986) first documented genetic differences in *Populus tremuloides* responses to O_3 from populations occurring in low or high O_3 areas. This work was later substantiated from similar results for a larger region (Berrang *et al.* 1989, 1991).

Recently, our team has established *P. tremuloides* clones of varying O_3 sensitivity in plantations designed to address (a) O_3 bioindication by foliar symptom development, (b) growth and yield under differing O_3 seasonal doses, and (c) competition between O_3 sensitive and tolerant genotypes under differing O_3 seasonal doses. The studies were established some five years ago in Rhinelander, Wisconsin (low O_3), Kalamazoo, Michigan (intermediate O_3), and Kenosha, Wisconsin (high O_3). These sites have proven very useful in showing the effects of O_3 on competitive interactions. Severe growth suppression and increased mortality among sensitive genotypes have been documented for the highest O_3 site within the first five years. We have also documented deterioration of the cuticle at the elevated ozone locations and linked this to increasing pest occurrence (Percy, Karnosky and Mankovska 1998). This would suggest that selection pressure is relatively higher than has previously been suggested (Taylor and Pitelka 1991) for the early development of densely populated aspen stands coming in after a fire, clearcut or other disturbance.

Heavily impacted by long-term atmospheric pollutants, the former Eastern bloc countries offer excellent study sites to examine the impact of atmospheric deposition on forest communities. These are sites of large-scale forest ecosystems where pollution loading has resulted in severe forest decline. At one such study point in central Slovakia (Mankovska *et al.* 1989), Karnosky, Mankovska and others are utilizing Müller-Starck's (Müller-Starck 1985) technique of matched pairs of pollution "sensitive" and "tolerant" European beech (*Fagus sylvatica*), Norway spruce (*Picea abies*), and silver fir (*Abies alba*) to compare gene frequencies and heterozygosity between sensitive and tolerant populations.

4.2. FACE technology

Scientists have long conducted atmospheric pollutant studies in laboratory, greenhouse or open-top chambers. These chambers have been very useful in characterizing early seedling development, basic physiology of pollution stress, and threshold levels for foliar and growth responses induced by pollution. Because of limited chamber size and because of the ever-present chamber effects, these techniques are not useful for longer term, larger scale research questions dealing with mature forest trees growing in natural plant communities.

Recently, a new technology has developed that allows for entire plant communities and trees of any size to be exposed to elevated CO_2 and/or O_3. Modified from the design originating in England (McLeod 1995), a large scale Free-Air CO_2 Enrichment (FACE) experiment (FACTS I) was established on a 20-year-old loblolly pine forest in North Carolina (DeLucia and Thomas 1995, Ellsworth *et al.* 1995). We have modified the FACTS I CO_2 distribution system to allow for dependable and reliable exposure of both elevated CO_2 and O_3 in an aggrading aspen/birch/maple stand in northern Wisconsin (FACTS II) (Karnosky *et al.* 1995). In this study, we are examining the relative CO_2 and/or O_3 responses (growth and physiology) via interspecific and intraspecific investigations. These studies should help the understanding of forest community responses to elevated levels of greenhouse gases, and they give us a window into what forests will be like in 50 years.

4.3. Molecular approaches

The field of molecular biology is rapidly evolving and many new tools are emerging that will be useful in studying the impacts of atmospheric pollutants on forest tree populations. For example, studies of pollution impact on genetic diversity and of population adaptability to pollution stress will likely be facilitated by emerging molecular marker techniques. While protein electrophoresis remains a useful tool for population biologists, techniques based on nucleic acids such as RFLP, RAPD, and AFLP and SSR (microsatellite) markers offer powerful new tools for analyzing germplasm (Powell *et al.* 1996), examining genetic divergence among populations (Parker *et al.* 1998), and utilizing marker-aided selection.

Acknowledgment

This research paper was supported, in part, by the U.S. Department of Energy (DE-FG02-95ER62125), the U.S. National Sciences Foundation (DBI-9601942; IBN-9652675), the U.S. Forest Service Global Change Program, the National Council of the Paper Industry for Air and Stream Improvement (NCASI), the Canadian Forest Service, the Slovakian Forest Research Institute, and Michigan Technological University.

Literature cited

Ayres, P.G., T.S. Gunasekera, M.S. Rasanayagam and N.D. Paul. 1996. Effects of UV-B radiation (280-320 nm) on foliar saprotrophs and pathogens. *In* J.C. Frankland, N. Mangan and G.M. Gadd (eds.): *Fungi and Environmental Change*. Cambridge University Press, Cambridge. 32-50.

Badiani, M., A.R. Paolacci, A. D'Annibale, F. Miglietta and A. Raschi. 1997. Can rising CO_2 alleviate oxidative risk for the plant cell? Testing the hypothesis under natural CO_2 environment. *In* A. Raschi, F. Miglietta, R. Tognetti, and P.R. van Gandingen (eds.): *Plant responses to elevated CO_2: Evidence from natural springs*. Cambridge University Press. United Kingdom. 221-241.

Barbo, D.N., A.H. Chappelka, G.L. Somers, M.S. Miller-Goodman and K. Stolte. 1998. Diversity of an early successional plant community as influenced by ozone. *New Phytol.* 138:653-662.

Barnola, J.M., M. Anklin, J. Porheron, D. Raynaud, J. Schwander and B.T.I. Stauffer. 1995. CO_2 evolution during the last millennium as recorded by Antarctic and Greenland ice. *Tellus* B 47:264-272.

Bazzaz, F.A. and K.D.M. McConnaughay. 1992. Plant-plant interactions in elevated CO_2 environments. *Aust. J. Bot.* 40:547-563.

Benoit, L.F., J.M. Skelly, L.D. Moore and L.S. Dochinger. 1983. The influence of ozone on *Pinus strobus* L. pollen germination. *Can. J. For. Res.* 13:184-187.

Berrang, P.C., D.F. Karnosky and J.P. Bennett. 1989. Natural selection for ozone tolerance in *Populus tremuloides* II. Field verification. *Can. J. For. Res.* 19:519-522.

Berrang, P.C., D.F. Karnosky and J.P. Bennett. 1991. Natural selection for ozone tolerance in *Populus tremuloides*: an evaluation of nationwide trends. *Can. J. For. Res.* 21:1091-1097.

Berrang, P.C., D.F. Karnosky, R.A. Mickler and J.P. Bennett. 1986. Natural selection for ozone tolerance in *Populus tremuloides*. *Can. J. For. Res.* 16:1214-1216.

Binder, W.D. and S. J. L'Hirondelle. 1996. Ultraviolet radiation effects on some British Columbia conifers: 1. Engelmann spruce from five elevations. Abstract IUFRO Meeting for Population, Ecological and Conservation Genetics, University of British Columbia, Vancouver, August 5-9, 1996.

Bornmann, J.F. and A.H. Teramura. 1993. Effects of ultraviolet-B radiation on terrestrial plants. *In* Young, A.R., Bjorn, L.O., Moan, J. And W. Nultsch (eds.): *Environmental UV Photobiology*. Plenum Press, New York. 427-471

Bornmann, J.F. and T.C. Vogelmann. 1988. Penetration of blue and UV radiation measured by fiber optics in spruce and fir needles. *Physiol. Plant.* 72: 699-705.

Ceulemans, R. and M. Mosseau. 1994. Effects of elevated atmospheric CO_2 on woody plants: a review. *New Phytol.* 127:425-446.

Cha, G. 1997. The impacts of climate change on potential natural vegetation distribution. *J. For. Res.* 2:147-152.

Chameides, W.L., P.S. Kasibhatla, J. Yienger and H. Levy II. 1995. Growth of continental-scale metro-agro-plexes, regional ozone pollution, and world food production. *Science* 264:74-77.

Chappelka, A.H. and L. J. Samuelson. 1998. Ambient ozone effects on forest trees of the eastern United States: A Review. *New Phytol.* 139:91-108.

Day, T.A., Martin, G. and T.C. Vogelmann. 1993. Penetration of UV-B radiation in foliage: evidence that the epidermis behaves as a non-uniform filter. *Plant, Cell and Environ.* 16: 735-741.

Day, T.A., T.C. Vogelmann and E.H. DeLucia. 1992. Are some plant life forms more effective than others in screening out ultraviolet-B radiation? *Oceol.* 92: 513-519.

DeLucia, E.H. and R.B. Thomas. 1995. Homeostatic adjustment of loblolly pine to CO_2 enrichment in a forest ecosystem. *In* J. Elwood (ed.): DOE Program for Ecosystem Research: Research Summaries 11-12.

DeLucia, E.H., T.A. Day and T.C. Vogelmann. 1992. Ultraviolet-B and visible light penetration into needles of two species of subalpine conifers during foliar development. *Plant, Cell and Environ.* 15: 921-929.

Dochinger, L.S. and C.E. Seliskar. 1970. Air pollution and the chlorotic dwarf disease of eastern white pine. *For. Sci.* 16:46-55.

Eamus, D. and P.G. Jarvis. 1989. The direct effects of increase in the global atmospheric CO_2 concentration on natural and commercial temperate forests. *Advances in Ecol. Res.* 19:1-55.

Ellsworth, D.S., R. Oren, C. Huang, N. Phillips, and G.R. Hendrey. 1995. Leaf and canopy responses to elevated CO_2 in a pine forest under free-air CO_2 enrichment. *Oecologia* 104:139-146.

Feder, W.A. and R. Shrier. 1990. Combination of UV-B and ozone reduced pollen tube growth more than either stress alone. *Environ. Exp. Bot.* 30:451-454.

Finlayson-Pitts, B.J. and J.N. Pitts, Jr. 1997. Tropospheric air pollution: Ozone, airborne toxics, polycyclic aromatic hydrocarbons, and particulates. *Science* 276:1045-1051.

Gordon, D.C., K.E. Percy and R.T. Riding. 1998a. Effects of UV-B radiation on epicuticular wax production and chemical composition of four Picea species. *New Phytol.* 138: 441-449.

Gordon, D.C., K.E. Percy and R.T. Riding. 1998b. Effect of enhanced UV-B radiation on adaxial leaf surface micromorphology and epicuticular wax biosynthesis of sugar maple. *Chemosphere* 36:853-858.

Hall, J.P., L. Magasi, L. Carlson, K. Stolte, E. Niebla, M.L. Bauer, C.E. Gonzalez-Vicente and T. Hernándes-Tejeda. 1996. Health of North American Forests. North American Forestry Commission Report. Natural Resources Canada. Ottawa. 66 pp.

Hogsett, W.E., J.E. Weber, D. Tingey, A. Herstrom, E.H. Lee and J.A. Laurence. 1997. An approach for characterizing tropospheric ozone risk to forests. *Environ. Management* 21:105-120.

Karnosky, D.F. 1977. Evidence for genetic control of response to sulfur dioxide and ozone in *Populus tremuloides* Michx. *Can. J. For. Res.* 7:435-436.

Karnosky, D.F. 1981. Changes in eastern white pine stands related to air pollution stress. *Mitteilungen der Forstlichen Bundesversuchsanstalt* Wien 137:41-45.

Karnosky, D.F., Z.E. Gagnon, D.D. Reed and J.A. Witter 1996a. Growth and biomass allocation of symptomatic and asymptomatic *Populus tremuloides* clones in response to seasonal ozone exposures. *Can. J. For. Res.* 22:1785-1788.

Karnosky, D.F., Z.E. Gagnon, R.E. Dickson, M.D. Coleman, E.H. Lee and J.G. Isebrands. 1996b. Changes in growth, leaf abscission, and biomass associated with seasonable tropospheric ozone exposure of *Populus tremuloides* clones and seedlings. *Can. J. For. Res.* 16:23-27.

Karnosky, D.F., G.K. Podila, Z. Gagnon, P. Pechter, A. Akkapeddi, Y. Sheng, D.E. Riemenschneider, M.D. Coleman, R.E. Dickson, and J.G. Isebrands. 1998. Genetic control of responses to interacting tropospheric ozone and CO_2 in *Populus tremuloides*. Chemosphere 36: 807-812.

Karnosky, D., K.S. Pregitzer, M. Kubiske, D.D. Reed, J.G. Isebrands, R.L. Lindroth, D.R. Zak, G.E. Host and G. Hendrey. 1995. Forest-atmosphere carbon transfer and storage-II: Interacting effects of elevated CO_2 and O_3 on aspen forest ecosystems. *In* J. Elwood (ed.): DOE *Program for Ecosystem Research*: Research Summaries. 28-29.

Keeling, C.D., T.P. Whort, M. Wahlen and J. vander Plicht. 1995. Interannual extremes in the rate of rise of atmospheric carbon dioxide since 1980. *Nature* 375:666-670.

Kelly, M. 1990. Halting global warming. *In* J. Leggett (ed.): *Global Warming*, The Greenpeace Report. Oxford University Press, Oxford.

Kickert, R.N. and S.V. Krupa. 1990. Forest responses to tropospheric ozone and global change: An analysis. *Environ. Pollut.* 68:29-65.

Körner, C. 1995. Toward a better experimental basis for upscaling plant responses to elevated CO_2 and climate warning. *Plant, Cell and Environment* 18:1101-1110.

Kossuth, S.V. and R.H. Biggs. 1981. Ultraviolet-B radiation effects on early seedling growth of Pinacea species. *Can. J. For. Res.* 11: 243-248.

Kull, O., A. Sober, M.D. Coleman, R.E. Dickson, J.G. Isebrands, Z. Gagnon and D.F. Karnosky. 1996. Photosynthetic response of aspen clones to simultaneous exposures of ozone and CO_2. *Can. J. For. Res.* 16:639-648.

Loucks, O.L. 1994. Sun-scald on 1993 foliage of white pine (*Pinus strobus*) linked to UV-B radiation in Ohio and Ontario. *In* K. Heidorn and B. Torrie (eds.): Proceedings International Conference on Ozone Depletion and Ultraviolet radiation: Preparing for the Impacts. The Skies Above Foundation, Victoria. April 27-29, 1994. 39-47.

Lloyd, J. and G.D. Farquhar. 1996. The CO_2 dependence of photosynthesis, plant-growth responses to elevated atmospheric CO_2 concentrations and their interaction with soil nutrient status. I. General principles and forest ecosystems. *Funct. Ecol.* 10:4-32.

Madronich, S., R.L. McKenzie, M.M. Caldwell and L.O. Bjorn. 1995. Changes in ultraviolet radiation reaching the earth's surface. *Ambio* 24: 143-152.

Mackerness, S. A-H., B.R. Jordan and B. Thomas. 1997. UV-B effects on the expression of genes encoding proteins involved in photosynthesis. *In* P. Lumsden (ed.): *Plants and UV-B: Responses to Environmental Change*. Cambridge University Press, Cambridge. 113-134

Mankovska, B., S. Huttunen and R. Peura. 1989. The effect of air pollution from the Krompachy and Rudnany smelters on *Picea abies* Karst. *Ecologia* (CSSR) 8:49-58.

Manning, W.J. and A.V. Tiedemann. 1995. Climate change: Potential effects of increased atmospheric carbon dioxide (CO_2), ozone (O_3), and ultraviolet-B (UV-B) radiation on plant diseases. *Environ. Pollut.* 88: 219-245.

Martin, P.H. 1996. Will forest preserves protect temperate and boreal biodiversity from climate change? *For. Ecol. and Management* 85:335-341.

McLeod, A.R. 1995. Open-air fumigation of field crops: criteria and design for a new experimental system. *Plant Cell Environ.* 18:215-225.

McLeod, A.R. and K.K. Newsham. 1997. Impacts of elevated UV-B radiation on forest ecosystems. *In* P. Lumsden (ed.): *Plants and UV-B: Responses to Environmental Change*. Cambridge University Press, Cambridge. 247-281

Miao, S.L., P.M. Wayne and F.A. Bazzaz. 1992. Elevated CO_2 differentially alters the responses of co-occurring birch and maple seedlings to a moisture gradient. *Oecologia* 90:300-304.

Midgley, G.F., S.J.E. Wand and C.F. Musil. 1998. Repeated exposure to enhanced UV-B radiation in successive generations increases developmental instability (leaf fluctuating asymmetry) in a desert annual. *Plant, Cell and Environ.* 21: 437-442.

Miller, P.R. 1973. Oxidant-induced community change in a mixed conifer forest. *In* Air pollution damage to vegetation. *Advances in Chemistry Series* No. 122:101-117.

Miller, P.R. 1993. Response of forests to ozone in a changing atmospheric environment. *Angew.* Bot. 67:42-46.

Miller, P.R., J.R. Parmeter, O.C. Taylor and E.A. Cardiff. 1963. Ozone injury to foliage of *Pinus ponderosa*. *Phytopathology* 53:1072-1076.

Miller, P.R., K.W. Stolte, D.M. Duriscoe and J. Pronos. 1996. Evaluating ozone air pollution effects on pines in the western United States. USDA Forest Service Pacific Southwest Research Station General Technical Report PSW-GTR-155. 79 pp.

Mooney, H.A., B.G. Drake, R.J. Luxmoore, W.C. Oechel and L.F. Pitelka. 1991. Predicting ecosystem responses to elevated CO_2 concentrations. *BioScience* 41:96-1104.

Mueller-Starck, G. 1985. Genetic differences between "tolerant" and "sensitive" beeches (*Fagus sylvatica* L.) in an environmental stressed adult forest stand. *Silvae Genet.* 34:241-247.

Musil, C.F. 1995. Differential effects of elevated ultraviolet-B radiation on the photochemical and reproductive performance of dicotyledonous and monoctyledonous arid-environment ephemerals. *Plant, Cell and Environ.* 18: 844-854.

Musil, C.F. 1996. Accumulated effect of elevated UV-B radiation over multiple generationsof the arid-environment annual *Dimorphotheca sinuata DC.* (*Asteraceae*). *Plant, Cell and Environ.* 19: 1017-1027.

Naidu, S.L., J.H. Sullivan, A.H. Teramura and E.H. DeLucia. 1993. The effects of ultraviolet-B radiation on photosynthesis of different-aged needles in field-grown loblolly pine. *Tree Physiol.* 12: 151-162.

Newsham, K.K., M.N.R. Low, P.D. Greenslade, A.R. McLeod and B.A. Emmett. 1997. UV-B radiation influences the abundance and distribution of phylloplane fungi on *Quercus robur*. *New Phytol.*

Parker, P.G., A.A. Snow, M.D. Schug, G.C. Booton, and P.A. Fuerst. 1998. What molecules can tell us about populations: Choosing and using a molecular marker. *Ecology* 79:361-382.

Percy, K.E. and D.C. Gordon. 1998. Ultraviolet (UV-B) radiation and Canada's forests: State of science and risk. *J. Air Waste Mgmt. Assoc.* (In Press)

Percy, K.E. and S.I. Cameron. 1997. Forests. *In* Wardle, D.I., Kerr, J.B., McElroy, C.T. and D.R. Francis (eds.): Ozone Science: A Canadian perspective on the changing ozone layer. CARD 97-3, Environment Canada, Toronto. 97-101

Percy, K.E., D.F. Karnosky and B. Mankovska. 1998. Epicuticular wax chemical composition as a bioindicator of bredisposing ozone injury in three aspen clones growing along a natural gradient. *In* IUFRO 18th International Meeting for Specialists in Air Pollution Effects on Forest Ecosystems Forest Growth Responses to the Pollution Climate of the 21st Century. (In Press)

Peterson, D.L., M.J. Arbaugh and L. Robinson. 1989. Ozone injury and growth trends of ponderosa pine in the Sierra Nevadas. *In* R.K. Olson and A.S. Lefohn (eds.): Effects of air pollution on western forests. Transaction Series No. 16. Pittsburgh, PA. Air and Waste Management Association :293-308.

Powell, W., M. Morgante, C. Andre, M. Hanafey, J. Vogel, S. Tingey, and A. Rafalski. 1996. The comparison of RFLP, RAPD, AFLP and SSR (microsatellite) markers for germplasm analysis. *Molecular Breeding* 2:225-238.

Pye, J.M. 1988. Impact of ozone on the growth and yield of trees: A review. *J. Environ. Qual.* 17:347-360.

Pyle, J.A. 1997. Global ozone depletion: observations and theory. *In* P. Lumsden (ed.): *Plants and UV-B: Responses to Environmental Change.* Cambridge University Press, Cambridge. 3-30

Schnitzler, J.-P., T.P. Jungblut, C. Feicht, M. Kofferlein, P. Hutzler, U. Heinzmann, E. Schmelzer, D. Ernst, C. Langbartels and H. Sandermann, Jr. 1996. Tissue localization of UV-B screening pigments and of chalcone synthase mRNA in needles of Scots pine seedlings. *New Phytol.* 132: 247-258.

SCOPE. 1992. Effects of Increased ultraviolet radiation on biological systems. Scientific Committee on Problems of the Environment (SCOPE), Paris.

Shaltsa, H., E. Verykokidou, C. Harvala, G. Karabourniotis and Y. Manetas. 1994. UV-B protective potential and flavonoid content of leaf hairs of *Quercus ilex*. *Phytochem.* 37: 987-990.

Sheng, Y., G.K. Podila and D.F. Karnosky. 1997. Differences in O_3-induced superoxide dismutase and glutathione antioxidant expression in O_3 tolerant and sensitive trembling aspen (*Populus tremuloides* Michx.) clones. *Forest Gen.* 4:25-33.

Sullivan, J.H. and A.H. Teramura. 1988. Effects of ultraviolet-B irradiation on seedling growth in the *Pinacea*. *Amer. J. Bot.* 75: 225-230.

Sullivan, J.H. and A.H. Teramura. 1989. The effects of ultraviolet-B radiation in loblolly pine. 1. growth, photosynthesis and pigment production in greenhouse grown seedlings. *Physiol. Plant.* 77: 202-207.

Sullivan, J.H. and A.H. Teramura. 1992. The effects of ultraviolet-B radiation on loblolly pine: 2. Growth of field grown seedlings. *Trees Struct. and Function* 6:115-120.

Sullivan, J.H., A.H. Teramura and L.R. Dillenburg. 1994. Growth and photosynthetic responses of field-grown sweetgum (*Liquidambar stryaciflua: Hamamelidaceae*) seedlings to UV-B radiation. *Amer. J. Bot.* 81: 826-832.

Sullivan, J.H., B.W. Howells, C.T. Ruhland and T.A. Day. 1996. Changes in leaf expansion and epidermal screening effectiveness in *Liquidambar stryaciflua* and *Pinus taeda* in response to UV-B radiation. *Physiol. Plant.* 98: 349-357.

Taylor, G.E., Jr. and L.F. Pitelka. 1991. Genetic diversity of plant populations and the role of air pollution. *In* J.R. Barker and D.T. Tingey (eds.): *Air pollution effects on biodiversity*. Van Nostrand Reinhold. New York. 111-130.

Taylor, R.M., A.K. Tobin and C.M. Bray. 1997. DNA damage and repair in plants. *In* P. Lumsden (ed.): *Plants and UV-B: Responses to Environmental Change*. Cambridge University Press, Cambridge. 54-93

Wang, D., D.F. Karnosky and F.H. Bormann. 1986. Effects of ambient ozone on the productivity of *Populus tremuloides* Michx. grown under field conditions. *Can. J. For. Res.* 16:47-55.

Woodward, F.I., G.B. Thompson and I.F. McKee. 1991. The effects of elevated concentrations of carbon dioxide on individual plants, populations, communities and ecosystems. *Ann. Bot.* 67 (Supplement 1):23-38.

Yakimchuck, R. and J. Hoddinott. 1994. The influence of ultraviolet-B light and carbon dioxide enrichment on the growth and physiology of seedlings of three conifer species. *Can. J. For. Res.* 24: 1-8.

Zeuthen, J., T.N. Mikkelsen, G. Paludan-Muller and H. Ro-Poulsen. 1997. Effects of increased UV-B radiation and elevated levels of tropospheric ozone on physiological processes in European beech (*Fagus sylvatica*). *Physiol. Plant.* 100: 281-290.

Sexual reproduction response of boreal trees to a changing environment

Veikko Koski

Introduction

The most pessimistic predictions of abrupt destruction of boreal forests caused by climate warming can be ignored, fortunately. Provenance trials invariably show that trees not only tolerate slight warming but even benefit from a slightly warmer climate (e.g. Beuker 1996, Mátyás 1997). We may therefore conclude that current boreal forests are not directly endangered by climate change. However, this does not mean we should be indifference to possible consequences of changes in climate and other atmospheric factors. The crucial question is: Will the boreal forests be able to retain their reproductive and adaptive potential?

Firstly, it is a question about the regeneration of current stands at the end of their rotation age. Secondly, and most profoundly; can the sexual reproductive system maintain potential genetic diversity? Profuse reproductive potential combined with high rate of outcrossing and recombination are necessary conditions in adapting to new environmental conditions. Disturbances to the reproductive cycle affect these factors, resulting in harmful consequences. A well known example of the consequences of disturbances to reproductive cycles is that of predatory birds, the numbers of which have declined as a result of offspring death related to the accumulation of pollutants.

The predicted climate warming may be expected to benefit flowering and regeneration. Warm and dry summers do promote flowering, whereas low temperature and energy balances are limiting factors (e.g. Despland and Houle 1997). However, there is also a potential risk of increased frost injury. Several of the stages in the reproductive cycle of trees are vulnerable to frost injury. If thaw periods during the winter disturb the frost resistance of flower bud, meiosis and microsporogenesis may be disturbed, as reported in an early study by Eriksson (1968) on *Larix spp.*. Late frosts damage female strobili of *Picea abies* and *Pinus sylvestris.* Visible damage to male catkins of early flowering species, such as *Populus tremula* and *Alnus incana,* can be observed after capricious winters. Occasional failure of flowering and seed set is not disastrous to perennial plants that can produce offspring numerous times during their life span. However, if exceptional events become frequent, the risk of failure may grow significantly.

C. Mátyás (ed.), Forest Genetics and Sustainability, Volume 63, 125–130.

The potential risk is not distributed evenly. Some species are more vulnerable than others and some populations and individuals within species are more susceptible than others. From the viewpoint of genetic diversity and species diversity it is necessary to investigate how serious the potential risks are. The aim of this paper is to discuss the biological aspects of sexual reproduction in a changing environment.

1. Potential risks

The reproductive cycle of boreal trees requires more than one growing season, and consequently the reproductive organs are exposed to at least one winter period. The female reproductive structures of some species have to tolerate three or even four winters. No systematic surveys have been conducted on the injury and vulnerability of various stages by species. In heavily polluted areas, pollution probably causes direct injury, and possibly indirect disturbance via effects on the annual rhythm and frost hardiness. In the following, I will illustrate the matter with some specific examples.

Pinus sylvestris

The flower induction and the early development of the primordia take place during the latter part of the summer, previous to the flowering. In some exceptional years a chill period in September is followed by a long warm spell, which can result in the untimely development of some flower buds. However, flowering is quite normal the following year. Male meiosis and microsporogenesis take place in early spring, and pollen shedding in early summer. Female flowers continue their development in late spring and reach the receptive stage in early summer. After dormancy release, the ontogenic development is strictly temperature dependent. In some years late frosts cause injury on receptive female strobili. In the event of early warm spells and late frosts, the loss of female strobili can be quite significant. Low but non-lethal temperatures during the flower development retard flowering (Chung 1981), which may affect the mating pattern.

Another critical period is towards the end of the female cycle. It has been observed, in both seed orchards and normal stands, that seeds sometimes loose their viability during the winter. Thaw periods with rainfall increase the moisture of cones, and if temperatures again rapidly drop, the frost damage to seeds can be considerable.

Picea abies

The flower induction and early development of flower buds take place in late summer of the previous year. The ontogenetic development proceeds somewhat longer than in pine, but the gametophyte stage does not begin until the early spring of the flowering year. Both microsporogenesis and female meiosis and subsequent development of female gametophyte take place prior to flowering proper. As the processes take place earlier than at *Pinus sylvestris*, the exposure of spruce to late frost is quite common. Frost damage on receptive female strobili is rather common. In our pilot study (data not shown) on female strobili of *Picea abies* the critical temperature was found to be -3...-4 °C. No actual observations on the other stages of the reproductive cycle have been published.

Larix spp.

The seed production of larch is irregular and very poor in most years. This is especially true for continental species when transferred to more maritime climates. Low pollen production and fertility of pollen is perhaps the main limiting factor. The microsporogenesis of *Larix* is quite peculiar; it starts during the previous autumn, overwintering is at a prophase of meiosis, and meiosis continues very early in next spring. For continental species, the winter dormancy and frost resistance are often broken during thaw periods, and the pollen mother cells deharden. Eriksson (1968) reported frost injury on pollen mother cells as the main reason for poor seed crops of Asiatic larches in Sweden. Climate warming may reduce the frequency of seed production of larch in the Nordic countries further, and even the decline of seed crops on original sites.

Populus tremula

The species is an example of early flowering broadleaved species. In fact it mainly colonizes sites by means of root suckers, and the importance of sexual reproduction may be ignored. However, in terms of genetic diversity and adaptive potential to new environment conditions, sexual reproduction is important. The flower buds are formed during the previous growing season, and they continue their development very early. Warm periods during late winter result in the growth of male catkins, and failure of viable pollen formation spoils seed production.

These few examples indicate that early flowering species would be more susceptible to injury, particularly pollen and female flower production. However, systematic and empirical observations are very rare. The normal pathway of flowering and embryogeny of most boreal species has been described quite well (see e.g. compilation by Owens and Blake 1985), but hardly anything is known about frost tolerance and response to pollution during various stages of reproduction.

The crucial question is whether the dehardening and dormancy release of flower buds are more strictly temperature-dependent than those of vegetative buds? If yes, the results obtained from experiments on vegetative processes must be carefully considered.

Table 1. Examples of disturbances in sexual reproduction

Species	Potential risk	Observed damage	Reference
Pinus sylvesrtris	late frost injury on female strobili frost injury on wet seed in winter	disturbance and damage in transferred seed orchards loss of viability in nature and manipulated environment	Chung 1981, p 35 Savonen *et al*, 1995 Tanaka 1980
Picea abies	frost injury on female strobili, disturbances in microsporogenesis	loss of female strobili in seed orchards	unpublished experience
Larix sibirica *L. decidua* *L. leptolepis*	frost injury on male flower buds	disturbances in male sporogenesis, lack of viable pollen	Eriksson, 1968
Alnus incana	injury on male catkins	desiccation of male catkins	unpublished observation
Populus tremula	frost injury during microsporogenesis	non-viable pollen	experience of controlled crossings

Reproductive organs are tender and fragile during the proper flowering period. Tiny pistils of broadleaved trees are seemingly susceptible to pollutants. Experimental tests have shown harmful effects of SO_2 exposure on pollen germination (Karnosky 1974, Scholz et al. 1985). Conspicuous variation in pollen germination among trees was observed for *Pinus sylvestris* after exposure to 170 µg and to 270µg SO_2/ m^3 (Scholz et al. op cit.).

At the moment, these facts should be considered only as potential risks. Not until we can verify real risks are we justified in drawing any genetical conclusions. My hypothesis is that, for some species and in some regions, the threat is real. However, it is not an all-or-nothing situation; a good seed crop at the individual and population level has a certain probability. Risk analysis by means of simulations with actual temperature data and data with expected change can be done, as has been with vegetative bud phenology (e.g. Häkkinen and Hari 1988 and Hänninen 1991). The problem is, how much does the probability of failure increase, and where does the critical value lie .

2. Genetical consequences

The worst scenario for a species in a certain area is the loss of sexual reproductive potential, which would lead to a gradual disappearance of natural populations. This kind of scenario may seem quite unlikely, but we must keep in mind some exotic species in experimental plantations, which thrive well but do not regenerate. This scenario in a hundred years' perspective would lead to a change in species composition.

A more moderate impact on reproductive processes would screen the parents, perhaps even gametes. If only a small fraction of trees contribute to sexual reproduction, it would influence the flowering ecology and subsequently reduce the amount of sound seed even more than the proportion of parents. Selection for fertility would be quite strong and perhaps it would increase sexual asymmetry.

Fertility selection would lead to a considerable decrease in the effective population size of the current generation, which would have profound effects on the population structure. The lowered supply of non-related pollen enhances self pollination and subsequent inbreeding. It is impossible to predict how persistent this kind of pattern would be over generations. It may be just a „bottle-neck" incident that gives rise to an adapted population and normal reproductive biology after one or two generations, but the process can be more degenerative. We do not know the genetic background of flowering and its relations to vegetative performance.

Strong fertility selection must cause bias in gene frequencies. Decline of the effective population size and increasing inbreeding rates inevitably reduce the rate of outcrossing. Together, the phenomena would lead to a narrower gene pool and lower frequency of recombination. In terms of conservation of genetic diversity and adaptive potential of populations, such a direction would be undesirable. As a rule, the genetic systems of forest trees are well buffered and effective, and therefore I am not concerned of slight disturbances. However, there is a risk for minimal breeding populations. Due to economic constraints and based on theoretical ideal models, the suggested number of trees is sometimes quite small. What happens, if only one half of the trees selected for their vegetative performance are fertile?

Once more, I want to emphasize that the potential risks mentioned above are only speculative. There is clear need for empirical research into the critical stages of tree sexual reproduction. The starting points are there - the normal progress has been thoroughly described in many classical studies on flowering and embryology. Mature provenance trials provide excellent experimental material. There will be no universal solutions. Each species and environment respectively is a case of its own. As there are surprisingly few studies reported in literature, I make the following suggestions for research.

- Investigate if the regulation of the annual cycle of reproductive organs differs from that of vegetative parts.
- Determine vulnerable stages of the reproductive cycle of each species.

- Determine the threshold temperature during critical stages experimentally.
- Determine the location of the critical period in an ecophysiological time scale.
- Calculate the probability of frost damage with current and predicted temperature data.
- Assess genetic consequences of decreased reproductive success.

Literature cited

Beuker, E. 1996. Implications of climate adaptability in provenance trials with scots pine and norway spruce in Finland for the possible effects of climate warming. Ph.D. Diss. University of Joensuu, Faculty of Forestry.

Chung, M-S. 1981. Flowering characteristics of *Pinus sylvestris* L. with special emphasis on the reproductive adaptation to local temperature factor. *Acta Forestalia Fennica* 169:1-69.

Despland, E. and Houle, G. 1997. Climate influences on growth and reproduction of *Pinus banksiana (Pinaceae)* at the limit of the species distribution in eastern North America. *American Journal of Botany* 84:928-937.

Eriksson, G. 1968. Temperature responses of pollen mother cells in *Larix* and its importance for pollen formation. *Studia Forestalia Suecica* 63:1-131.

Häkkinen, R. and Hari, P. 1988. The efficiency of time and temperature driven regulation principle in plants at the beginning of the active period. *Silva Fennica* 22:163-170.

Hänninen, H. 1991. Does climatic warming increase the risk of frost damage in northern trees? *Plant Cell Environment* 14:449-454.

Karnosky, D. F. and G. R. Stairs, 1974. The effects of SO_2 on *in vitro* forest trees pollen germination and tube elongation. *J. Environm. Quality* 3:406-409.

Mátyás, Cs. (ed.) 1997. Perspectives of forest genetics and tree breeding in a changing world. IUFRO World Series Vol. 6, Vienna.

Oleksyn, J., Chalupka, W., Tjoelker, M.G. and Reich, P.B. 1992. Geographic origin of *Pinus sylvestris* populations influences the effects of air pollution on flowering and growth. *Water, Soil, and Air Pollution* 62:201-212.

Owens, J.N. and Blake, M. D. 1985. Forest tree seed production. A review of the literature and recommendations for future research.. Information report PI-X-53. Petawawa National Forestry Institute, Canadian Forestry Service.

Savonen, E.-M., Pulkkinen, M. and Leinonen, K. 1995. Frost resistance of overwintering Scots pine seeds. Abstracts of Poster Presentations. Fifth International Workshop on Seeds. The University of Reading. 11.-15.9.1995

Scholz, F, Vornweg, A. and Stephan, B.R. 1985. Wirkungen von Luftunreinigungen auf die Pollenkeimung von *Forstarchiv* 56 :121-124.

Tanaka, Y. 1980. Effect of freezing temperatures on noble fir seed during cone storage. Proceedings of the international symposium on forest tree seed storage, Petawawa National Forestry Institute, Chalk River, Ontario, Canada, 23.-27.9.1980.

5. SUSTAINED MANAGEMENT OF GENETIC RESOURCES – THE IMPACT OF FORESTRY

Conserving genetic diversity of forest trees in managed landscapes

Timothy J.B. Boyle

Introduction

With tropical forests estimated to contain more than 50% of all terrestrial biodiversity (Myers 1988), the ways in which these forests are used can have very significant implications for global biodiversity, and the direct and indirect benefits resulting from its conservation and use. From the perspective of genetic diversity, the large numbers of species obviously also implies very high genetic diversity supported by molecular studies (e.g. Loveless 1992). As these forests are so critical for the survival and welfare of so many people in the tropics, conservation of biodiversity and/or genetic diversity must be compatible with continued economic benefits being derived from the forests. It is necessary to achieve a balance between these potentially conflicting demands to ensure that all facets of sustainability - ecological, economic, and social are met. Such a balance requires an understanding of the impacts of harvesting on the genetic resources of tropical forests.

1. Harvesting and disturbance

The debate about mankind's role in nature - whether as an integral part of natural systems, or separate from them, continues (e.g. Visser 1992). In this paper, in order to restrict the topic under discussion, the terminology used treats human activities as distinct from natural processes. Thus, "disturbance" is used to mean the impact of human activities on forests, as distinct from the consequences of "natural" events associated with wind, water, or natural fire.

Human activities in forests have the effect of modifying the size and age structure of the tree species, and potentially altering genetic structure and

C. Mátyás (ed.), Forest Genetics and Sustainability, Volume 63, 131–146.

levels of genetic diversity. The most dramatic human impact on forests comes from large-scale conversion to other land uses. This has occurred historically throughout huge areas of Europe, and much of North America, and at the present time is impacting substantial areas of tropical forests in South America, Africa and Asia. The impact of land conversion on genetic diversity of forest species is obviously enormous, resulting as it does in dramatically smaller population sizes, and increased isolation of populations. For the purposes of this paper, attention will mainly be focussed on those human activities which impact the forest without changing the vegetation cover. Such activities usually involve the extraction, or harvesting of products from the forest. The variety of products harvested from tropical forests is immense: wood, fruit and other foods, medicinal plants, construction materials, and many more. Commercial harvesting of timber through logging is a major form of disturbance in many areas.

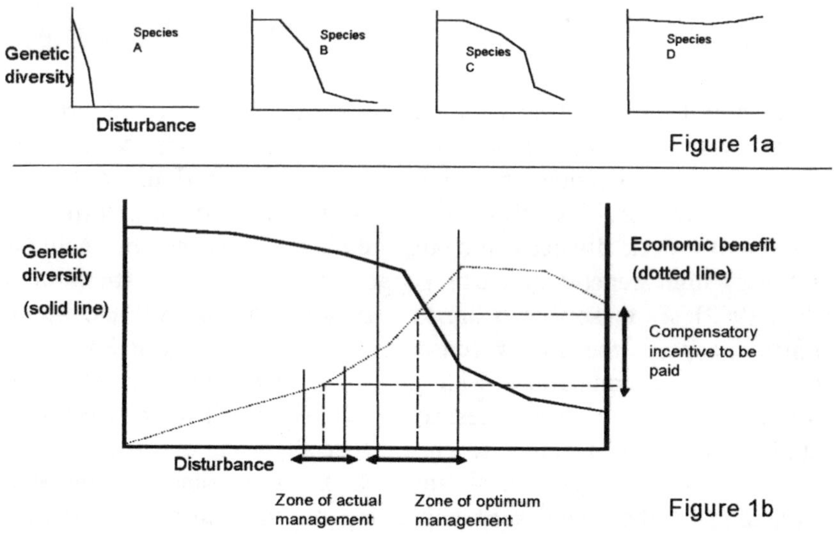

Figure 1. In Figure 1a, possible relationships between intensity of disturbance (x axis) and genetic diversity (y axis) are shown for four possible "model" species. Species A is highly sensitive to disturbance, and can only be conserved in strictly protected areas. Species D is insensitive to disturbance (or may actually benefit), while Species B and C exhibit different buffering capacity and threshold tolerance levels. However, only through the multidisciplinary research that yields Figure 1b can the consequences of altering disturbance intensities be fully understood, and socially sensitive conservation strategies designed. Disturbance is measured in relative, rather than absolute terms, using an undisturbed location as the reference point.

Figure 1 shows a generalized relationship between economic benefits from the forest and genetic diversity of its component species. As intensity of use increases, genetic diversity of many species will be affected, with some species unable to tolerate disturbance becoming locally extinct. The optimum level of management in economic terms may be substantially more intensive than the optimum for conservation of diversity. If, for political/policy reasons, a greater effort is considered necessary to conserve biodiversity (either at the genetic or species level), the need for compensation is inescapable, and research is required to develop an approach for establishing appropriate levels of compensation. Numerous examples show (Fairhead and Leach 1993; McNeely 1993; Pinedo-Vasquez and Padoch 1993) that unless local people and other affected forest users can gain economic benefits from protected areas, or are adequately compensated for their loss, there is little likelihood that effective long-term protected area conservation can be achieved.

Consequently, the forest manager – whether a large industrial organization, community group, or individual, needs to consider the impact of forest practices on genetic aspects of sustainability. Some of most common forms of human interventions in forests, and their impacts on genetic diversity, are outlined below.

1.1. Logging

Logging directly impacts the genetic resources of the commercial species through changes in their population age and density distributions. Non-commercial species, both plant and animal, will also be affected, either due to intentional or accidental damage during the logging process, or due to the resulting changes in environmental conditions. Forest-dependent animal species will also be affected.

1.2. Grazing

Grazing of livestock is a very common practice, especially in tropical seasonal forests. Grazing will have a thinning effect on regenerating vegetation and changes the population density of grazed species. Since grazing may also compact soils and alter stand structure, it may also induce significant environmental changes. It could also affect populations of forest-dwelling herbivores.

1.3. Fire

Intense fire inflicts heavy mortality and reduces population sizes. Although small fires may differentially affect species or genotypes, devastating fires can potentially kill all individuals. Although much attention is currently focussed on "slash-and-burn" agriculture in the tropics, this agricultural system is often practiced sustainably where traditional approaches still apply. Where large-scale in-migration occurs spontaneously or with government encouragement, or where the policy is to change land use from forestry to other uses, prescribed burning can be extensive and cause widespread damage. However, this type of activity fall under the category of land conversion which, as mentioned earlier, is not a focus of this paper as the impacts are obvious and catastrophic.

1.4. Harvesting non-timber products

Harvesting parts of trees or shrubs may involve their reproductive structures such as fruit or seed, and directly reduce the effective size of the pool of reproducing parents. Fruit and seed are often harvested very intensively, resulting in severe genetic impacts. When whole plants are harvested, the effects of reduced population size will be genetically significant. Harvesting of animals (i.e., hunting and fishing), may have similarly severe impacts as harvesting of whole plants. Fuelwood, including charcoal, is often classified as a non-timber forest product. As for other NTFP's , traditional systems of collection may have been sustainable, but social and policy changes may result in serious problems. In-migration or in situ population growth may result in dramatically larger settlements, which may lead to non-sustainable levels of production, or government programmes to settle previously nomadic groups may also lead to non-sustainable levels of harvesting, as exploitation which was previously spread over a large area becomes concentrated in a much small area (Murali et al. 1996; Uma Shanker et al. 1996).

There are an increasing number of studies of the impacts of human activities on genetic diversity, including several in this volume (e.g. Wickneswari et al.; Glaubitz and Moran, this volume). While most such studies focus on a single species, or a small group of similar species, it is important to recognize that impacts will vary according to the life history strategies of the species and the type of product harvested. For example, at Huay Kha Khaeng in Thailand, the impact of timber harvesting has been studied for two dipterocarp species, *Shorea siamea* and *Dipterocarpus obtusifolius* that are contrasting, both in terms of ecology, and economic uses. *Shorea siamea* is cut by local people for construction timber. The species is predominantly pollinated by a species of weak-flying bee. Along a gradient of increasing disturbance, the density of mature trees falls by a factor of 10, while at the same time the density of flowers per tree increases dramatically. There appears to be a threshold point where the distance between adjacent flowering trees exceeds the distance normally covered by the pollinator (Ghazoul *et al.* 1998). Consequently reproductive success and outcrossing rates decline rapidly and there is a parallel decrease in genetic diversity, both among the embryos and among mature individuals (Chaisurisri *et al.* 1998). In contrast, outcrossing rates for *Dipterocarpus macrocarpus* are constant along the gradient, as tests have shown it to be almost completely self-sterile. However, reproductive success increases on the more disturbed sites. It appears as though this is due to the effective control of fires within the wildlife sanctuary, which results in much higher losses of flowers and fruit to insects and other pests.

The most disruptive influence of harvesting is likely to derive from harvesting of reproductive structures. An example of this comes from southern India, where fruits of *Phyllanthus emblica* are harvested for food and as the base for many traditional medicines. In two locations, samples collected from areas subject to three different intensities of collection, and separated by only a few kilometres, showed distinct differences in characters such as germination percentage, seedling vigour, and seed predation (e.g. Murali *et al.* 1997). The samples collected from the most highly disturbed site had the lowest germination percent and vigour and the highest rates of seed predation. Plotting weightings of seedling multilocus allozyme phenotypes on the first two principle components of allozyme variation also showed clear separation of seedlings from the three intensities of collection. While the differentiation in seedling allozyme phenotypes could simply be due to isolation by distance, the reproductive strategy of *Phyllanthus emblica*, involving wind pollination and animal seed dispersal, suggests that this is unlikely to be the case.

2. Effects on Genetic Dynamics

The impacts of human activities on genetic parameters, as illustrated in the examples above, are attributable to changes in the basic evolutionary processes (Figure 2). These processes are: random genetic drift, selection, migration, mating and mutation. Mutation is excluded from consideration since we are primarily concerned with relatively short term processes of less than ten generations and with effects of forest interventions that are not likely to significantly change mutation rates.

All these evolutionary processes affect the levels and distribution of genetic variation and the present state of the genetic resource is a result of their joint effects. Variation within populations may be increased by migration if populations were originally distinct, but at the expense of variation between populations and possibly with decreased fitness. The size of the recipient population may be increased but the adaptability of the resultant genetic variation is dependent on the forms of selection and migration that also occur.

It is also clear that the present condition is not necessarily an optimum state for adaptation to future conditions. However, regardless of any difficulties inherent in defining an optimum condition - so-called undisturbed forests may not exist - we assume that if the processes that maintain the levels and distribution of genetic variation remain intact, then the biological capabilities for adaptability are expected to be in at least as good a condition in the immediate future as they are now. The exact levels and distribution of genetic variation would be expected to have changed during evolutionary history, but our central concern is for the processes and the structures necessary to maintain those processes, and not for the present distribution of variation as an end in itself.

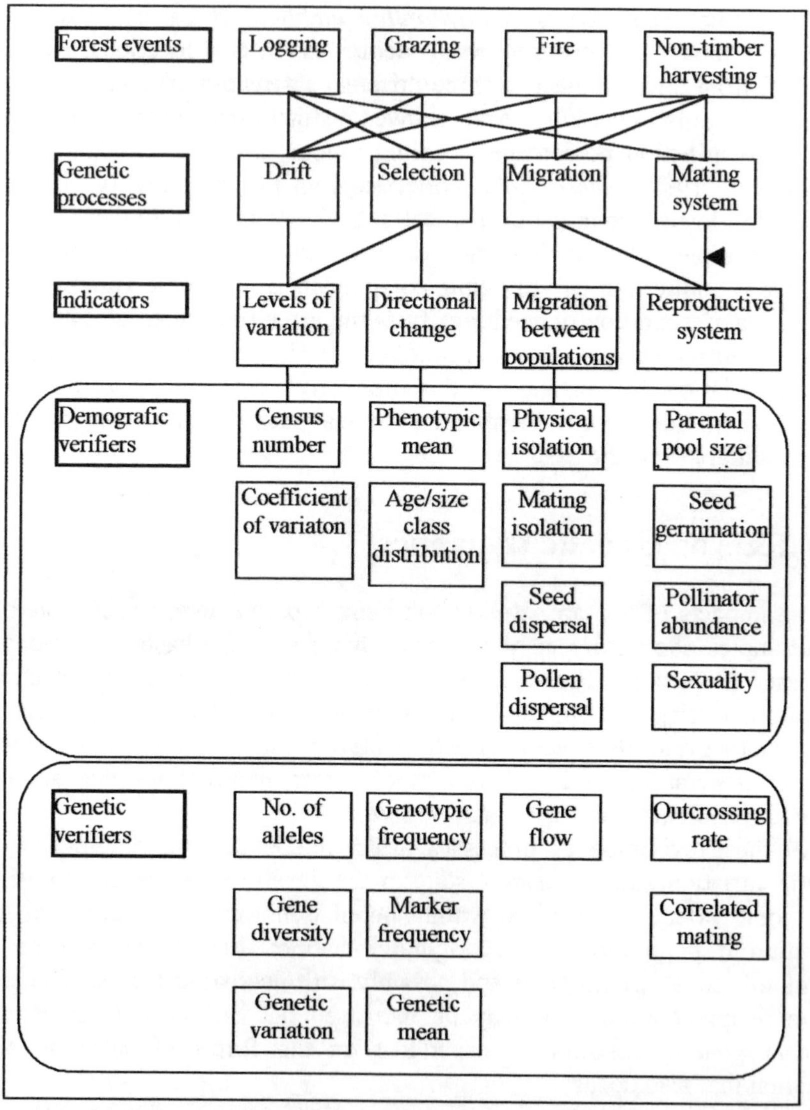

Figure 2. Relationship of forest events to genetic processes and their indicators.

In the absence of any other information, it has been proposed that population sizes of several hundred or a few thousand are sufficient to maintain quantitative genetic variation and to save even rare alleles indefinitely (Franklin 1980). For species with random mating and without interpopulation divergence, simple counts of reproductive adults may be sufficient for assessing genetic viability, but such rules are imprecise, and are of limited value for the complex structure of many forest tree populations. Local populations may be much smaller and transient, and genetic divergence among populations may be an important adaptive feature of the species to conserve. Especially in the species-rich tropics, most species are widely dispersed and contiguous populations of large size are unusual. Therefore, it is necessary not only to measure population size parameters, but to monitor the effects of forest practices on adaptive genetic processes.

The effects of various types of biotic and abiotic changes that are evident at the forest stand level usually affect, or are the result of, more than one of the genetic processes. For example, changes in the abiotic environment can change relative genotypic viabilities but can also affect pollinator behaviour and other inter-species interactions. Because biotic interactions, such as in pollinator-plant behaviour, are generally more complex in the tropics than in temperate ecosystems, the indirect effects of several, simultaneous environmental changes are likely to be more complex and to operate on a variety of time scales.

2.1. Drift

Random genetic drift is characteristic of small populations, and especially of populations undergoing fluctuations in size with frequent reductions to small numbers (Hartl and Clark 1989). Forest operations or fire may effectively reduce population sizes, leaving few individuals contributing seed or pollen. In addition, sources of migration may be cut off, and the effectiveness of pollinators may be reduced. Frequently, the number of mating individuals and their reproductive output may be much smaller than the total number of adults or number of flowers. Furthermore, inequities in mating frequencies and non-random mating generally further reduce effective population size. If the reproductive population remains small, the random drift of alleles will severely reduce genetic variation that may be significant for future adaptability for several generations.

2.2. Selection

Selection can simultaneously reduce genetic variation, and change population averages (Hartl and Clark 1989). Both of these shifts may influence future adaptability. Events that lead to genetically significant selection may be unintentional, for example, a single event such as harvest, could result in changes in several traits, not only those considered desirable for harvest. Thus, traits such as juvenile growth rates, reproduction age, and pollinator behaviour, can be affected by logging as well as by the timing and design of harvests. Furthermore, selection on one trait may also induce genetically correlated changes in other traits and genetic processes. Hence, even simple selection events may have multiple effects on the distribution and dynamics of genetic variation that may affect present and future viability and reproduction.

2.3. Migration among populations

The form and rates of migration mediate the genetic divergence between partially independent populations of a species (i.e., meta-populations) and affects the distribution of genetic variation. The creation of forest islands by fires or creation of patch openings within continuous forests, can change migration rates and the relative sizes and patterns of openings vs. forest cover can strongly influence genetic divergence. An additional factor is the proximity of an adequate recruitment pool of immigrants.

2.4. Mating system

The levels of heterozygosity, the relative importance of vegetative propagation, and the exposure of mutants to selection, are all affected by changes in the mating system (Hartl and Clark 1989). Forest practices such as logging, thinning, or conversion, can change population density, pollen and seed dispersal rates and dispersal mechanisms, and timing of flowering, which change the mating system. Populations that are usually large, but suddenly undergo severe reductions may also be exposed to inbreeding depression and loss of viability, reducing their immediate capacity for survival.

3. Criteria and indicators

Criteria and indicators are tools for collecting and organizing information in a way that is useful to assist in evaluating and implementing sustainable forest management. They have proved to be a popular assessment tool, but this popularity has, in itself, led to some confusion resulting from the profusion of "jargon" terms associated with the large number of proposed systems of Criteria and Indicators.

The assessment of sustainability, including conservation of biodiversity, is essential at both national and local scales. Local-scale assessment is needed because it is the management decisions made in individual forest management units that ultimately determine the fate of forest biodiversity. However, comprehensive assessment is not possible solely at a local scale. For example, the establishment of a system of "protected areas" is an essential for conservation purposes as one aspect of sustainable forest management. It is neither necessary nor possible for each forest management unit to include protected areas, so national scale Criteria and Indicators are required to provide an assessment of such factors.

National-level Criteria and Indicators, which summarize the status of forest management over the entirety of a country's forest area, have been developed under a number of international initiatives. Several systems of Criteria and Indicators which reflect sustainability at a local level, corresponding with individual forest management units, have been developed by organizations such as the Rainforest Alliance, the Soil Association, and the African Timber Organization. Yet Prabhu et al. (1996) concluded that the entire question of biodiversity assessment has been avoided in the sets of criteria and indicators proposed to date, instead the emphasis has been to ensure that management conforms with good forest stewardship. This is an understandable and pragmatic first step towards the development of a sustainability assessment tool. Nevertheless the question how forest management actually impacts on biodiversity still begs answer.

"Sustainability" in the context of conserving forest biodiversity must be considered in relation to a forest management plan. The first requirement for sustainability is therefore the existence of an articulated management plan. Key questions are:

- is the new level of biodiversity stable or continuing to decrease (or increase); and
- is the new level adequate to support all ecological processes (i.e. above a critical threshold level)?

These questions must consider multiple scales both because the definition of biodiversity requires this, and because the processes affecting biodiversity operate at different scales.

Criteria and Indicators are most powerful if they can be used as a tool in "adaptive" management, by which forest managers (be they corporations, communities, or individuals) can regulate the quality of their forest management, and adapt management practices accordingly. This implies that that members of the assessment team will be not be "scientific experts", and that conservation of biodiversity will need to be assessed in only a few (± 10) person-days. These considerations dictate that the most important characteristic of an effective indicator will be the practicality of assessment in a very short period. This is a serious constraint for assessment of "conservation", which implies a need to consider temporal dynamics, and for the assessment of genetic diversity, which is an extremely difficult concept for non-specialists. It is largely a consequence of this constraint that Prabhu et al. (1996) concluded, on the basis of field trials in different parts of the world, that most of the currently proposed biodiversity Criteria and Indicators were ineffective.

Stork *et al.* (1997), building on concepts originally developed by Namkoong et al. (1996), suggested that changes in biodiversity may be assessed indirectly through assessment of the processes that maintain and generate biodiversity. For example, successful pollination is essential for plant species that reproduce sexually. Changes in pollination success will indicate changes in the diversity, abundance or behaviour of pollinating species as well as the structure of future plant (and animal) communities.

Table 1. Genetic consequences of different types of utilization

	Drift	Direct selection	Indirect selection	Mating system	Gene flow
Logging - commercial species	X	X		X	
Logging- non-commercial species	X		X	X	
Grazing	X				
Fire	X				X
NTFP - reproductive		X		X	X
NTFP - non-reproductive		X			
NTFP - whole individual	X		X		

In developing genetic criteria, indicators, and verifiers, it is important to emphasize that the conservation of evolutionary processes is a necessary biological criterion by which the sustainability of forests can be judged. To conserve those processes, Namkoong *et al.* (1996) defined four genetic indicators that are necessary for sustainability. For each indicator, they listed several ways to assess the status and dynamics of the indicator, through measures termed "verifiers". Two types of verifiers were described, those that can be obtained by field estimation of demographic parameters and those requiring laboratory analysis of genetic parameters.

Namkoong *et al.* (1996) proposed that a single criterion for conservation of genetic resources could be expressed as "The conservation of the processes that maintain genetic variation". Different types of human activities are likely to affect several of the processes that maintain genetic variation as indicated in Table 1. The processes of genetic change have effects that can be measured in the level and structure of genetic variation and in the processes themselves. These measures can serve as indicators of whether genetic variation and the processes of evolution are being maintained. The four indicators of the genetic processes that were proposed include:

1. Level of genetic variation;
2. Directional change in gene or genotypic frequencies;
3. Gene migration between populations;
4. Mating system processes.

Generally more than one indicators need to be assessed for each of the genetic processes. To be most useful to forest managers, the effects of forest-level events on genetic processes can be combined with the indicators of those processes as given in Table 1, to provide managers with tools to warn of changes in genetic processes which arise from particular management practices or by accident. This is shown in Table 2, where indicators are directly linked to forest practices and other forest level events. Table 2 allows forest managers to determine which genetic indicators need to be assessed based only on their knowledge of forest practices in the forest management unit. Thus, genetic expertise is unnecessary for the assessment of genetic sustainability.

For example, if logging is taking place, the following indicators need to be assessed:

- Levels of genetic diversity are maintained;
- There is no directional change in genic/genotypic frequencies; and
- There are no changes in the mating system.

Table 2. Indicators relevant to different forest operations.

	Indicators			
Modes of exploitation	Levels of genetic diversity are maintained	There is no directional change in genic/genotypic frequencies	There are no changes in gene flow/migration	There are no changes in the mating system
Logging - commercial species	X	X		X
Logging- non-commercial species	X	X		X
Grazing	X			
Fire	X		X	X
NTFP - reproductive	X	X	X	X
NTFP - non-reproductive	X	X		
NTFP - whole individual	X	X		

For each of the indicators, the verifiers may be parameters of the genetic system that directly reflect genetic processes or other factors of the reproductive system. For example, genetic parameters such as gene frequencies can serve as measures of the level of genetic variation, and differences in frequencies can be used to estimate population differentiation. These data can also be analyzed to estimate parameters of the mating system, such as the outcrossing rate, selection effects, and migration rates. Alternatively, verifiers may be estimates of demographic statistics from which we can infer genetic processes. For example, changes in population sizes, and age class structure may reflect changes in the levels and distribution of genetic variation as well as the mating system. However, it is important to note that measures based on censuses are likely to be significantly biased. Methods to overcome this problem have been proposed, for example by Lindgren et al. (1996), who proposed "status number" as a measure of the effective size of the breeding population, being calculated as half the inverse of the average coancestry. The distinction between genetic and demographic verifiers reflect the point made by Brown *et al.* (1997) that the choice of verifier will always be a trade-off between high information content (generally genetic verifiers) and ease of measurement (generally demographic verifiers). Verifiers proposed by Namkoong *et al.* (1996) are shown in Table 3.

Table 3. Indicators and verifiers of conservation of genetic diversity (from Namkoong *et al.* 1996)

Indicator	Demographic verifiers	Genetic verifiers
1: LEVELS OF VARIATION		
	1.D.1 Census number of sexually mature individuals This number accounts for all individuals expected to be sexually mature at present time, according to their age or size.	1.G.1 Number of alleles Allelic richness is drastically affected by changes in populations sizes, and could be assessed on a reduced number of loci that are polymorphic in the reference population.
	1.D.2 Census number of reproducing individuals If fecundity and fertility data can be obtained, then this number corresponds to the subset of 1D1 that is expected to contribute to the next generation.	1.G.2 Gene diversity If multi-allelic and codominant markers are not available, then levels of diversity could be assessed at anonymous and randomly distributed markers
	1.D.3 Coefficient of phenotypic variation Assessments of *in situ* phenotypic variation of important adaptive traits will provide some crude indication of genetic variation of the same traits.	1.G.3 Genetic variation Additive variance of adaptive important traits could be estimated in common garden experiments.
2: DIRECTIONAL CHANGE IN ALLELE OR GENOTYPE FREQUENCIES		
	2.D.1 Phenotypic Shifts. If phenotypic mean differences between a reference population and the affected population are noticed, then a genotypic change may have occurred that may cause a change in the economic or ecological sustainability of the population.	2.G.1. Genotypic Frequency Shifts. Differences in genotypic frequency distributions among cohorts may be measured as changes in heterozygosity
	2.D.2. Age/Size Class Shifts. If differences occur in the age or size class distribution exist between a reference population and the affected population, genic level changes may also be occurring in age related selection.	2.G.2 Marker Frequency Shifts. In some cases, selection will result in directional changes in genotypic frequencies at some marker loci among generations or from a reference population
	2.D.3. Environmental Shifts. If differences in the environment that affect either viability or reproduction are caused by forest level events, then selection may be so strong as to affect population viability.	2.G.3 Genetic Mean Shifts. Most direct but most costly of time and effort is the establishment of a genetic estimation experiment analyzing differences in traits of adaptive significance.

Table 3, continued

3: MIGRATION AMONG POPULATIONS		
	3.D.1. Physical Isolation. Reductions in degree of physical isolation by loss of neighbors or by easing of physical or biotic barriers to pollen or seed migration.	3.G.1 Gene Flow. Gene flow can best be estimated by tracing neutral marker loci and estimating statistics such as Fst, Gst, d, Dj, and theta statistics.
	3.D.2. Mating Isolation. Breaking mating isolation by changes in age class distribution and by loss of neighbors that may increase the frequency of mating among distant parents.	
	3.D.3. Seed Dispersal. Changes in seed dispersal distances due to physical or biotic factors affecting seed dispersal vectors.	
	3.D.4. Pollen Dispersal. For plant species, the effective distance of pollen dispersal may be increased by factors increasing pollen vector distances either by changes in the pollen vector or in inter-individual distances.	
4: REPRODUCTIVE SYSTEM		
	4.D.1. Parental pool size Modification of the density of potential parents indicates possible changes in future mating success	4.G.1. Outcrossing rate Modification of the outcrossing rate reflects change in the potential inbreeding, and possible increases in inbreeding depression.
	4.D.2. Seed Germination Modification of the percentage of seed germination (or of filled seed count) relative to the reference population, can reflect changes in mating success.	4.G.2. Correlated mating Even if the outcrossing rate is not modified, modification of the gene flow within a population can strongly affect the genetic makeup of the next generation.
	4.D.3. Pollinator Abundance Modification of pollinator abundance or composition during the mating season indicates potential changes in the male gamete pool.	
	4.D.4. Sexuality Modification of the resources allocated to male and female functions portends changes in mating success.	

4. Prediction at non-sampled points

There is increasing recognition of the need to deal with forest management issues on a landscape-wide basis. This is in contrast with most research and monitoring which has tended to be at a stand level. Biodiversity must also be dealt with on a landscape-wide basis. There are a number of reasons for this. First, there are processes that contribute to the long term viability of animal species in a landscape that are relatively large-scaled. For example, many animal species occur in landscapes as disjunct populations. Their persistence and genetic viability are affected by the meta-population dynamics that determine the extinction of individuals in a patch and the dispersal of individuals between patches (for example, Lindenmayer and Possingham 1994).

A further and critical premise of landscape ecology is that landscapes are not environmentally homogeneous. Rather, they comprise often surprisingly broad and complex environmental gradients, which exert a fundamental control on the nature and distribution of biodiversity in a landscape.

Conventional forestry tends to smooth out the inherent environmental heterogeneity found in landscapes by planning and managing to uniform, "lowest common denominator" guidelines. There is widespread recognition that such an approach is neither necessary, nor optimal, but methods need to be developed to facilitate more flexible and environmentally sensitive management.

The need for accurate predictive information for forest management planning, including the location and shape of conservation areas, is vital. The scale of operations of concessions in Kalimantan, for example, makes detailed biodiversity surveying over more than a fraction of the area impractical. The ability to predict species occurrence and habitat type from environmental information that can be obtained from readily available data, GIS and remote sensing would be a valuable management tool. In addition such prediction would also be of use in developing sampling designs for further biodiversity studies and in PSP location for growth and yield research. Elevation is known to be an important factor in determining species composition, but locally topography may also have significant influence on the distribution and availability of ecological resources and resultant biodiversity of an area. There is a need therefore to undertake biodiversity surveys inclusive of topographic processes. The results will be more robust, relevant to management, and have greater applicability.

5. Conclusions

Conservation of genetic diversity in managed landscapes requires both an understanding of the impacts of human activities on genetic diversity and an ability for non-specialist forest managers to measure and monitor genetic diversity under an adaptive management regime. Our understanding of the impacts of human activities is expanding as more empirical studies are undertaken. However, results from these studies also indicate that many factors related to life history strategies will determine the actual impact for any given species. In order to generate greater predictive power for the benefit of forest managers, systems analysis offers a valuable approach to understanding complexity and generating more accurate predictions. Systems analysis can integrate existing tools, such as spatial modelling, with results from empirical studies of the type described in this paper and others in this volume.

The second requirement, a practical and efficient tool for measurement and monitoring is offered through criteria, indicators and verifiers. More work is required to refine verifiers and to undertand the relationship between ecological or demographic factors and genetic processes. The relationships described by the matrices in Tables 1, and 2, are very crude and their refinement to include levels of intensity would be very useful in a decision support system. More precise surrogate measures can be developed and critical levels of parameters need to be established. Namkoong *et al.* (1996) assumed independence among the factors and hence had to set conservative, independent critical values. It can be imagined that factors can interact in such ways that high levels of one indicator can compensate for low levels of another. The joint factor evaluation of risk and interpretation of information from multiple indicators need to be developed.

Literature cited

Chaisurisri, K., Wungplong, P., Liewlaksaneeyanawin, C., and Boyle, T.J.B. 1999: Impacts of disturbance on genetic diversity of some forest species in Thailand. In preparation.

Fairhead, J. and Leach, M. 1993. Contested Forests: Modern Conservation and Historical Land Use of Guinea's Ziama Reserve. *Working paper 7. Connaissance et Organisation Locales Agro-ecologiques (COLA),* Kissidougou (Guinea).

Franklin, I.R. 1980. Evolutionary change in small populations. Pages 135-149 *In* Soulé, M.E. and Wilcox, B.A. (eds.): Conservation Biology: An evolutionary-Ecological Perspective. Sinauer Assocs., Sunderland, Mass., USA.

Ghazoul, J., Liston, K.A. and Boyle, T.J.B. 1998. Disturbance-induced density-dependent reproductive success in a tropical forest tree. *Journal of Ecology* 86(3): 462-473.

Hartl, D.L. and Clark, A.G. 1989. Principles of Population Genetics. Sinauer Associates, Sunderland, Mass.

Hutchinson M.F. 1991. Climate analysis in data sparse regions. Pages 55-71 *In* Muchow, R.C. and Bellemay, J.A. (eds.): Climatic risk in crop production. CAB International, Wallingford.

Lindenmayer D.B. and Possingham H.P. 1994. The risk of extinction: ranking management options for Leadbeater's Possum using Population Viability Analysis. Centre for Resource and Environmental Studies, Australian National University, Canberra, 204 pp.

Lindgren, D., Jefferson, P., and Gea. L. 1996. Loss of Genetic Diversity Monitored by Status Number. *Silvae Genetica* 45: 52-59.

Loveless, M.D. 1992. Isozyme variation in tropical trees: patterns of genetic organization. Pages 67-94 *In* Adams, W.T., Strauss, S.H., Copes, D.L. and Griffin, A.R. (eds.): *Population Genetics of Forest Trees*. Kluwer Academic, Dordrecht.

McNeely, J.A. 1993. Economic Incentives for Conserving Biodiversity: Lessons from Africa. *Ambio* 22: 144-150.

Murali, K.S., Uma Shanker, Uma Shaanker, R., Ganeshaiah, K.N. and Bawa, K.S. 1996. Extraction of non-timber forest products in the forests of Biligiri Rangan Hills, India. 2. Impact of NTFP extraction on regeneration, population structure, and species composition. *Economic Botany* 50: 252-269.

Myers, N. 1988. Tropical forests and their species: Going, going ...? *In* Wilson, E.O. (ed.): *Biodiversity*. National Academy Press, Washington, D.C. 28-35

Namkoong, G., Boyle, T.J.B., Gregorius, H.R., Joly, H.I., Savolainen, O., Ratnam, W. and Young, A.J. 1996. Criteria and Indicators for Impacts of Forest Management on Genetic Variation. Testing Criteria and Indicators for Sustainable Management of Forests. *CIFOR Working Paper* 10. Bogor, Indonesia

Nix H.A. 1982. Environmental determinants and evolution in Terra Australis. Pages 47-65 *In* Evolution of the flora and fauna of arid Australia. Peacock Publications, Sydney.

Pinedo-Vasquez, M. and Padoch, C. 1993. Community and Governmental Experiences Protecting Biodiversity in the Lowland Peruvian Amazon. Pages 199-211 *In* Potter, C.S., Cohen, J.I. and Janczewski, D. (eds.): Perspectives on Biodiversity: Case Studies of Genetic Resource Conservation and Development. AAAS Press, Washington, DC.

Prabhu, R., Colfer, C.J.P., Venkateswarlu, P., Tan, L.C., Soekmadi, R., Wollenberg, E. 1996. Testing criteria and indicators for sustainable management of forests: final report of phase I. CIFOR, Indonesia.

Stork, N.E., Boyle, T.J.B, Dale, V., Eeley, H., Finegan, B., Lawes, M., Manokaran, N., Prabhu, R. and Soberon, J. 1997. Criteria and Indicators for Assessing the Sustainability of Forest Management: Conservation of Biodiversity. *CIFOR Working Paper 17*. Center for International Forestry Research, Bogor, Indonesia.

Uma Shankar, Murali, K.S., Uma Shaanker, R., Ganeshaiah, K.N. and Bawa, K.S. 1996. Extraction of non-timber forest products in the forests of biligiri rangan hills, India. 3. Productivity, Extraction and prospects of sustainable harvest of amla *Phyllanthus emblica* (Euphorbiaceae) *Economic Botany* 50: 270-279

Visser, M.B.H. 1992. Environmental ethics - a case for survival. *In* Yap, S.K. and Lee, S.W. (eds.): *In harmony with nature*. Malayan Nature Society, Kuala Lumpur. 534-544.

World Resources Institute, United Nations Environment Programme, United Nations Development Programme and The World Bank. 1996: *World Resources 1996–97*. World Resources Institute, Washington, D.C.

Plantations vs. naturally regenerated forests: Risks and stability in using bred, cloned, or biotech products

Gene Namkoong

Introduction

There are at least four questions addressed in the title of this paper which can be considered for discussion:

- are the materials derived from breeding any more or less stable in trait expression than materials derived from uncontrolled regeneration;
- are cloning or other so-called biotechnology products any more or less stable than seed propagules;
- are there risks in their uses that go beyond fears of instability, and
- are naturally regenerated forests any more or less risky or stable than planted forests?

The first three questions involve effects expressed at the individual tree level or at the population level of operation. Since we lack evidence that either breeding or propagation techniques affect the stability of individual tree trait expression, I will only consider stability and risk at the level of the population. I also distinguish between the two kinds of problems, stability as a problem of variation in trait expression such as disease susceptibility, and risk as a problem of probability of an event such as an insect epizootic.

The title of this session carries an implication that naturally regenerated forests are more stable and less risky than planted forests, an implication based on an unexamined assumption of the optimality of „naturalness" (Thomas, J. W. 1993). I do not wish to further propagate that logical fallacy, but wish to consider the bases on which reasonable fears exist in developing forests from plantings of materials derived by breeding and propagated by clonal technologies. By examining the kinds of problems that such breeding and propagation technologies can feasibly generate, I hope to conclude this paper by identifying some real issues and ways of addressing them. As will become evident, my proposition is that technology is not the problem. Rather, technology offers opportunities for changing forests that can increase or decrease risks and decrease or increase stability, but technology must be part of the solution for achieving biologically feasible social objectives.

C. Mátyás (ed.), Forest Genetics and Sustainability, Volume 63, 147–154.

1. Stability

I assume that trait expression in individual trees on any specific site is not affected by the act of choosing the parents, whether it is we who design the choice or we allow chance to do the choosing. That is, the act of choosing two parents instead of not choosing is assumed to have no influence on how those parents impart traits to their progeny. The influence of parental choice lies in which particular parents are chosen, if they have certain levels of heritable traits, and the numbers of such matings that may be made.

It is conceivable that certain parental combinations could result in progeny that are less stable in trait expression, but except in the case of severe inbreeding, there is no evidence that much variation exists in the phenotypic variance among individuals within families. What may vary is that some families exhibit greater phenotypic response ranges to variations in planting site environments. This kind of genotype x environment interaction is known to exist but is not discernable by observation of the parents. Thus, increased or decreased sensitivity to environmental variations may result from selective breeding, but there seems to be no *a priori* reason for suggesting that selection for growth rate, for example, necessarily decreases stability. In fact, this kind of effect can be estimated in breeding experiments and stability can then be increased by breeding (Namkoong *et al.*, 1992).

It is also possible when selecting for one trait that a second trait, which is genetically correlated, would respond to selective breeding with instability. However, the same arguments would hold for this case as for the above case, and this is merely another opportunity for breeders to either increase or decrease stability.

Therefore, the effect of breeding does not pose any direct threat to stability, but instead offers some opportunities for improving how we would like trees to respond to environmental variations (Eriksson 1995). In one sense, however, breeding may pose an indirect threat to stability, if we define instability as a departure of predicted from realized performance (Yanchuk, personal communication). Since the number of sampled families affects the variance of the mean of those families, then breeding small numbers of families increases the variance of prediction, and hence decreases that measure of stability. When breeding a single family, regardless of how stable its own performance is over any set of environments, the mean performance may not be well predicted, especially if small tests are used to estimate mean performance. If several families are chosen for breeding, their mean performance is better predicted than that of any one of them, and hence the breeder may wish to increase the number selected, to better stabilize the predicted mean performance.

A separate issue is whether the act of cloning or of using embryogenesis techniques makes phenotypic expression any less stable. Evidence is still being gathered on any induced instability at the individual clone level. However, many of the forest tree experiences with these techniques have indicated that families vary widely in their adaptability to propagation by somatic cell manipulations (Park *et al.* 1993). While it is not clear that this results in instability, it is clear that there is a selectivity that is imposed by the technique itself, in addition to selection for economic or other objectives. This may hopefully be uncorrelated with performance levels but raises the problem of decreasing the number of individuals and families that can be propagated (Högberg *et al.* 1998). The genetic variance in regenerated forests would be decreased by any form of family and within family selection for a trait that is extraneous to economic gain, with its associated problems of sample size. Improvements in technology and in breeding may yet solve these problems.

2. Risks

The concern is that any change in the average performance of a population, or in its variances and covariances, would increase the chances of an adverse event such as an epizootic. It is possible by breeding to directly increase the exposure of a stand to some debilitating influence, for example, if breeding for increased growth also increases the stand's susceptibility to frost damage, or its visibility to insects. It is also possible to indirectly increase risks of catastrophic failure by decreasing the genetic variance in a stand as a result of breeding or propagation constraints.

It seems obvious that there are some threats to forests that can be predicted and directly included as objectives for breeding enhancement. Resistance against or adaptability to some climatic or edaphic stresses can be directly improved by selection (White 1987) and would be obvious candidates for a breeding program. Problems with known biotic agents may be more difficult to incorporate into a breeding program due to the evolving nature of the agent itself which then requires that the co-evolutionary system be managed rather than just breeding for immediate resistance (Namkoong 1991). But these problems are merely questions of adding resistance traits to the multiple trait selection objectives that are usually included in any kind of plant or animal breeding program. Thus, if problems can be reasonably anticipated, they become breeding problems and the efficacy of including those response phenotypes in forests can be a technical problem of multiple trait selection.

The main open question is whether by breeding, we unknowingly increase susceptibility to genetic problems that are induced by the breeding or propagation operation itself. There are two problems to consider; are selected populations more susceptible to unknown future threats, and are the probabilities of failure increased by having fewer genotypes present in a stand than might exist if no breeding or propagation technologies were applied? Our concern is that some presently unknown threat will arise that are subject to genetic improvement, but for which we inadvertently decrease the population's capacity to respond. A suspicion exists that we increase stand susceptibility to these unknown problems by changing tree physiology or by limiting the number of genotypes that may buffer a population. That is, we directly increase stability by changing trait performance, or reduce the variance needed by the population to respond to that future challenge.

The factors that determine whether a random decrease in the number of different genotypes will lead to decreased stand viability through loss of a resistant allele are the frequency of susceptible genes in the population, and the level of stand mortality that can be tolerated (Roberds and Bishir 1997 Roberds *et al.* 1990). The result was anticipated earlier by others and can be briefly summarized as whether the probability of stand failure is increased or decreased by greater genetic sample size. If the frequency of susceptible genotypes is higher than the acceptable level of stand mortality, then increasing the number of random genotypes merely ensures that a high level of susceptibility will exist in the stand. If the frequency of susceptible genotypes is low, then increasing the number of genotypes past a few does little to further increase the probability of stand success. The range of parameters for susceptibility vs. stand failure, where increased genotypic sampling may help, is small. Therefore, the effects of breeding that could increase genetic susceptibility, including selective propagation techniques on reducing genotypic sampling, are not detrimental to stand safety.

The greater problem for reduced genotypic sample sizes lies in the recurrent breeding program where an increased level of inbreeding and a loss of genetic variance may be experienced in future breeding generations. This however, is a breeding problem that is well recognized by tree breeders (Namkoong *et al.* 1988) and involves the methods that are available for managing the genetically effective population size. For breeding populations the number of parents in populations and the number and diversity of populations can be used to control the total genetic variance available for forest managers to use in placing or deploying genotypes into plantations. The genetic structure and population size of the breeding population should be managed for long term gains, which can be quite different for the populations of trees that is planted.

The useful variance for population response exists in both the breeding as well as in the planted population. The breeding population can be managed to have a variance structure to allow for future responses, even if the planted population is limited. In comparing either of these two genetically controlled populations to genetically uncontrolled forest regeneration, several factors can increase genetic sources of risk. Natural regeneration depends on the vagaries of which parents are particularly fecund at the time and place of regeneration. The genetically effective population size can be a small fraction of the total number of adult trees in a stand, whereas breeding can control and equalize the distribution of parental representation in the progeny pool.

As suggested elsewhere (see for example Koshy *et al.* 1998) the genetic variance among all the members of a well structured set of multiple breeding populations can be larger than the genetic variance of a single large population and can be larger than the genetic variance of a population that is subject to uncontrolled pollination and seed dispersal.

3. Naturally regenerated vs. planted forests

The assumed ideal of a naturally regenerated forest being maximally stable, maximally diverse, and minimally susceptible or maximally resistant to unanticipated threats, is a commonly held paradigm. The three properties are assumed to be simultaneously achieved by the normal processes of forest regeneration without human intervention. Since such naturally regenerated forests follow the dynamics of evolution and are subject to the usual forces of selection, migration, mutation, and drift, the assumption requires that the joint effect of all four forces is the same effect on all three properties. While it is now quite obvious that there are no forests that have been entirely free of human influence, there may be strong tendencies for the debilitation of the three properties to be exacerbated with increases in human interventions. At one extreme, it would seem obvious that extensive stands composed of a single clone would be susceptible to any single biotic or climatic challenge to which it may be especially vulnerable. There is no genetic variance either in the stand or in the pool of genotypes available for regeneration, and the accidents of drift would have fixed the particular alleles of that genotype as the only ones available. Mutation is too slow to bring useful levels of variation into the population for many generations, and only reproduction by selfing would generate limited levels of recombination.

By this paradigm, there is a single scale of quality of all three properties that declines with increasing intensity of genetic interventions such as tree breeding. It follows that breeding not only makes trees more susceptible to unanticipated threats, it also reduces the parental gene pool, and propagation technologies further reduce the number of genotypes, so that evolutionary capacities are reduced.

However, the paradigm is fundamentally flawed in the assumption that all three properties are highly correlated, that all four forces of evolution act similarly, and that gene management can act in only one dimension. It has been shown that stability and diversity are not identical (May 1974). While it also seems clear that some levels of species diversity increase resource utilization (Tilman *et al.* 1997) there is a limit to species packing and to the kinds of species interactions that lead to stability of any sort. Therefore, maximizing diversity does not logically lead to maximizing stability.

It can also be argued that stability can be achieved by many means, one of which is to reduce the number of species and genotypes involved in destabilizing interactions, or select those that can provide mutualistic interactions. While this is possible, it is by no means obvious that natural regeneration would necessarily lead to such selection. Natural regeneration would be dependent on the pool of species and genotypes that are available at particular times in the historical development of stands and that is subject to past migrations and sources of genotypes which are in turn partially dependent on the ecological history of glaciation and climatic constraints. Thus, the present distribution of species and genotypes at best, lags any selective optima. The stands we may see today or that we can estimate for the recent past would rarely if ever reflect an optimum condition for current or future conditions.

In addition, the forces of evolution themselves operate on different time and spatial scales and can rarely if ever be considered to be in equilibrium or at an optimum state (Eriksson *et al.* 1993). Species and genotypes cannot be considered to be in an adapted state, but only in an adaptable condition. Therefore, there is no foundation for believing that the three properties of stability, diversity, and resistance or resilience can simultaneously exist in any one standing forest. In every generation of trees, the processes of adaptation must be jointly affected on the basis of their history.

4. Conclusion

If natural regeneration cannot be held as an ideal method to achieve an ideal state, we also have reasonable concerns about breeding for limited and short-term gain objectives with very small genotypic diversity. If genetic management is limited to only one tool such as the intensity of selection for multiple traits, then the gene manager can choose a best compromise between selection objectives and selection intensity. However, the forces of evolution are complex and not uniform in their scales or effects. The genetic architecture of species depends on the effects of the forces on many partially independent gene loci, and each locus can be considered to respond to different forces. That is, some loci have been under strong selection pressure, and others under less intensive present pressure. Since each locus also has a distribution of frequencies among populations, the frequency distribution of alleles under strong vs. no selection would have to differ. This level of complexity implies that the natural distribution of alleles is heterogeneous, and there is no single ideal state or ideal distribution.

Among the tools that are used in tree breeding, selection for different objectives can include selection for traits of value for future adaptational value. Breeders also have the capacity to create multiple levels of evolution, which is not directly available to naturally regenerated forests. By using managed levels of gene migration between diversely selected populations, we can create heterogeneous distributions of alleles and can plant them in designed patterns in forests independent of their origin or physical location. The breeding populations can be controlled and maintained separately from the standing forests, each with alleles that can be either the same among the breeding population units or divergent. They can then be deployed in standing forests independent of their history, and in patterns that can include higher or lower levels of genetic variation within stands, or can be uniform within stands, but divergent between stands or different over a time series of replacement stands.

Thus, the technologies of breeding and propagation offer opportunities for managing the genetics and evolution of forests that can either increase or decrease risks over that of natural regeneration. Technology is neither the problem nor the solution, it is a tool that we have available that can be used well or poorly.

Literature cited

Eriksson, G. 1995. Which traits should be used to guide sampling for gene resources? *In* Baradat, Ph., W. T. Adams and G. Müller-Starck (eds.): *Population genetics and conservation of forest trees*. 349-358.

Eriksson, G., G. Namkoong and J.H. Roberds 1993. Dynamic gene conservation for uncertain futures. *For. Ecol. And Mgmt*. 62:15-37.

Höberg, K. A., Ekberg, I., Norell, L. and von Arnold, S. 1998. Integration of somatic embryogenesis in a tree breeding programme – a case study with *Picea abies*. *Can. J. For. Res*. 28:1536-1545.

Koshy, Mathew P., Namkoong, G. and Roberds J. H. 1998. Genetic variance in the F2 generation of divergently selected parents. *Theor. and Appl. Genet*. Vol. 97:990-993.

May, R. M. 1973. Qualitative stability in model ecosystems. *Ecology* 54: 638-641.

Namkoong, G. , Kang, H. C., and Brouard, J. S. 1988. *Tree breeding: Principles and strategies*. Springer-Verlag, New York, 179 pp.

Namkoong, G., Jonsson, A., and Eriksson, G. 1992. Genetic variation in nutrient response functions. *Theor. and Appl. Genet.* 85:165-172.

Namkoong, G. 1991. Maintaining genetic diversity in breeding for resistance in forest trees. Annu. Rev. *Phytopathol*. 29:325-342.

Park, Y. S., Pond, S. E. and Bonga, J. M. 1993. Initiation of somatic embryogenesis in white spruce (*Picea glauca*): genetic control, culture treatment effects, and implications for tree breeding. *Theor. and Appl. Genet*. 86:427-436.

Roberds, J. H., Namkoong, G., and Tore Skroppa 1990. Genetic analysis of risk in clonal populations. *Theor. and Appl. Genet.* 79:841-848.

Roberds, J. H., and Bishir, J. 1997. Risk analysis in clonal forestry. *Can. J. For. Res*. 27:425-432.

Tilman, D. Knops, J., Wedin, D. Reich, P., Ritchie, M. and Siemann, E. 1997. The influence of functional diversity and composition of ecosystem processes. *Science* 277: 1300-1302.

Thomas, J. W. 1993. Addressing the retention of biodiversity in national forests in the United States. *In* Fenger, M. A., Miller E. H., Johnson J. A. and Williams E. J. R. (eds.): *Our Living Legacy: Proceedings of a Symposium on Biological Diversity*. Royal British Columbia Museum, Victoria, B. C. 47-63.

White, T. L. 1987. Drought tolerance of southwestern Oregon Douglas-fir. *Forest Science* 33(2): 283-293.

Impacts of industrial forestry on genetic diversity of temperate forest trees

Yousry A. El-Kassaby

Introduction

Domestication of forest tree species can be characterized as taking genotypes from their wild, undomesticated state through a series of sampling and selection stages, each with the potential of reducing genetic variation and in the same time resulting in genetically superior plantations (El-Kassaby and Namkoong 1994,1995). This domestication follows a systematic process by which the populations involved descend from large collections to screening and testing, and finally to advanced breeding with small sets of parents before incorporation in production populations (i.e., seed orchards).

The amount of available genetic data related to the study of forest tree domestication is very limited. Furthermore, published studies were devoted mainly to the genetic consequences of domestication, rather than to the evaluation of the process itself. In this paper, we review genetic diversity studies of artificial and natural regeneration methods with special reference to temperate zone forests.

1. Artificial Regeneration

1.1. Tree-improvement programs

1.1.1. Phenotypic selection

The contemporary state of the genetic variation of wild populations is a result of its history and evolution that has responded to mutation, migration and inbreeding, recombination events, and various forms of natural and human-influenced selection. The level of genetic variation of the resource can be evaluated and characterized by morphological, physiological, and quantitative (fitness) traits, as well as biochemical and molecular markers. Comparing the genetic diversity and its organization between natural populations and the product of selection or breeding programs through phenotypic and genotypic selection (i.e., breeding and production populations) ideally requires the utilization if a combination of these attributes (i.e., morphological, adaptive, fitness, and biochemical and molecular markers) due to the apparent limitation if using only one type. In

C. Mátyás (ed.), Forest Genetics and Sustainability, Volume 63, 155–169.

this review, a combination of these attributes has been used due to their ease and suitability to specific steps of the process.

Table 1. Heterozygosity parameter comparison between phenotypic selection (seed orchards) and natural populations.

Species	Population	Heterozygosity parameters		
		% polymorphic loci	# of alleles/ locus	Mean expected heterozygosity
Norway spruce[1]	Seed orchard	88.0	2.4	0.320
	Natural populations	100.0	2.5	0.230
Western red cedar[2]	Seed orchard	11.1	1.2	0.058
	Natural populations	11.1	1.1	0.055
Sitka spruce[3]	Seed orchard	100.0	2.8	0.229
	Natural populations	66.9	1.8	0.183
Douglas-fir[4]	Seed orchard	62.5	2.3	0.172
	Natural populations	52.6	2.1	0.171
"Interior"spruce[5]	Seed orchard	64.7	2.4	0.194
	Natural populations	64.7	2.7	0.189

[1]Bergmann and Ruetz 1991, [2]El-Kassaby et. al. 1993b, [3]Chaisurisri and El-Kassaby 1994, [4]El-Kassaby and Ritland 1996a,b, [5]*Picea glauca* × *engelmannii,* Stoehr and El-Kassaby 1997

To evaluate genetic variability, several heterozygosity parameters (number of alleles per locus, percentage of polymorphic loci, and mean expected and/or observed heterozygosity) of first-generation seed orchards with their counterpart natural populations were conducted for several species (Table 1). In general, the heterozygosity parameter comparisons between the seed orchards and their corresponding natural populations indicated that the amount of genetic variation present in natural populations was retained or increased during the phenotypic selection step (Table 1). Chaisurisri and El-Kassaby (1994) and El-Kassaby and Ritland (1995b) attributed the difference between the sampling methods of natural populations (collection of several trees from one location) and individual plus trees (one or few trees per location) for the maintenance of most of the variation observed in natural populations. Sampling of natural populations is usually restricted to a relatively small area (i.e., the local gene pool); on the other hand, the selection of large numbers of plus trees from several populations represents a situation whereby the sampled trees for breeding purposes may include other alleles not captured by more restricted sampling. It is also interesting to note that the frequently observed reduction of genetic diversity in agricultural crop plants was not observed in these studies. This may be due to the fact that conifers, with their known high genetic variability, are still in their natural state and that early domestication is not effective in causing appreciable reduction to genetic variability. Similar results were also reported on white spruce (Cheliak *et al.* 1988), however, this study's main emphasis was on the presence and/or absence of rare alleles. Cheliak *et al.*

(1988) indicated also that the heterozygosity estimates were similar between the two samples (phenotypic selection and natural populations).

1.1.2. Breeding vs. production populations

Most seed orchards are established with a smaller set of the parent trees present in the breeding populations, thus estimates of the level of genetic variability present in these smaller population production populations is very important in the tracking of genetic variability. This type of comparison sheds light on how tree improvement could affect the genetic variability and, more importantly, it provides guidelines for the implementation of various breeding and conservation strategies.

Table 2. Comparision of Douglas-fir heterozygosity parameters: (a) natural, breeding and production, populations and (b) first- and second-generation seed orchards (El-Kassaby and Ritland 1996a,b).

	Population	Heterozygosity parameter		
		% polymorphic loci	# of alleles/locus	Mean expected heterozygosity
a	Natural	53	2.14	0.171
	Breeding	65	2.50	0.176
	Production	65	2.65	0.167
b	1st gen. S.O.	63	2.28	0.172
	2nd gen. S.O.	56	2.25	0.163

In Douglas-fir, El-Kassaby and Ritland (1995a,b) monitored the genetic variability (i.e., heterozygosity) through the process of tree improvement programs. Additionally, first- and second-generation seed orchard populations were evaluated and compared (Table 2). 12 first- and 3 second-generation orchards were analysed. This study has demonstrated the ability of phenotypic selection to capture most of the genetic diversity in the species' natural range, confirming the expectation of many tree breeders. The comparison between the first- and second-generation seed orchards showed only a slight reduction in % polymorphic loci and mean heterozygosity. However, the average number of alleles per locus indicated permanent loss of some rare or localized alleles. It should be noted that rare alleles do not contribute much to heterozygosity. The slight differences in genetic variability observed between the first- and second-generation seed orchard populations might disappear when more second-generation seed orchards are established and included in the analysis.

1.2. Seed orchards

Seed orchards represent the deployment tools of breeding programs through the consistent delivery of genetically improved seed. Panmictic equilibrium (random mating) is achieved only when seed orchards function as closed, perfect populations. Reaching this goal in seed orchard populations requires that the entire seed orchard's population (clones or seedlings) are in reproductive synchrony and contribute similarly to the gene pool (i.e., reproductive output equality). Additionally, the satisfaction of this condition assumes lack of or minimal levels of: 1. outside pollen contamination, 2. inbreeding (selfing and related mating), and 3. selection among the various clones' gametes. In most cases seed orchards do not fulfil these conditions due to several biological factors. The following section lists these factors.

1.2.1. Reproductive phenology

It is well known that both time and temperature, summarized as heat sum, have a significant effect on reproductive phenology (Worrall 1983). Because seed orchards consist of several genotypes that are usually adapted to different environmental conditions due to their different geographic origin, with various heat-sum requirements for bud burst, differences in reproductive bud phenology will occur. Thus, the orchard clones form several temporally isolated breeding sub-populations, increasing the rate of inbreeding, reducing panmixis, and exposing the early and late "flowering" classes to higher levels of non-orchard pollen contamination (see El-Kassaby 1989, for review). The dynamics of reproductive phenology in seed orchards of several species were investigated, and, with the exception of western redcedar (El-Kassaby 1992), all displayed an extended pollination season and subdivision of the seed orchard population into several temporally isolated breeding sub-populations. This temporal isolation also affected seed yield, rate of pollen contamination, and mating pattern (selfing rate and level of correlated matings).

Differences in reproductive phenology also affect the gametic contribution by every clone to the resultant seed crop. A high seed- or pollen- producing clone in the early or late phenology classes could contribute the same as or less than the amount contributed by a low or medium reproductive-output clone during the peak of pollination (El-Kassaby and Askew 1991, Roberds *et al.* 1991).

The frequency and timing of receptive females and pollen-shedding males also affect the rate of pollen contamination. If background pollen level is high when early or late phenology classes are receptive, most of the receptive trees will be pollinated by background pollen (El-Kassaby *et al.* 1988). This, in effect, will reduce the observed rate of inbreeding (El-Kassaby and Ritland 1986), consequently increasing the effective population size (i.e., increasing the genetic variability). However, pollen contamination is often associated with reduction of the expected genetic gain (Askew 1986), as well as increased maladaptation, especially if the orchard is located outside of the seed zone it is planned for.

1.2.2. Reproductive output

The maintenance of similar allelic frequencies between a seed orchard and its seed crop is dependent on the presence of balanced gametic production among the orchard's clones and absence of effective selection involving the alleles studied. Parental-balance estimates of several seed orchards have revealed that the parental contribution to cone/seed crops and pollen varied substantially among clones (see El-Kassaby 1995, for review). In most cases, a small proportion of clones contributed the majority of male and/or female gametes, thus reducing the expected genetic variation and producing crops with unpredictable allelic frequencies (Muona and Harju 1989, Roberds *et al.* 1991, El-Kassaby and Ritland 1992, Nakamura and Wheeler 1992).

1.2.3. Inbreeding

Reported estimates of outcrossing in natural coniferous populations are usually high for most species (t>0.9; Adams and Birkes 1991). High levels of inbreeding in seed orchards could be detrimental due to the non-competitive plantings commonly practiced in nursery production and plantation establishment in most forestry programs. In general, outcrossing rate estimates obtained from seed orchards are higher than those reported for natural stands of the same species, suggesting that seed orchard designs (i.e., the physical placement of ramets of the same clone, individuals of the same full- or half-sib families, or clones with common ancestry within the seed orchard) affect the rate of outcrossing (Rudin *et al.* 1986, Shaw and Allard 1982). Recently, the use of clonal rows has been proposed and used as an efficient seed orchard design for ease of management activities and flexibility in deployment of new clones. These types of seed orchard designs require aggressive pollen management practices due to the presence of genetically identical individuals in close proximity. Results from a clonal row and random seed orchard design comparison in a western hemlock seed orchard have indicated the presence of higher rate of selfing and correlated matings in clonal-row than in the random design (El-Kassaby 1997).

1.3. Seed biology

1.3.1. Germination parameters

Seed orchard managers endeavour to maximize the genetic variability in their seed crops. These crops are usually collected, extracted, and stored in bulked seedlots. Bulked seedlots are in most cases composed of unknown proportions of seed from several seed parents. Treating a collection of seed produced from different parents (genotypes) with one uniform treatment during extraction, germination, and storage assumes that all seeds will respond to these various treatments in a uniform fashion. However, published work on the genetics of conifer seed size, dormancy, and germination have demonstrated that genotypic differences represent a major component of the total variation observed for these attributes (Chaisurisri *et al.* 1992, El-Kassaby *et al.* 1992). Therefore, it is expected that any uniform treatment such as dormancy-breaking treatment applied to bulked seedlots will produce different responses due to the presence of different genetic backgrounds in the seedlots. Furthermore, germination studies have demonstrated that germinative speed is under strong genetic control, thus different genotypes will germinate according to their genetic background, resulting in greater variation during nursery emergence.

1.3.2. Seed storage

Seed viability is usually highest at the time of cone harvest and declines with age. In agricultural crops, it has been demonstrated that genotypes differ in the degree of seed viability loss during seed storage (Delouche and Baskin 1973). If conifer seed crops are to be collected, extracted, and stored on a bulk basis, then seed viability losses by specific genotypes cannot be traced, and a reduction in the total genetic diversity passes undetected (El-Kassaby 1992). Our work on viability loss caused by simulated long-term storage in Sitka spruce (Chaisurisri *et al.* 1993), mountain hemlock (El-Kassaby and Edwards 1998), Douglas-fir, western hemlock, white spruce and lodgepole pine (unpublished results) has demonstrated that the rate of germination decline varied among genotypes, indicating differential viability losses. These differences alter the genetic makeup of stored, bulked, seedlots from collection to sowing.

1.4. Seedling production

It is expected that the genetic diversity present in seed orchards and their seedlots is to be translated to plantations through nursery-grown plants. In container nurseries, maximizing the number of seedlings that meet stock requirements and minimizing the number of empty cavities and the culling of seedlings that do not meet "standards" are important, if container nurseries are to reach their maximal economic potential. Reducing the number of empty cavities is usually achieved by multiple sowing to ensure at least one germinant per cavity. Cavities with multiple germinants, on the other hand, require thinning. If the seedlots used are bulked, then some of the biological factors discussed above, such as differences in germination parameters (percent and speed), may influence the effects of these practices on the genetic representation of various parents. Thus, it would be of interest to evaluate the assumption of genetic similarity between seedlots and their seedling crops.

El-Kassaby and Thomson (1995) evaluated the impact of management practices commonly used in container nurseries (thinning and culling) on the genetic diversity (i.e., family representation) of seedling crops. Douglas-fir seed germination, thinning, and culling contributed 66, 20, and 14%, respectively, to the variation in family representation. They also reported the presence of unintentional directional selection during the seedling production phase. To avoid this, the authors proposed the use of either individual seed sowing or multiple sowing on an individual family basis.

Table 3. Percent variation corresponding to the effect of germination, thinning, survival, and culling comparing 1-, 2- and 3-seeds/cavity and single-family-block experiment nursery studies for Douglas-fir and western hemlock

	Proportion contribution			
	Germination	Thinning	Survival	Culling
Douglas-fir				
1-seed / cavity	0.65	--[1]	0.04	0.31
2-seeds / cavity	0.09	0.91	0.00	0.00
3-seeds / cavity	0.04	0.95	0.00	0.01
Single family block	0.20	--[1]	0.18	0.62
Western Hemlock				
1-seed / cavity	0.19	--[1]	0.12	0.69
2-seeds / cavity	0.06	0.83	0.01	0.10
3-seeds / cavity	0.07	0.85	0.01	0.07
Single family block	0.08	--[1]	0.00	0.92

[1]no thinning was conducted

El-Kassaby (unpublished) followed these recommendations and conducted two studies on Douglas-fir and western hemlock seedlots that were represented with an equal contribution of seeds from several seed orchard parents. In each species experiment, seeds were sown in four different arrangements (single-, two-, and three seeds per cavity (seeds within cavity were randomly selected among the 15 families) and family blocks (seeds within block belonged to one family)). This experimental design allowed for the comparison between inter- and intra-family competition. Families were compared based on: 1) changes in their rank order from emergence to post-thinning and post-culling status, and 2) relative performance from seed contribution to seedling production. Changes were observed in both experiments (Table 3). It was determined that seed germination, germinant thinning and culling affected seedling production per source, again indicating the presence of several consecutive bottlenecks in the process. Family sowing with culling standards that recognize the growth differences among families in the nursery and single seed sowing, after understanding the inter-/intra-family competition, were recommended for seedling production to maintain genetic diversity.

Stoehr and El-Kassaby (1997) have evaluated the cumulative effect of phenotypic selection, breeding, and seed and seedling production. In this study, the level of genetic diversity of nine natural stands of "interior spruce" (*Picea glauca x engelmannii*) representing one breeding zone was assessed and used as a benchmark for comparison of the levels of genetic variation detected in: 1) 100 clones representing the production and breeding populations of this breeding zone; 2) a bulked seedlot from the same seed orchard; and 3) a sample of 120 seedlings from a plantation established from a seedling crop produced from the same seedlot. The results demonstrated that the genetic diversity parameters of the natural populations did not differ from those of the seed orchard population, confirming our earlier observations (Figure 1). Additionally, a steady decline (represented by divergence in the Figure) was observed when the evaluation was extended to the seedlot and the seedling crop. This study clearly demonstrated that the domestication process could cause genetic erosion and calls for vigilant monitoring.

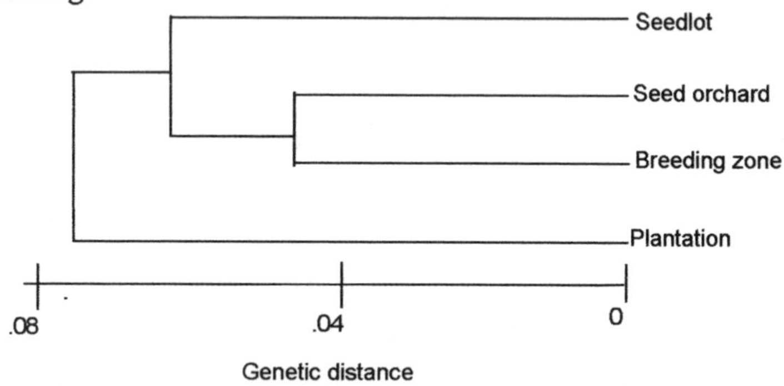

Figure 1. Dendrogram depicting the genetic relationship among the four "interior spruce" test populations (breeding zone, seed orchard, seedlot, and plantation) (after Stoehr and El-Kassaby 1997).

1.5. Plantation management

Commercial thinning, a common plantation management activity, is currently being practiced by the coastal forest industry in British Columbia and in the Pacific Northwest of the United States. Commercial thinning provides an opportunity to significantly alter the species/genetic diversity of the thinned stand, thus rigorous evaluation of the impact of commercial thinning on tree species representation and their genetic diversity is required.

El-Kassaby (unpublished) conducted a study to evaluate the impact of commercial thinning on the biological diversity (species/genetic) of forest trees on two Douglas-fir plantations. The study plantations' average age is 60 years. In the study plots, all tree species were stem mapped and vegetative buds or foliar tissue from every individual tree were collected for biochemical fingerprinting using isozyme analysis. The genetic data provided a baseline for the degree of genetic variation within and among the various species in the study plot. A commercial thinning treatment was applied to the study plots and the thinned trees were identified. The genetic data for the thinned trees were removed from the pre-thinning data, and a new assessment of the level of genetic variation within and among the various species was determined (Table 4).

Table 4. Number of loci, total number of alleles, number of alleles per locus (N_a), proportion of polymorphic loci ($PLP_{95\%}$), expected heterozygosity (H_e), and allelic loss for western redcedar (Cwr), western hemlock (Hw), Douglas-fir (Df), western white pine (Pw), and Pacific silver fir (Ba) before and after commercial thinning in two Douglas-fir plantations (SD's are given in parentheses).

Plantation	Species	Thinning	# of loci	# of alleles	N_a	PLP_{95}	H_e	Allelic loss
A	Cwr	Before	21	27	1.29(0.12)	19.05	0.068(0.039)	--
		After	21	27	1.29(0.12)	23.80	0.087(0.041)	--
	Hw	Before	20	31	1.55(0.15)	15.00	0.051(0.026)	--
		After	20	24	1.20(0.09)	15.00	0.043(0.027)	7
	Df	Before	22	56	2.55(0.18)	54.55	0.153(0.031)	--
		After	22	55	2.50(0.19)	59.10	0.158(0.033)	1
B	Df	Before	22	48	2.2(0.2)	54.5	0.164(0.035)	--
		After	22	48	2.2(0.2)	50.0	0.163(0.036)	--
	Hw	Before	15	23	1.5(0.2)	13.3	0.041(0.023)	--
		After	15	18	1.2(0.1)	20.0	0.045(0.026)	5
	Cwr	Before	14	14	1.0(0.0)	00.0	0.000(0.000)	--
		After	14	14	1.0(0.0)	00.0	0.000(0.000)	--
	Pw	Before	15	19	1.3(0.1)	26.7	0.081(0.040)	--
		After	15	17	1.1(0.1)	13.3	0.067(0.045)	2
	Ba	Before	11	11	1.0(0.0)	00.0	0.000(0.000)	--
		After	11	11	1.0(0.0)	00.0	0.000(0.000)	--

The results indicated that the species composition differed from the original plantation. For example, site A was originally planted with Douglas-fir (i.e., 100%), however, at the time of thinning the species composition was 46%, 39%, and 15%, Douglas-fir, western hemlock, and western red cedar, respectively. The thinning altered the species composition to 76%, 17%, and 7%, Douglas-fir, western hemlock, and western red cedar, respectively. Species diversity estimates showed reduction from 1.01 to 0.70 (Shannon index: Pielou (1975)). It is noteworthy to mention that all three species remained after the thinning. Estimates of heterozygosity parameters for the two sites (A and B) before and after thinning are presented in Table 4. In general, average number of alleles per locus, percent polymorphic loci, and expected heterozygosities did not differ significantly. However, thinning resulted in a loss of 8 and 7 alleles for sites A and B, respectively (Table 4).

2. Natural regeneration

2.1. Alternative silvicultural systems

Detailed knowledge and understanding of the mating system in natural and artificial plant populations are fundamental prerequisites for understanding their genetic architecture and genetic variation potential. Recently, several areas of research into plant mating systems, such as the extent of consanguineous mating (Ritland and El-Kassaby 1985), and similarity of paternity among siblings (Ellstrand and Marshall 1986; Meagher 1986) or correlated matings (Ritland 1989) have received increased attention as other facets of the mating system (Ritland and El-Kassaby 1995). While most of the mating systems studies have focused on the population's mating system estimation (see Schemske and Lande 1985, for review), variation among individuals' selfing rate and the extent of correlated mating (i.e., sibs' genetic relatedness) in an individual progeny array have proven to be revealing and more informative. For example, constant outcrossing rate estimates over three years were observed for a western white pine population. However, individual outcrossing rates varied within (among trees) and among years (for the same trees), revealing the existence of consistent inbreeders and outcrossers (El-Kassaby *et al.* 1987, 1993a). The mating system in a population is influenced by genetic and environmental factors (Clegg 1980). Population density represents one of the environmental factors potentially influencing the mating pattern in forest tree populations. Positive relationships between population density and outcrossing rate were reported for ponderosa pine (Farris and Mitton 1984), tamarack (Knowles et al. 1987), and Engelmann spruce (Shea 1987), but not for coastal Douglas- fir (Neale and Adams 1985a), Jeffrey pine (Furnier and Adams 1986), subalpine fir (Shea 1987), and Norway spruce (Morgante et al. 1991). It should be emphasized, however, that density should be considered in concert with other factors, such as reproductive ecology (El-Kassaby and Davidson 1991), reproductive phenology (El-Kassaby and Ritland 1986b; El-Kassaby et al. 1988), fecundity (Schoen and Stewart 1986, 1987; Denti and Schoen 1988), and the distance between mates (Erickson and Adams 1989; Adams et al. 1992).

Different regeneration systems such as clear cut, patch cut, seed tree (green tree), and shelterwood with a residual old growth reserve in the same forest type provide rare opportunities for genetic evaluation of the mating system and the genetic variability.

The Montane Alternative Silvicultural System (MASS) (Arnott and Beese 1997), with its replicated experimental design, represents a unique opportunity for such evaluation. The project compares four silvicultural treatments:

1. Clearcut

2. Patch out (patch size: 1.5 to 2 ha)
3. Green tree retention (minimum of 25 trees/ha left)
4. Shelterwood (approx. 200 stems per ha over 17.5 cm DBH).

Genetic evaluation of these four (clear cut, patch cut, green tree, and shelterwood) regeneration systems, specifically assessment of heterozygosity parameters of adult trees and their natural regeneration and the rate of selfing, mating among relatives (the difference between single-locus (ts), multilocus (tm), outcrossing rate estimates (Ritland and El-Kassaby 1985), and the extent of correlated mating (Ritland 1989), have been conducted and were compared to that of the control old growth forest. Generally, no large differences were detected between the two life-cycle stages (Table 5).

Table 5. Number of loci, number of alleles per locus N_a, % polymorphic loci (PLP95%), expected heterozygosity H_e for Pacific silver fir (PSF) and western hemlock (Hw) adult and offspring (natural regeneration) in the MASS site (SD's are given in parentheses).

Species	Population	# of loci	N_a	PLP_{95}	H_e
PSF	Adults	22	2.1(0.2)	31.8	0.085(0.030)
	Offspring	22	2.2(0.2)	27.3	0.071(0.027)
Hw	Adults	21	2.0(0.2)	42.9	0.093(0.026)
	Offspring	21	2.4(0.2)	33.3	0.109(0.030)

These results are similar to those reported by Neale and Adams (1985b) from shelterwood regeneration of Douglas-fir. However, these evaluations should take into account the breeding system and its impact on the level of inbreeding and genetic homogeneity of regeneration.

Evaluation of the mating system of these "alternative" silvicultural treatments (patch cut, green tree, and shelterwood) and comparing their results with that of the old growth "control" produced interesting results (Table 6). This evaluation was conducted on Pacific silver fir and similar evaluation on western hemlock is in progress. The old growth sample of Pacific silver fir produced a significant level of inbreeding (tm) one that is similar or higher than that reported by Davidson and El-Kassaby (1997) for the same species from several other old growth populations. The Green tree treatment was the only treatment that produced inbreeding values lower than that observed for the old growth and were much lower than that reported for natural populations (Davidson and El-Kassaby 1997). The low density of the leave tree treatment may have been the cause for this observed difference. Higher estimates of inbreeding were observed for Scots pine (Rudin *et al.* 1997, Yazdani *et al.* 1985, 1989) and western larch (El-Kassaby and Jaquish 1996). The patch cut and shelterwood treatments produced higher estimates than either the old growth and green tree systems. It is hypothesised that the higher density of leave trees in the shelterwood and the edge effect in the patch cut were the reasons for the relatively higher estimates of outcrossing rates in these two treatments. However, Neale and Adams (1985b) did not observe any change in the inbreeding level of Douglas-fir between their shelterwood and control treatments. Similar results were observed by El-Kassaby and Stoehr (unpublished results) from another study on Douglas-fir with different densities of leave trees.

Table 6. Population estimates of single-locus (ts), multilocus (tm), and correlated paternity (rp) estimates for Pacific silver fir in the MASS site populations using a sample size of 10 seeds/tree (SD's are given in parentheses).

Estimate	Old Growth	Patch Cut	Shelterwood	Green Tree
t_s[1]	0.650 (0.054)	0.707 (0.074)	0.739 (0.070)	0.582 (0.097)
t_m	0.737 (0.059)	0.753 (0.074)	0.883 (0.080)	0.637 (0.105)
r_p	0.355 (0.087)	0.278 (0.018)	0.616 (0.196)	0.652 (0.194)

[1]Single-locus minimum variance mean.

The four comparisons indicated that high level of mating among relatives is occurring in these treated stands. In all comparisons the multilocus estimates were higher than the single-locus estimates, confirming our observation. These results are common among most conifers and usually produce lower estimates of inbreeding.

Correlated-mating results were very surprising. The level of correlated paternity in the old growth was high and reached 36% (Table 6). The patch cut with its probable edge effect produced a lower estimate; however, this estimate is very high. The two remaining treatments produced fairly high levels of correlated paternity and confirm that few trees are participating in the mating as males. These correlated paternity estimates are of great importance if natural regeneration will be the method of choice. In other words, significant levels of inbreeding, high estimates of mating among relatives and extremely high correlated paternity are the characteristics of potential seeds produced from the remaining trees for regeneration. These estimates of correlated paternity were much higher that that reported by El-Kassaby and Jaquish (1990) in their seed tree study on western larch.

In conclusion, the present paper illustrates the necessity of understanding the biological ramifications of the various management practices to avoid the effect of some particular practices that may have long-term consequences on genetic diversity.

Literature cited

Adams, W.T. and D.S. Birkes. 1991. Estimating mating patterns in forest tree populations. *In* Fineschi, S., M.E. Malvolti, F. Cannata, and H.H. Hattemer, (eds.): Proc. Internat. Workshop on Plant Biology, Biochemical Markers in Population Genetics of Forest Trees. Institute for Agroforestry of the National Research Council of Italy. Porano-Orvieto, Italy. 157-172.

Adams W.T., A.R. Griffin, and G.F. Moran. 1992. Using paternity analysis to measure effective pollen dispersal in plant populations. Am. Nat. 140:762-780.

Arnott, J.T. and W.J. Beese. 1997. Alternatives to clearcutting in BC Coastal Montane Forests. For. Chron. 73:670-678. Accessible website: http://www.pfc.cfs.nrcan.gc.ca/-www_user/lgalbraith/mass/mass.html

Askew, G.R. 1986. Implications of non-synchronous flowering in clonal conifer seed orchards. *In* IUFRO Conference: A joint meeting of working parties on breeding theory, progeny testing and seed orchards. 182-191.

Bergmann, F. and W. Ruetz. 1991. Isozyme genetic variation and heterozygosity in random tree samples and selected orchard clones from the same Norway spruce populations. *For. Ecol. Manage*. 46:39-47.

Chaisurisri, K. and Y.A. El-Kassaby. 1994. Genetic diversity in a seed production population vs. natural populations of Sitka spruce. *Biodiversity and Conservation*. 3:512-523.

Chaisurisri, K., D.G.W. Edwards, and Y.A. El-Kassaby. 1992. Genetic control of seed size and germination in Sitka spruce. *Silvae Genet*. 41:348-355.

Chaisurisri, K., D.G.W. Edwards, and Y.A. El-Kassaby. 1993. Accelerating aging in Sitka spruce seed. *Silvae Genet*. 42: 303-308.

Cheliak, W. M., G. Hurray and I. A. Pitek 1988. Genetic effect of phenotypic selection in white spruce. For. Ecol. Manage. 24: 139-149.

Clegg, M. T. 1980. Measuring plant mating systems. Bio Science. 30: 814-818.

Davidson, R. and Y. A. El-Kassaby. 1997. Genetic diversity and gene conservation of Pacific silver fir (*Abies amabilis*) on Vancouver Island, British Columbia. *For. Genetics* 4:85-98.

Delouche, J.C. and C.C. Baskin. 1973. Accelerated aging techniques for predicting the relative storability of seed lots. *Seed Sci. & Technol*. 1:427-452.

Denti, D. and D.J. Schoen. 1988. Self-fertilization rates in white spruce: effect of pollen and seed production. *J. Hered*. 79:284-288.

El-Kassaby, Y.A. 1989. Genetics of seed orchards: expectations and realities. *In* Proc. of the 20th South. For. Tree Improve. Conf., June, 1989. Charleston, South Carolina, USA. 87-109

El-Kassaby, Y.A. 1992. Domestication and genetic diversity - should we be concerned? *For. Chron*. 68:687-700.

El-Kassaby, Y.A. 1995. Evaluation of tree-improvement delivery system: factors affecting genetic potential. *Tree Physiology* 15:545-550.

El-Kassaby, Y.A. 1997. Clonal-row and random seed orchard designs: comparison of mating pattern and seed yield. *In* Proc. of the 24th South. For. Tree Improve. Conf., June, 1997. Orlando, Florida. 371-373.

El-Kassaby, Y.A. and G.R. Askew. 1991. The relation between reproductive phenology and output in determining the gametic pool profile in a Douglas-fir seed orchard. *For. Sci*. 37:827-835.

El-Kassaby, Y.A. and R. Davidson. 1991. Impact of pollination environment manipulation on the apparent outcrossing rate in a Douglas-fir seed orchard. *Heredity* 66:55-59.

El-Kassaby, Y.A. and B. Jaquish. 1996. Population density and mating pattern in western larch. *J. Hered*. 87:438-443.

El-Kassaby, Y.A. and Edwards, D.G.W. 1998. Genetic control of germination and the effects of accelerated aging in mountain hemlock seeds and its relevance to gene conservation. *For. Ecol. Manage*. 112: 203-211.

El-Kassaby, Y.A. and G. Namkoong. 1994. Impact of forest management practices on genetic diversity and its conservation. *In* Drysdale R.M., S.E.T. John, and A.C. Yapa, (eds.): Proc. International Symposium on Genetic Conservation and Production of Tropical Forest Tree

Seed ASEAN-CANADA Forest Tree Seed Centre Project, Muak-Lek, Saraburi, Thailand. 205-213.

El-Kassaby, Y.A and G. Namkoong. 1995. Genetic diversity of forest tree plantations: consequences of domestication. *In* IUFRO XX World Congress (Caring for the Forest: Research in a Changing World), August 1995, Tampere, Finland.Congress Report, Vol. II:218-227.

El-Kassaby, Y.A. and K. Ritland. 1986. The relation of outcrossing and contamination to reproductive phenology and supplemental mass pollination in a Douglas-fir seed orchard. *Silvae Genet.* 35:240-244.

El-Kassaby, Y.A. and K. Ritland. 1992. Frequency-dependent male reproductive success in a polycross of Douglas-fir. *Theor. Appl. Genet.* 83:752-758.

El-Kassaby, Y.A. and K. Ritland. 1996a. Genetic variation in low elevation Douglas-fir of British Columbia and its relevance to gene conservation. *Biodiversity and Conservation* 5:779-794.

El-Kassaby, Y.A. and K. Ritland. 1996b. Impact of selection and breeding on the genetic diversity in Douglas-fir. *Biodiversity and Conservation* 5:795-813.

El-Kassaby, Y.A. and A.J. Thomson. 1995. Effect of seed biology and common seedling nursery production practices on parental contribution rank and relative performance. *For. Sci.* 42:228-235.

El-Kassaby, Y.A., D.G.W. Edwards and D.W. Taylor. 1992. Genetic control of germination parameters in Douglas-fir and its importance for domestication. *Silvae Genet.* 41:48-54.

El-Kassaby, Y.A., M.D. Meagher, J. Parkinson, and F.T. Portlock. 1987. Allozyme inheritance, heterozygosity and outcrossing rate among *Pinus monticola* near Ladysmith, British Columbia. *Heredity* 58:173-181.

El-Kassaby, Y.A., M.D. Meagher, and R. Davidson. 1993. Temporal variation in the outcrossing rate in a natural stand of western white pine. *Silvae Genet.* 42:131-135.

El-Kassaby, Y.A., K. Ritland, A.M.K. Fashler, and W.J.B. Devitt. 1988. The role of reproductive phenology upon the mating structure of a Douglas-fir seed orchard. *Silvae Genet.* 37:76-82.

El-Kassaby, Y.A., J. Russell and K. Ritland. 1993b. Mixed-mating in an experimental population of western redcedar, *Thuja plicata*. *J. Hered.* 85:227-231.

Ellstrand, N.C. and D.L. Marshall. 1986. Patterns of multiple paternity in populations of *Raphanus sativus*. *Evolution* 40:837-842.

Erickson, V.J. and W.T. Adams. 1989. Mating success in a coastal Douglas-fir seed orchard as affected by distance and floral phenology. *Can. J. For. Res.* 19:1248-1255.

Farris, M.A. and J.B. Mitton. 1984. Population density, outcrossing rate, and heterozygote superiority in ponderosa pine. *Evolution* 38:1151-1154.

Furnier, G.R. and W.T. Adams. 1986. Mating system in natural populations of Jeffrey pine. *Amer. J. Bot.* 73:1022-1008.

Knowles, P., G.R. Furnier, M.A. Aleksiuk, and D.J. Perry. 1987. Significant levels of self-fertilization in natural populations of tamarack. *Can. J. Bot.* 65:1087-1091.

Meagher, T.R. 1986. Analysis of paternity within a natural population of *Chamaelirium luteum*. I. Identification of most-likely male parents. *Am. Nat.* 128:199-215.

Morgante, M., G.G. Vendrain, and P. Rossi. 1991. Effects of stand density on outcrossing rate in two Norway spruce (*Picea abies*) populations. *Can. J. Bot.* 69:2704-2708.

Muona, O. and A. Harju. 1989. Effective population sizes, genetic variability, and mating system in natural stands and seed orchards of *Pinus sylvestris*. *Silvae Genet.* 38:221-228.

Nakamura, R.R. and N.C. Wheeler. 1992. Pollen competition and parental success in Douglas-fir. *Evolution* 46:846-851.

Neale, D.B. and W.T. Adams. 1985. The mating system in natural and shelterwood stands of Douglas-fir. *Theor. Appl. Genet.* 71:201-207.

Neale, D.B. and W.T. Adams. 1985. Genetic implications of shelterwood regeneration of Douglas-fir in southwest Oregon. *For. Sci.* 31:995-1005.

Pielou, E.C. 1975. Ecological diversity. John Wiley and Sons. N.Y.

Ritland, K. 1989. Correlated matings in the partial selfer *Mimulus guttatus*. *Evolution* 43:848-859.

Ritland, K. and Y.A. El-Kassaby. 1985. The nature of inbreeding in a seed orchard of Douglas-fir as shown by an efficient multilocus model. *Theor. Appl. Genet.* 71:375-384.

Ritland, K. and Y.A. El-Kassaby. 1995. Evolutionary and ecological significance of plant mating systems. *In* Siddiqui K.A. (ed.): Inter. Symp. New Genetical Approaches to Crop Improvement Atomic Energy Agricultural Research Centre, Tandojam Sind, Pakistan (in press).

Roberds, J.H., S.T. Friedman and Y.A. El-Kassaby. 1991. Effective number of pollen parents in clonal seed orchards. *Theor. Appl. Genet.* 82:313-320.

Rudin, D., D. Lindgren, and M. Rasmusson. 1977. Inbreeding in a seed tree stand of *Pinus sylvestris* L. in northern Sweden. A study by the aid of the isozyme technique. Res. Notes by Dept. *For. Genet.* No. 25. 45pp.

Rudin, D., O. Muona, and R. Yazdani. 1986. Comparison of the mating system of *Pinus sylvestris* in natural stands and seed orchards. *Hereditas* 104:15-19.

Schemske, D.W. and R. Lande. 1985. The evolution of self-fertilization and inbreeding depression in plants. II. Empirical observations. *Evolution*. 39:41-52.

Schoen, D.J. and S.C. Stewart. 1986. Variation in male reproductive investment and male reproductive success in white spruce. *Evolution* 40:1109-1120.

Schoen, D.J. and S.C. Stewart. 1987. Variation in male fertilities and pairwise mating probabilities in *Picea glauca*. *Genetics* 116:141-152.

Shaw, D.V. and R.W. Allard. 1982. Estimation of outcrossing rates in Douglas-fir using isozyme markers. *Theor. Appl. Genet.* 62: 113-120.

Shea, K. L. 1987. Effects of population structure and cone production on outcrossing rates in Engelmann spruce and subalpine fir. *Evolution* 41:124-136.

Stoehr, M.U. and Y.A. El-Kassaby. 1997. Levels of genetic diversity at different stages of the domestication cycle on interior spruce in British Columbia. *Theor. Appl. Genet.* 94:83-90.

Worrall, J. 1983. Temperature - bud-burst relationships in amabilis and subalpine fir provenance tests replicated at different elevations. *Silvae Genet.* 32:203-209.

Yazdani, R., D. Lindgren, and D. Rudin. 1985. Gene dispersion and selfing frequency in a seed-tree stand of *Pinus sylvestris* (L.). *In* Proceedings of the meeting of the IUFRO Working party "Ecological and Population Genetics: held in Gottingen, Aug. 21-24, 1984, 138-154.

Yazdani, R., D. Lindgren, and S. Stewart. 1989. Gene dispersion within a population of *Pinus sylvestris* . *Scand. J. For. Res.* 4:295-306.

Impact of logging on genetic diversity in humid tropical forests

Wickneswari Ratnam, Chai T. Lee, Norwati Muhammad and Timothy J. B. Boyle

Introduction

Sustained management of natural forests depends on their ability to regenerate. Successful regeneration of forests accompanied by conservation of genetic diversity of its species is important for both short-term adaptation to environmental change and long-term impact on species and communities (Templeton 1995).

Maintenance of genetic diversity in forest tree populations that are undergoing population changes due to natural or human-induced events is seen to be instrumental for adaptability and continued evolution (Müller-Starck 1985; Gregorius *et al.* 1985; Ledig 1988; Namkoong 1991; Müller-Starck *et al.* 1992; Bush and Smouse 1992). Hence, assessment of regeneration capacity and genetic diversity would be useful for developing sustainable forest management guidelines and effective conservation strategies.

The aims of the present study are to assess the impact of selective logging on some genetic diversity parameters describing genetic diversity of adult trees and to compare the respective parameters between unlogged and adjacent stands regenerated naturally after cutting more than 40 years earlier.

1. Impact of logging on genetic diversity

Logging in the production forests in Malaysia is managed under two harvesting systems i.e. the 30-year Selective Management System (SMS) cutting cycle and the 55-year Malayan Uniform System (MUS) cutting cycle (Wyatt-Smith 1963; Thang 1987). In the SMS practice, selective felling is carried out where the cutting limit for dipterocarp species should not be less than 50 cm diameter at breast height (dbh) and for non-dipterocarp species the cutting limit should not be less than 45cm dbh. In this practice, sufficient adolescent trees (i.e. at least 32 commercial trees/ha or equivalent should be left after logging (Thang 1988). In the MUS practice, all commercial trees not less than 45cm dbh are allowed to be logged if there are sufficient regenerants of the species.

C. Mátyás (ed.), Forest Genetics and Sustainability, Volume 63, 171–181.

Two case studies on impact of logging on genetic diversity parameters of plant species are reported here. Genetic structure of plant species with different economic importance and life history strategies were examined in two humid tropical forest types; a ridge forest and a mixed dipterocarp lowland forest which had been logged less than a year ago and more than 40 years ago respectively. About 30 mature individuals, each of 5-6 species from each forest type were randomly sampled for isozyme and randomly amplified polymorphic DNA (RAPD) analysis.

1.1. Tropical ridge forest

Genetic diversity measures for five tropical rainforest species in Compartment 48 of Serting Tambahan Forest Reserve, Negeri Sembilan (2° 50' N, 102° 15'E) were compared between before logging and immediately after logging gene pools of adult individuals. This forest management unit (FMU) was logged in late 1996 under the Selective Management System (SMS). The size of the FMU is about 140 ha. The elevation of the FMU ranges from 400-700m. Genetic diversity of three emergent/canopy tree species, viz. *Parkia speciosa, Scaphium macropodum, Shorea leprosula* and a climbing palm, *Daemonorops verticillaris* were analysed using 4-13 isozyme loci and a pteridophyte, *Tectaria singaporeana* using 27 RAPD loci in the FMU (Wickneswari *et al.* 1997a). Leaves or inner bark tissues were used for isozyme and RAPD analysis. Details on the biology of the species are given in Table 1.

The reduction in basal area of the FMU after logging was about 56%. The number of individuals that could be sampled for each species was reduced after logging by more than 50% for *S. leprosula* and *T. singaporeana* (Table 2). Genetic diversity (expected heterozygosity (He) for isozyme data and Shannon Diversity Index (H) for RAPD data) was reduced by 5.0-23.4% after logging for all species examined. Generally, the order of reduction in number of individuals corresponded with the order of reduction in genetic diversity (He or H and loss in total of alleles) (Table 2). However, the magnitude of change in numbers did not result in a similar magnitude of change in expected heterozygosity for *P. speciosa, S. macropodum, S. leprosula and D. verticillaris* or in Shannon's Diversity Index for T. singaporeana indicating high heterogeneity among individuals within the FMU.

Reduction in expected heterozygosity due to logging was highest *in S. leprosula* (23.4%) followed by *D. verticillaris* (9.4%), *S. macropodum* (6.0%) and *P. speciosa* (5.0%). Shannon's Diversity Index for *T. singaporeana* before logging was 0.487 and after logging 0.443 with a reduction of 9.0%. Loss of alleles for the different species ranged from 7.7-25.0% after logging; lowest for *S. macropodum* and highest for *S. leprosula*. Appreciable loss in alleles was observed in *S. leprosula* with the low frequency class, 0.25>p=0.01 being most affected (Table 3). Among the non-timber species examined, the climbing palm, *D. verticillaris* exhibited the highest loss of alleles (13.8%) with also the low frequency class of 0.25>p=0.01 being most affected (Table 4). Buchert *et al.* (1997) also reported appreciable losses of alleles (about 25%) in *Pinus strobus* (white pine) in Ontario immediately after harvesting.

Table 1. Biology of species investigated in Serting Tambahan F.R. and Pasoh F. R.

Species Family Local name	Position in forest strata	Pollinator	Seed dipersal agent	Economic importance
Parkia speciosa Leguminosae petai	canopy tree	bats	birds and mammals	edible fruits, medical value in treating diabetes
Scaphium macropodum Sterculiaceae kembang semangkok jantan	emergent tree	insects	birds and mammals	commercial timber; plywood and veneer
Shorea leprosula Dipterocarpaceae meranti tembaga	emergent tree	insects	wind	commercial timber; structural purposes, furniture
Garcinia malaccensis Guttiferae manggis hutan	understorey tree	apomitic	birds and mammals	edible fruits
Daemonorops verticillaris Palmae rotan sabung	climbing palm	Insect?, wind?	?	Low quality rattan
Labisia pumila Myrsinaceae kacip fatimah	herbaceous shrub	?	? (sucker roots present)	local medicinal plant
Tectaria singaporeana Tectaria group -	ground fern	asexual	wind, water	?

? not known yet

The loss in total number of alleles (7.7-25.0%) was higher than the loss in expected heterozygosity (5.0-23.4%) for all species investigated indicating most probably loss in homozygous genotypes. All species with the exception of *T. singaporeana* showed an increase in observed heterozygosity after logging. Buchert *et al.* (1997) also reported a similar phenomenon for white pine where a 12% increase in observed heterozygosity was observed after harvesting.

Buchert *et al.* (1997) reported that the appreciable losses of alleles in white pine in Ontario due to harvesting could imply that gene pools of naturally regenerated progeny stands may be quite different from the original parental stands. They recommended silvicultural practices to ensure that gene pools of remaining pristine old-growth stands be reconstituted in the regenerating stands. The observed changes in genetic diversity of *P.*

speciosa, S. macropodum, S. leprosula, D. verticillaris and *T. singaporeana* due to logging may also imply that gene pools of naturally regenerated progeny stands after logging under the SMS may be slightly different from the original parental stands. As these genetic changes are very gradual, its consequences may only be observable in the third or fourth rotation coupes of the forest management unit. Hence, post-felling silvicultural practices should include enrichment planting of affected species with the threatened genotypes.

Table 2. Estimates of genetic diversity within populations for each species. (For explanation of symbols see table 5.)

	Measures of Genetic Diversity						Reduction after logging (%)		
Population	N	A	A_e	P	H_o	H_e	in H_e	in N	in total no. of alleles
P. speciosa									
before	12.3	2.1 ±0.5	1.28	55.6	0.156±0.084	0.218±0.089			
after	9.3	1.9 ±0.3	1.26	55.6	0.188±0.085	0.207±0.074	5.0	24	10.5
S. macropodum									
before	29.0	3.3 ±0.6	1.71	100	0.348±0.081	0.415±0.097			
after	19.5	3.0 ±0.4	1.64	100	0.399±0.116	0.390±0.105	6.0	33	7.7
S. leprosula									
before	9.2	2.2 ±0.3	1.72	81.8	0.258±0.079	0.418±0.073			
after	3.4	1.6 ±0.2	1.47	63.6	0.318±0.107	0.320±0.079	23.4	63	25.0
D. verticillaris									
before	19.2	2.2 ±0.4	1.34	61.5	0.209±0.062	0.254±0.070			
after	10.8	1.9 ±0.3	1.30	53.8	0.238±0.070	0.230±0.066	9.4	44	13.8
T. singaporeana									
before	33	-	-	-	0.487*	-			
after	15	-	-	-	0.443*	-	9.0	55	12.2

Table 3. Distribution of alleles in allele frequency classes for *Shorea leprosula* in a ridge forest management unit before and immediately after logging.

	Number of alleles (percentage of total no. of alleles)		**Loss of alleles (%)**
Allele frequency class	**Before logging**	**After logging**	
Total no. alleles	24	18	25.0
Four classes:			
High $p\geq0.75$	3(12.5)	8(44.4)	- 166.7
Intermediate $0.75> p\geq 0.25$	16(66.7)	10(55.6)	37.5
Low $0.25> p\geq0.01$	5(20.8)	0(0)	100
Rare $p<0.01$	0(0)	0(0)	0
Two classes:			
Common $p\geq0.05$	24(100.0)	18(100.0)	25.0
Rare $p<0.05$	0(0)	0(0)	0

Commercial timber species of low abundance in a FMU are more likely to undergo higher loss in genetic diversity than commercial species of high abundance. This was observed for *S. leprosula* which occurred sporadically in the FMU where the loss in genetic diversity was 23.4%, whereas *S. macropodum* which was more abundant experienced a loss of 6.0% only. Hence, there is a need to examine regeneration status for timber species of low abundance before permitting logging of such species.

Among the non-timber species examined, the climbing palm, *D. verticillaris* exhibited the highest loss of alleles (13.8%) most probably due to death of individuals following logging damage to supporting trees and being trampled during extraction of felled trees. Loss of alleles can lead to loss of adaptability (Gregorius et at. 1985) and environmental fitness (Müller-Starck 1985, Raddi *et al.* 1994). The existence of *D. verticillaris* may be further threatened as it is a dioecious species requiring sufficient individuals for natural regeneration. As it is quite common to allow extraction of palm species before logging is carried out, this prelogging practice needs to be reexamined to prevent further genetic erosion of these climbing plant species.

Table 4. Distribution of alleles in allele frequency classes for *Daemonorops verticillaris* in a ridge forest management unit before and immediately after logging.

	Number of alleles (percentage of total no. of alleles)		**Loss of alleles (%)**
Allele frequency class	**Before logging**	**After logging**	
Total no. alleles	29	25	13.8
Four classes:			
High $p \geq 0.75$	9(31.0)	9(36.0)	0
Intermediate $0.75 > p \geq 0.25$	7(24.2)	7(28.0)	0
Low $0.25 > p \geq 0.01$	13(44.8)	9(36.0)	30.8
Rare $p < 0.01$	0(0)	0(0)	0
Two classes:			
Common $p \geq 0.05$	24(82.8)	24(96.0)	0
Rare $p < 0.05$	5(17.2)	1(4.0)	80

Thus, logging caused loss in genetic diversity of adolescent timber and non-timber tree species in the FMU. However, commercial timber species of low abundance in an FMU are more likely to undergo higher losses in genetic diversity than commercial species of high abundance. Logging also caused loss of genetic diversity of climbing palms and ferns in the FMU. The loss in genetic diversity measured in this study are immediate losses after logging. These immediate losses in genetic diversity may be compensated by good seed or seedling bank in the FMU or nearby undisturbed gene pools.

Table 5. Estimates of genetic diversity measures for each species in the three FMUs.

Measures of Genetic Diversity*								
Species	FMU	N	A	A_e	P	H_o	H_e	F
P. speciosa	Unlogged stand	27.9	2.1 ±0.5	1.15	55.6	0.127 ±0.055	0.149 ±0.061	0.148
	Regenerated stand 1	32.0	2.1 ±0.4	1.18	55.6	0.213 ±0.084	0.184 ±0.071	-0.158
	Regenerated stand 2	32.2	2.2 ±0.5	1.28	55.6	0.223 ±0.077	0.249 ±0.080	0.104
S. macropodum	Unlogged stand	26.5	3.5 ±0.9	1.62	100	0.395 ±0.132	0.381 ±0.112	-0.037
	Regenerated stand 1	10.0	3.0 ±0.4	2.18	100	0.457 ±0.120	0.541 ±0.106	0.155
	Regenerated stand 2	14.3	3.0 ±0.8	1.99	75	0.411 ±0.159	0.498 ±0.166	0.175
S. leprosula	Unlogged stand	27.3	2.5 ±0.3	1.61	90.0	0.349 ±0.065	0.380 ±0.060	0.082
	Regenerated stand 1	30.4	2.6 ±0.3	1.68	90.0	0.331 ±0.083	0.405 ±0.063	0.183
	Regenerated stand 2	34.9	2.7 ±0.3	1.65	90.0	0.372 ±0.066	0.393 ±0.054	0.053
G. malaccensis	Unlogged stand	28.3	2.4 ±0.7	1.41	62.5	0.262 ±0.081	0.290 ±0.091	0.097
	Regenerated stand 1	20.3	2.3 ±0.6	1.38	62.5	0.299 ±0.107	0.273 ±0.095	-0.095
	Regenerated stand 2	25.5	2.3 ±0.6	1.37	62.5	0.267 ±0.084	0.272 ±0.084	0.018
D.verticillaris	Unlogged stand	31.9	2.5 ±0.4	1.39	61.5	0.214 ±0.054	0.280 ±0.070	0.236
	Regenerated stand 1	22.7	2.4 ±0.4	1.42	61.5	0.184 ±0.062	0.295 ±0.077	0.376
	Regenerated stand 2	13.6	2.2 ±0.3	1.42	61.5	0.197 ±0.053	0.300 ±0.079	0.343
L. pumila	Unlogged stand	36.8	2.0 ±0.4	1.30	55.6	0.222 ±0.111	0.231 ±0.082	0.039
	Regenerated stand 1	30.3	2.2 ±0.4	1.35	55.6	0.283 ±0.098	0.257 ±0.085	-0.101
	Regenerated stand 2	29.3	2.0 ±0.4	1.20	55.6	0.225 ±0.117	0.164 ±0.074	-0.372

*N - mean sample size per locus; A - mean number of alleles per locus; A_e - effective number of alleles per locus; P - percentage of loci polymorphic; H_o - observed heterozygosity; H_e - expected heterozygosity and F - mean fixation index ($1-H_o/H_e$). Figures with ± sign are standard deviations.

1.2. Lowland mixed dipterocarp forest

Genetic diversity of six lowland mixed dipterocarp forest species with different life history strategies from two adjacent regenerated stands logged in 1951 and 1955 in comparison with an adjacent unlogged stand in Pasoh

Forest Reserve were estimated using isozyme markers. Details on the biology of the species are given in Table 1.

Appanah and Mohd. Rasol Abd. Manaf (1990) reported sufficient regeneration after logging in a dipterocarp forest managed under the Selective Management System (SMS) of harvesting practice in Malaysia. However, Nicholson (1958) and Wyatt-Smith and Foenander (1962) reported that damage to residuals (small to large tree classes in lowland forests managed under the Malayan Uniform System varied from 28-45%. Notwithstanding this, they found sufficient undamaged adolescent trees (about 8-33 per acre), which could ensure regeneration of the forests. In this study, we observed good regeneration of the two logged stands with increases in tree density (for trees ≥ 10 cm dbh) of 18.3-28.7% compared to the unlogged stand after a single logging event. However, the mean basal area (for trees ≥ 10 cm dbh) in the regenerated stands was reduced by 13.5-40.7% compared to the unlogged stand, with the large tree diameter class showing the largest reduction, indicating that the logged stands have not matured yet.

Genetic diversity parameters were estimated using 10 isozyme loci for *S. leprosula*, 8 for *G. malaccensis*, 9 for *L. pumila* and *P. speciosa*, 13 for *D. verticillaris* and 4 for *S. macropodum* (Wickneswari *et al.* 1997b.) The measures of genetic diversity are summarised in Table 5. Mean genetic distances between stands for each species varied from 0 for *G. malaccensis* to 0.062 for *S. macropodum* (Table 6). There was no clear effect of logging on genetic diversity measures of the species investigated in the regenerated stands as substructuring of populations was detected *in P. speciosa, S. macropodum, S. leprosula* and *L. pumila*. The observed differences in allele frequencies for these four species in the three stands may have prevailed before logging or may be an effect of logging.

Table 6. Nei's (1978) mean genetic distances between stands for each species.

Species	Mean Genetic Distance (D)
P. speciosa	0.009
S. macropodum	0.062
S. leprosula	0.016
G. malaccensis	0.000
D. verticillaris	0.002
L. pumila	0.020

Effective number of alleles and expected heterozygosity for the timber species, *S. macropodum* and *S. leprosula* were generally higher in both the regenerated stands compared to the unlogged stand. Effective number of alleles and expected heterozygosity for the non-exploited timber species, *P. speciosa* and the climbing palm, *D. verticillaris* was also higher in both the regenerated stands compared to the unlogged stand. However, the understorey species, *G. malaccensis* and the herbaceous plant, *L. pumila* showed a decrease in mean number of effective alleles and heterozygosity in both the regenerated stands for the former and in regenerated stand 2 only for the latter compared to the unlogged stand. Fixation indices were generally low for all species with the exception of *D. verticillaris* which exhibited an excess of homozygotes in all three stands even though it is an obligate outbreeder.

There was no significant loss or gain of alleles in the species investigated from both regenerated stands compared to the unlogged stand. However, for *D. verticillaris* where the sampled populations in the three stands were very closely related (D = 0.002), a 12.5% loss in total number of alleles was detected in regenerated stand 2, compared to the unlogged stand (Table 7).

Table 7. Distribution of alleles in allele frequency classes for *Daemonorops verticillaris* in unlogged and regenerated forest management units in Pasoh F.R.

	Number of alleles(percentage of total no. of alleles)		
Allele frequency class	**Unlogged stand**	**Regenerated stand 1**	**Regenerated stand 2**
Total no. alleles	32	31	28
Four classes:			
High $p \geq 0.75$	8(25.0)	7(22.6)	7(25.0)
Intermediate $0.75 > p \geq 0.25$	8(25.0)	12(38.7)	12(42.9)
Low $0.25 > p \geq 0.01$	16(50.0)	12(38.7)	9(32.1)
Rare $p < 0.01$	0(0)	0(0)	0(0)
Two classes:			
Common $p \geq 0.05$	25(78.1)	27(87.1)	26(92.9)
Rare $p < 0.05$	7(21.9)	4(12.9)	2(7.1)

Generally, the early logging activities in Pasoh Forest Reserve did not cause adverse changes in genetic diversity of *P. speciosa, S. macropodum, S. leprosula, G. malaccensis, D. verticillaris* and *L. pumila*. In fact, an appreciable increase in both observed and expected heterozygosity was observed for *P. speciosa* and *S. macropodum* in the regenerated stands, indicating most probably good gene flow in the species as seeds of both species are dispersed by birds and mammals. Loss in alleles (14%) was detected in *S. macropodum* in both regenerated stands compared to the unlogged stand. As substructuring of populations was detected in this species, the difference in allelic composition cannot be clearly attributed to logging damage. 3-13% loss in alleles was detected in *D. verticillaris* in the regenerated stands compared to the unlogged stand which may be attributed to logging damage and exploitation of its canes.

2. Impact of logging on the mating system

2.1. Lowland mixed dipterocarp forest

Seeds of *S. leprosula* were collected from 5 trees in each stand with the exception of the unlogged stand where only one tree was fruiting. Seeds were germinated and about 25 germinated embryos per tree were assayed for electrophoretic variation at 7 isozyme loci viz. GPI-1, GDH-1, IDH-1, MDH-2, PGM-1, UGP-1 and UGP-2. Maximum likelihood estimation for multilocus outcrossing rate was computed using MLTR (multilocus mating system program version 0.9 with expectation-maximization method and 200 bootstraps - Ritland 1990).

Table 8. Estimates of multilocus outcrossing rates (t_m) for *Shorea leprosula* from the three Forest Management Units in Pasoh Forest Reserve.

FMU	No. of mother trees	Mean no. of progenies assayed per mother tree	No. of loci assayed	t_m (S.D.)
Unlogged stand	1	22.0	7	0.806 ± 0.143
Regenerated stand 1	5	24.7	7	0.929 ± 0.065
Regenerated stand 2	5	24.7	7	1.000 ± 0.006

Outcrossing rates for *S. leprosula* were high in all three stands indicating no inbreeding (Table 8). Manokaran *et al.* (1992) reported *S. leprosula* to be the most dynamic dipterocarp tree species with high mortality (17%) compensated by high recruitment (10%). High heterozygosity (He = 0.380-0.405) and high outcrossing rates (tm = 0.806-1.000) were detected for this species in this study corroborating the above finding. Comparison of outcrossing rates between the regenerated stands and the unlogged stand could not be made because the estimate of outcrossing rate in the unlogged stand was from one tree only. It is quite common to find low numbers of dipterocarp trees to fruit in undisturbed forests compared to disturbed forests especially in plantations and arboreta (Fox 1972; Kochummen 1961; Tang 1978; Lee 1980; Appanah and Mohd. Rasol Abd. Manaf 1990).

Kitamura *et al.* (1994) also reported no significant difference in outcrossing rates between primary forest and secondary forest, logged selectively 20 years ago for *Dryobalanops aromatica* (Dipterocarpaceae) in Brunei. They concluded that high levels of outcrossing rates were maintained by high flowering density regardless of size structure or the topography of the habitat. However, Murawski *et al.* (1994) reported reduction of outcrossing rates in *Shorea megistophylla* (Dipterocarpaceae) in a forest logged selectively about 20 years ago in Sri Lanka, which indicates inbreeding.

3. Conclusion

A single logging event in a lowland dipterocarp forest in Peninsular Malaysia in the 1950s did not cause adverse changes in genetic diversity of timber and non-timber species. More lowland forest types need to be assessed for impact of logging and other accruing disturbances especially fire on genetic diversity and population structure of both commercial and non-commercial plant species. Immediate loss in genetic diversity was detected in adolescent and non-timber tree species in a ridge forest in Peninsular Malaysia after a single logging event in 1996. These immediate losses in genetic diversity may be compensated by good seed or seedling bank in the forest management unit or nearby undisturbed gene pools. Hence, it would be useful to assess the genetic diversity of seed crops and regenerants in the FMU and adult trees in nearby FMUs. As commercial timber species of low abundance are more likely to undergo higher losses in genetic diversity than commercial species of high abundance, the current harvesting guidelines on cutting limit for commercial timber species and number of adolescent trees to be left after logging need to be reconsidered.

Acknowledgement

The authors would like to thank the Centre for International Forestry Research (CIFOR) and the International Plant Genetic Resources Institute (IPGRI) for providing a research grant to the author to carry out this study.

Literature cited

Appanah, S. and Mohd. Rasol Abd. Manaf. 1990. Smaller trees can fruit in logged dipterocarp forests. *Journal of Tropical Forest Science*, vol. 3, no. 1, pp. 80-87.

Buchert, G. P., Rajora, O. P., Hood, J. V. and Dancik, B. P. 1997. Effects of harvesting on genetic diversity in old-growth eastern white pine in Ontario, Canada. *Conservation Biology*, vol. 11, no. 3, pp. 747-758.

Bush, R.M. and Smouse, P.E. 1992. Evidence for the adaptive significance of allozymes in forest trees. *New Forests*, vol. 6, pp. 179-196.

Fox, J.E.D. 1968. Logging damage and the influence of climber cutting prior to logging in the lowland dipterocarp forest of Sabah. *Malaysian Forester*, vol. XXXI, no. 4, pp. 326-347.

Gregorius, H.R., Hattemer, T.T., Bergmann, F. and Müller-Starck, G. 1985. Unweltbelastung und Anpassungsfähigkeit von Baumpopulationen. *Silvae Genetica*, vol. 34, pp. 230-241.

Hamrick, J.L., Godt, M.J.W. and Sherman-Broyles. 1992. Factors influencing levels of genetic diversity in woody plant species. *New Forests*, vol.6, pp. 95-124.

Kitamura, K., Mohamad Yusof, A.R., Ochiai O. and Yoshimaru, H. 1994. Estimation of outcrossing rate on *Dryobalanops aromatica* Gaertn. F. in primary and secondary forests in Brunei, Borneo, Southeast Asia. *Plant Species Biology*, vol.9, pp. 37-41.

Kochummen, K. M. 1961. Precocious flowering (paedogenesis) in Dipterocarpaceae. *Malayan Forester*, vol. 24, p. 236.

Ledig, F.T. 1988. The conservation of diversity in forest trees. *BioScience*, vol. 38, pp. 471-479.

Lee, H.S. 1980. *Shorea pinanga* Burck flowers six after planting in research plot number 76. *Malaysian Forester*, vol. 43, pp. 251-253.

Manokaran, N., Abd. Rahman Kassim, Azman Hassan, Quah, E.S. and Chong, P.F. 1992. Short-term population dynamics of dipterocarp trees in a lowland rain forest in Peninsular Malaysia. *Journal of Tropical Forest Science*, vol. 5, no. 1, pp. 97-112.

Müller-Starck, G. 1985. Genetic differences between "tolerant" and "sensitive" beeches (*Fagus sylvatica* L. in an environmentally stressed adult forest stand. *Silvae Genetica*, vol. 34, pp. 241-247.

Müller-Starck, G., Baradat, P. and Bergmann, F. 1992. Genetic variation within European tree species. *New Forests*, vol. 6, pp. 23-47.

Murawski, D.A., Gunatilleke, I.A.U.N., Bawa, K.S. 1994. The effects of selective logging on inbreeding in *Shorea megistophylla* (Dipterocarpaceae) from Sri Lanka. *Conservation Biology*, vol. 8, no. 4, pp. 997-1002.

Namkoong, G. 1991. Biodiversity – issues in genetics, forestry and ethics. *Forestry Chronicle* 68, pp. 438-443.

Raddi, S.F., Stefanini, M., Camussi, A. and Giannini, R. 1994. Forest decline index and genetic variability in *Picea abies* (L. Karst.). *Forest Genetics*, vol.1, pp. 33-40.

Ritland, K. 1990. A series of FORTRAN computer programs for estimating plant mating systems. *Journal of Heredity*, vol. 81, pp. 235-237.

Tang, H.T. 1978. Age at first fruiting of *Shorea leprosula*. *Malaysian Forester*, vol. 41, pp. 294-296.

Templeton, A.R. 1995. Biodiversity at the molecular genetic level : experiences from disparate macroorganisms. *In* Hawksworth, D.L. (ed.): *Biodiversity. Measurement and Estimation.* Chapman & Hall, London. 59-64.

Thang, H.C. 1987. Forest management systems of tropical high forest, with special reference to Peninsular Malaysia. *Forest Ecology and Management*, vol. 21, pp. 3-20.

Thang, H.C. 1988. Selective management system: Concept and Practice. (Peninsular Malaysia. Forestry Department Headquarters, Kuala Lumpur. 19p.

Wickneswari,R., Lee, C.T., Norwati, M. and Boyle, T.J.B. 1997a. Immediate effects of logging on the genetic diversity of five tropical rainforest species in a ridge forest in Peninsular Malaysia. Paper presented at CIFOR Wrap-Up Workshop on Impact of Disturbance, Bangalore, India, August 1997.

Wickneswari,R., Lee, C.T., Norwati, M. and Tim, T.J.B. 1997b. Effects of logging on the genetic diversity of six tropical rainforest species in a regenerated mixed dipterocarp lowland forest in Peninsular Malaysia. Paper presented at CIFOR Wrap-Up Workshop on Impact of Disturbance, Bangalore, India, August 1997.

Wyatt-Smith, J. 1963. Manual of Malayan Silviculture for Inland Forests. Volume 1. Malayan Forest Records No. 23. Forest Research Institute, Kepong, Malaysia.

Genetic impacts of different silvicultural practices in native eucalypt forests

Jeffrey C. Glaubitz, Jimena Strk and Gavin F. Moran

Introduction

Native forests cover 20%, or 156 million hectares, of the Australian land mass (769 million hectares). Policies and strategies for the ecologically sustainable management of Australia's forests are outlined in the National Forest Policy Statement, which was jointly developed by the Commonwealth, State and Territory Governments (Commonwealth of Australia 1992). Conservation of the biological diversity in Australia's forests is one of the key national goals of this statement. It is well known that biological diversity consists of three components: ecosystem diversity, species diversity, and genetic diversity, with genetic diversity being the baseline component upon which the other two higher level components ultimately rest. Hence, the goal to conserve biological diversity in native forests by definition implies that genetic diversity should also be conserved. With regard to conservation of genetic and species diversity on managed native forests, the National Forest Policy Statement states: "reforestation of extensively managed native forest should be with species and provenances native to the area or an equivalent locality to help conserve species patterns and gene pools present prior to harvesting" (Commonwealth of Australia 1992). Hence, there is a mandate to conserve not only the levels of genetic diversity of species in a locality, but also the local character of that diversity.

Although there is a policy mandate to conserve genetic diversity in species in managed native forests in Australia, there is currently a lack of data on the impacts of management on genetic diversity levels or population genetic processes. This study was undertaken in response to this lack of data. The species of initial focus is *Eucalyptus sieberi* L. Johnson, or silvertop ash. This is the commercially most important tree species in the lowland forests of the East Gippsland region of the state of Victoria, where it is also the dominant species, and is harvested for sawlogs and/or wood chips (Boland et al. 1984, Cumming and Black 1993). It is in the East Gippsland portion of its range where *E. sieberi* achieves its best form and highest growth rates, with some stands attaining heights of over 45 metres (Boland *et al.* 1984). The Victorian Department of Natural Resources and Environment (DNRE) is conducting a large research project in East Gippsland (near Orbost) on alternative silvicultural systems in *E. sieberi*-dominated forests. The

C. Mátyás (ed.), Forest Genetics and Sustainability, Volume 63, 183–195.

Lowland Sclerophyll Forest Silvicultural Systems Project (SSP) was established "to test the hypothesis that a better balance between economic and environmental concerns could be achieved with systems other than clear-felling" (Squire 1990). The range of silvicultural treatments utilised on this site, the large amount of information collected in relation to the silvicultural studies, as well as convenient access make this experimental site an ideal location for our genetic diversity studies.

In this paper we describe in detail the design and implementation of this study on the impacts of silvicultural treatments on genetic diversity in *E. sieberi*. We also present results from a preliminary analysis of a subset of the data being collected. These results show no significant loss of genetic diversity due to silviculture. Implications of these results and considerations for future studies are discussed.

1. Materials and Methods

1.1. Field Site

On the approximately 400 hectare site the DNRE has tested eight silvicultural treatments, along with unharvested controls (Geary in prep.). These treatments represented points along two continuums: increasing gap size (from single-tree gap to 10 ha clear-fells) and decreasing overwood retention (from 50% overwood to none). Also, the DNRE examined two alternative methods of site preparation, burning and mechanical disturbance, on all of the eight treatments except for the 10 ha clear-cuts where only burning was used. Both the 4 ha and 10 ha clear-fells were regenerated by aerial sowing via helicopter using seed collected on the site itself at the time of harvest (12 trees in the case of *E. sieberi*). All other treatments were regenerated via seed-fall from retained trees (overwood and/or edge trees). The main result to date from the DNRE's silvicultural study is that although eucalypt regeneration was successfully achieved in all treatments, the best height growth was attained only on the treatments with the largest gap size and the least retained overwood (e.g., clear-felling or the seedtree system) (Lutze 1998).

1.2. Sampling strategy and treatments for the genetic diversity study

For the current genetic diversity study only three of the silvicultural treatments were chosen, large (10 ha) clear-cut, the seed tree system (10% overwood) with site preparation by burning, and the seed tree system with site preparation by mechanical disturbance. From each of these three treatments, two replicate coupes were selected for sampling of the regeneration. For the genetic diversity analysis, 100 saplings were sampled from each coupe. Sampling was done with a rectangular grid system, with 15 × 15 metre spacing in the 5 ha seed tree coupes and an 18 × 18 metre spacing in the large clear-fall coupes. At each sample point, leaves were sampled from the nearest dominant *E. sieberi* sapling. Dominant saplings (i.e., in height) were chosen because they were considered to be more likely to survive to adulthood, and hence are more likely to be representative of the eventual mature stand. They are also more likely to have originated from one of the earliest two cohorts of seedling regeneration, which have exhibited the highest survival rates (Faunt et al. in prep.). In laying out the sampling grid, care was taken to avoid getting close to the coupe boundaries, in order to minimise the genetic contribution of the bordering trees via seed. This was done because operational coupes tend to be much larger than the experimental coupes used here.

Two unharvested control stands were chosen in opposite corners of the site. These are assumed to be representative of the diversity present in *E. sieberi* across the entire site prior to harvest. These stands were predominantly even aged, and virtually every mature tree was sampled, i.e. a total of 100 trees from each control stand. Trees were sampled by shooting down a branch with a 0.308 calibre rifle – both leaves and fruit were collected from the sampled trees. All of the *E. sieberi* seed trees on the four seed tree coupes were sampled in a similar fashion (25 seed trees in total from the four seed tree coupes).

DNA was extracted from about 8 g of leaves (fresh weight) following a modified version of the procedure of Wagner et al. (1987).

1.3. RFLP markers

For single copy nuclear RFLP analysis, 4 μg of DNA was digested with the restriction endonucleases EcoRV, BglII, or DraI and electrophoresed through 0.8% agarose. Southern Blots were prepared by alkaline transfer (Reed and Mann 1985). Probes were radioactively labelled for hybridisation with [α32P]-dCTP to high specific activity via random primer extension (Sambrook et al. 1989). Thirty single copy nuclear RFLP probes were chosen for this study from the 40 employed in a study in *E. nitens* by Byrne *et al.* (1998). Fifteen of the probes selected for use in *E. sieberi* were from a *E. globulus* cDNA library, while the other 15 were from an *E. nitens* genomic library (Byrne et al. 1994). All except 7 of the 30 chosen probes have been placed on a genetic linkage map of *E. nitens* (Byrne *et al.* 1995). For each probe the best restriction enzyme was selected from those listed above based upon which enzyme gave the most clearly interpretable genetic polymorphisms. Results from 10 of the 30 RFLP loci are presented here (Table 1).

Table 1. 10 RFLPs and 2 microsatellites employed for preliminary analysis of genetic diversity in *Eucalyptus sieberi*.

Marker*	Type †	Source species	Total no. alleles	H_T ‡	F_{IS} §	HWE (G^2 probability)¶
c030B	RFLP (cDNA)	*E. globulus*	14	0.579	0.004	0.9892
c092E	RFLP (cDNA)		6	0.116	-0.049	0.9005
c135E	RFLP (cDNA)		14	0.635	0.031	0.9852
c288E	RFLP (cDNA)		8	0.400	0.009	0.8990
c395E	RFLP (cDNA)		13	0.739	-0.015	0.6280
g086E	RFLP (genomic)	*E. nitens*	12	0.483	0.023	0.9999
g121E	RFLP (genomic)		5	0.675	0.047	0.3153
g142E	RFLP (genomic)		2	0.304	-0.197	**0.0304**
g195E	RFLP (genomic)		3	0.010	-0.005	1.0000
g256E	RFLP (genomic)		7	0.094	-0.046	1.0000
Es115	Microsatellite $(ACAT)_8(AC)_{19}(TA)_5$	*E. sieberi*	23	0.876	-0.023	0.9990
Es211	Microsatellite (GA_{12})		13	0.852	-0.020	0.9598
Overall			120	0.480	-0.009	

* The last letter in the RFLP marker names refers to the restriction enzyme employed (B: BglII; E: EcoRV).

† All RFLP markers reveal single copy nuclear loci (Byrne et al. 1998). For the microsatellites, the repeat type and number of repeats observed in the cloned allele are given.

‡ H_T (total gene diversity) is calculated according to Nei (1973) over all of the 610 trees sampled.

§ F_{IS} (fixation index) is given for each locus in the Unharvested Control 1 (UC1) population (sample size of 100 trees). The overall F_{IS} is that over the 12 markers in the UC1 population and was calculated using (He - H_O)/ He.

¶ The probability values from G^2-tests for deviation from Hardy Weinberg equilibrium (HWE) genotype proportions in the unharvested control replicate 1 are given. Probability values less than 0.05 are shown in bold.

1.4. Microsatellite markers

For this study ten microsatellite markers were developed in *E. sieberi* from a non-enriched genomic library. Clones containing dinucleotide microsatellite motifs were identified via hybridisation of colony filters with poly(CA) and poly(GA) probes simultaneously. Putative positive clones were sequenced and primer pairs were designed from appropriate sequences with the aid of the computer program Primer, version 0.5 (Lincoln et al. 1991). Primer pairs were tested and PCR conditions optimised for each working pair. One of the primers for each working primer pair was re-synthesized with fluorescent labels to allow microsatellite analysis to be carried out on an ABI 310 Genetic Analyzer (Applied Biosystems). Results from 2 of the 10 microsatellite markers developed are presented (Table 1).

1.5. Data analysis

Preliminary analysis of data from 10 RFLP loci and 2 microsatellites was performed with the computer package POPGENE (Yeh *et al.* 1997). This preliminary analysis was conducted on 3 of the 4 treatments: on the unharvested trees from the two replicate control treatments, on the unharvested seed trees and the sapling regeneration from the two replicates of the seed tree system with site preparation by burning, and on the sapling regeneration from the two large clear-fall coupes with aerial re-sowing (610 trees in total). These two Seed Tree Burnt (SB) coupes each had 5 *E. sieberi* seed trees left behind to re-seed the coupes after logging. As RFLP and microsatellite assays are currently in progress on sapling regeneration from two seed tree coupes with site preparation by mechanical disturbance (Seed Tree Disturbed or SD), data will not be presented from this treatment.

2. Results

2.1. Overall level of genetic diversity

The levels of genetic variability detected by the 12 selected markers are shown in Table 1. The number of alleles detected at each locus over the 610 trees and the total gene diversity (expected heterozygosity) over these 610 trees are given. The RFLP markers indicate that there is a high level of genetic diversity in *E. sieberi*, with an average gene diversity across the 10 RFLPs of 0.404. As expected, the microsatellites display much greater variability than do the RFLPs (average gene diversity across the two microsatellites of 0.864).

2.2. Conformation of the marker genotypes to Hardy-Weinberg ratios

Table 1 also shows that genotypic ratios at most of the markers conform to Hardy-Weinberg Equilibrium (HWE) expectations in the Unharvested Control replicate 1 (UC1) population (similar results were found in Unharvested Control replicate 2 – data not shown). In spite of the fact that many of the RFLP markers and both of the microsatellite markers detected a number of rare alleles, only one of the loci (g142E, having only two alleles) showed significant deviation from HWE (excess of heterozygotes, $\alpha = 0.05$). As there were 12 tests, this sole deviation may be due to chance alone. However, g142E also showed a significant excess of heterozygotes in Unharvested Control replicate 2 (UC2) (data not shown). The latter result suggests that there may be an underlying biological explanation for the departure of g142E from HWE – for example, one of the alleles may be closely linked to a deleterious recessive allele.

2.3. Comparative genetic diversity analysis

Levels of genetic diversity of the three treatments are compared in Table 2. For this comparison, the two replicates of each treatment have been pooled and treated as single populations. Also, the 10 seed trees from Seed Tree Burnt replicate 1 (SB1) and Seed Tree Burnt replicate 2 (SB2) were pooled with the other unharvested trees. Regeneration on the Seed Tree Burnt treatment showed somewhat lower values for all three diversity measures given in the table, whereas the Clear Fell regeneration and the Unharvested Control trees had nearly identical values for all three measures. Although the lower values observed in the Seed Tree Burnt regeneration for all three measures are consistent, the differences are not large and are likely not statistically significant – proper statistical tests of the differences, utilising bootstrapping and/or permutation techniques, will be performed once the entire data set has been amassed. Differences in the number of alleles detected (N_a) in each of the three groups in Table 2 are entirely due to rare alleles. All of the 48 alleles that were not detected in all three groups were present at a frequency of less than 1% in the overall sample of 610 trees.

Table 2. Measures of genetic diversity for 3 silvicultural treatments in *E. sieberi* based on 12 loci (10 RFLPs and 2 microsatellites.)

Treatment	# Trees	N_a *	N_e †	H_e ‡
Unharvested Control and Seed Trees	210	8.00(96)	2.95	0.486
Clear Fell Regeneration	200	8.08(97)	3.00	0.480
Seed Tree Burnt Regeneration	200	7.25(87)	2.68	0.467

* N_a = Average number of alleles per locus (total number in parentheses)
† N_e = Average effective number of alleles (Hartl and Clark 1989, p. 125)
‡ H_e = Expected heterozygosity or gene diversity (Nei 1973)

Diversity measures from the six individual replicates (i.e., without pooling) are compared in Figure 1. The 10 seed trees from the Seed Tree Burnt (SB) replicates have been excluded from this analysis (i.e., the diversity estimates presented for the SB replicates are based upon the sapling regeneration only). In each bar graph in the figure the replicates are ranked in descending order. Here the differences in N_a are more pronounced, with the two Unharvested Control replicates (UC1 and UC2) having the second highest and the lowest N_a respectively (Figure 1A). Again the differences among the replicates in N_a are entirely due to rare alleles. Of the 69 alleles that were not detected in at least one of the replicates, 66 had an overall frequency (among all 610 trees) of less than 2%, and the remaining 3 were all less than 5% in frequency. Based upon the measures Effective Number of Alleles (N_e) and Expected Heterozygosity (H_e), the two Seed Tree Burnt replicates perform consistently poorly, and the Clear Fells perform well

(Figure 1B and 1C). Again however, the differences among the replicates are small and are not likely to be statistically significant. A paired T-test (Nei 1987) indicated that the lowest N_e , found in SB1, was not significantly different from that of UC2.

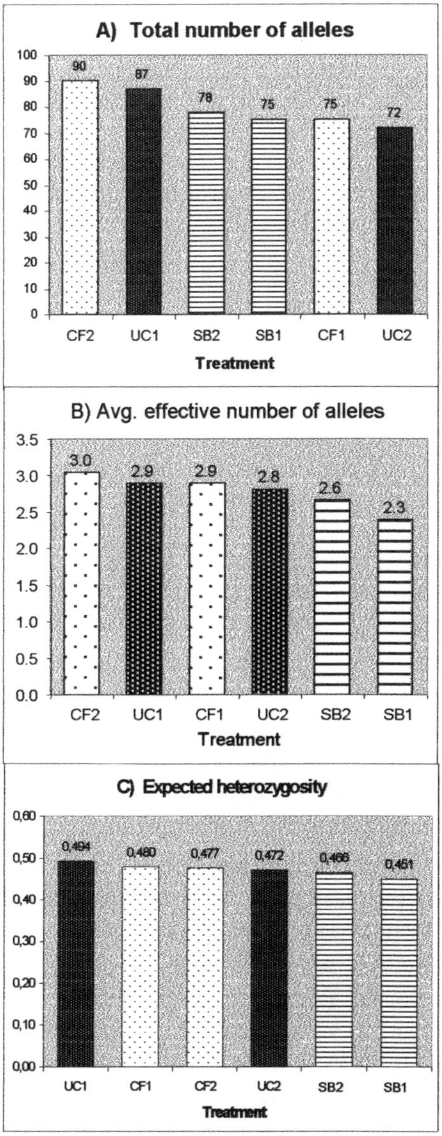

Figure 1. Comparative genetic diversity analysis of the 6 individual coupes. UC1: unharvested control replicate 1; UC2: unharvested control replicate 2; CF1: regeneration on clear-felled replicate 1; CF2: regeneration on clear-felled replicate 2; SB1: regeneration on seed tree burnt replicate 1; SB2: regeneration on seed tree burnt replicate 2. A: Comparison of the total number of alleles. B: Comparison of the average effective number of alleles. C: Comparison based on expected heterozygosity. Replicates are shown in rank order (highest to lowest) for each statistic.

The sampled trees were also pooled into two populations – the 210 unharvested trees vs. the 400 sapling regenerants. F_{ST} between these two populations was only 0.002, indicating that differences between the two generations accounted for only 0.2% of the total genetic variation.

2.4. Inbreeding

Fixation indices (F_{IS}) for each replicate over all the 12 markers are given in Table 3. All of these are close to zero indicating no marked deviations from HWE proportions in any of the replicates. Furthermore, there is no clear tendency for the fixation indices to be higher in the sapling regeneration, indicating that harvesting has not led to increased levels of inbreeding. These results do not suggest that there is a higher proportion of inbred individuals in stands at a younger developmental stage (i.e., in saplings vs. unharvested trees) as has been observed by Hardner and Potts (1997) in *E. regnans*

Muell. However, the sampling in this study of "dominant" saplings (with a higher probability of survival to maturity and most likely derived form the earliest cohorts of regeneration) may have influenced this observation, as it may have resulted in a tendency to sample non-inbred individuals with little or no inbreeding depression and thus better growth. However under the consideration that it is the genetic makeup of the eventual mature stands that develop from the sapling regeneration that is of most interest, the sampling strategy employed in this study seems appropriate.

Table 3. Fixation indices for two replicates of each of 3 silvicultural treatments of *E. sieberi*.

Replicate	Fixation index
Unharvested control 1	-0.009
Unharvested control 2	-0.044
Clear Felled 1	0.013
Clear Felled 2	-0.008
Seed tree burnt 1	-0.041
Seed tree burnt 2	0.003

2.5. Relationships among the replicates

The relationships among the replicates revealed by the 12 markers are summarised in a UPGMA dendrogram based upon Nei's (1972) genetic distance (Figure 2). The two clear-felled replicates (CF1 and CF2) cluster closest together and in turn form a group with the two unharvested control replicates (UC1 and UC2). The seed trees from the two Seed Tree Burnt replicates (SB1 and SB2) cluster together with the sapling regeneration on their respective coupes but away from each other and away from the other treatments, with SB1 showing the most pronounced divergence.

The two microsatellite markers had an average genetic distance among the 8 "populations" that was more than 9 times higher than that of the RFLPs (0.1776 vs. 0.0196 respectively). Accordingly, genetic distances from the 2 microsatellites dominate the dendrogram based upon all 12 loci – the dendrogram based upon the 2 microsatellites alone has identical topology (with longer branch lengths) as that based upon all 12 loci while the topology of the dendrogram derived from the 10 RFLPs alone is somewhat different (results not shown). It is likely that the topology of the latter, "RFLP-only" dendrogram is more sensitive to sampling error due to the much smaller genetic distances upon which it is based. Hence it appears that the microsatellites are more sensitive indicators of fine scale genetic structure than are the RFLPs.

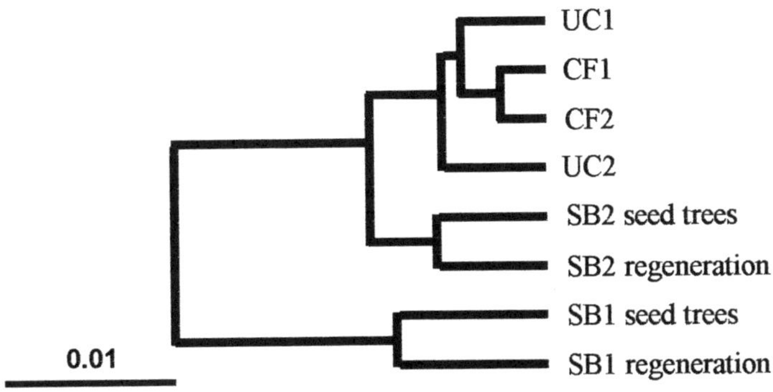

Figure 2. Cluster analysis of silvicultural treatments in *E. sieberi* based on Nei's (1972) genetic distance and the UPGMA algorithm. UC: Unharvested Control; CF: Clear-felled; SB: Seed Tree Burnt.

3. Discussion

Preliminary analysis of a subset of the data being collected in this study indicates little or no loss of genetic diversity in *E. sieberi* due to regeneration practices. In general, differences in diversity measures between the sapling regeneration and the unharvested controls are quite small and are not likely to be statistically significant. No loss of genetic diversity was observed due to clear-felling and only very slight reductions in the effective number of alleles (N_e) and in expected heterozygosity (H_e) were observed under the seed tree system with site preparation by burning.

As theory predicts that H_e is reduced by genetic drift at the rate of $1/(2 N_e)$ each generation (where N_e is the effective population size) (Frankel *et al.* 1995), it is likely that the bottlenecks imposed by the harvesting and regeneration systems were not severe enough to result in significant reductions in H_e over one generation. The true effective sizes of the populations from which regeneration on the sampled coupes were derived are quite likely much larger than the 5 seed trees left behind on each of the seed tree coupes or the 12 trees from which seed was collected for re-sowing of the clear falls. This is because a large number of pollen parents are likely to have contributed to the seed produced on each tree (see below). However, even an effective population size as small as 5 would only result in a 10% reduction in expected heterozygosity – a difference that may be difficult to detect without very large samples of individuals and loci.

The number of alleles (N_a), in contrast, should be much more sensitive to bottlenecks than is H_e (Nei et al. 1975, Leberg 1992, Frankel et al. 1995). Accordingly, genetic bottlenecks due to silviculture that are not severe enough to palpably affect H_e may instead be detected by significant reductions in N_a (Buchert et al. 1997). However, because of the high likelihood that alleles that are present in a population at low frequencies may not be included in a sample of that population (Marshall and Brown 1975, Gregorius 1980, Nei 1987, Namkoong 1988), apparent differences in N_a due to rare alleles alone should be interpreted with a great deal of caution. With the sample sizes employed in this study, failure to detect a given allele in a sample of the regeneration can only be interpreted as probable loss of that allele if it was relatively common (i.e., at a frequency of 5% or greater) in the unharvested trees in the first place. As the observed differences in N_a among replicates in this study were almost entirely due to alleles of less than 2% in overall frequency, they cannot be considered to be statistically or biologically significant. The fact that all alleles of greater than 5% overall frequency were present in all replicates indicates that there was no significant loss of genetic diversity due to silviculture.

However, genetic drift effects due to the seed tree regeneration method are apparent in the UPGMA dendrogram (Figure 2). The topology of this dendrogram is quite consistent with logical expectations, indicating that the genetic markers employed (particularly the microsatellites) are sensitive and accurate indicators of fine scale genetic structure. The two clear-cut replicates (CF1 and CF2) are genetically quite similar as these coupes were re-sowed with the same seed lot – collected from 12 *E. sieberi* trees from the site at the time of harvest. As these 12 trees were from scattered locations on the eastern half of the site (Geary in prep.), CF1 and CF2 are genetically closer to the Unharvested Control replicate 1 (UC1), located in the north-eastern corner of the site, than to the more spatially distant Unharvested Control replicate 2 (UC2) in the opposite corner. The positions in the dendrogram of the two replicates samples of Seed Tree Burnt regeneration ("SB1 regeneration" and "SB2 regeneration" in Figure 2) suggest that it is these two populations that have been most strongly affected by genetic drift ("SB1 regeneration" in particular). This follows from the fact that these populations were regenerated via natural seed fall from a small number of *E. sieberi* seed trees (only 5 seed trees on each coupe – "SB1 seed trees" and "SB2 seed trees" on the dendrogram), causing stochastic shifts in some allele frequencies. Dendrogram construction using a clustering algorithm (such as UPGMA) has hence facilitated the detection of subtle genetic drift effects that could not have been convincingly detected with genetic diversity measures alone.

The non-zero distances between the SB1 and SB2 seed trees and the regeneration upon their respective coupes most likely reflect the genetic contribution of pollen parents other than the seed trees. In addition, these genetic distances may also reflect contributions of other seed sources (e.g., edge trees and logging slash) as well as sampling error due to sampling the regeneration. There is no sampling error associated with the seed trees themselves, however, as in both cases ("SB1 seed trees" and "SB2 seed trees") the entire population of interest was examined.

On the whole, except for the subtle genetic drift effects due to the seed tree system apparent via cluster analysis, it appears that *E. sieberi* is quite resilient to the effects of logging in terms of genetic diversity. The local abundance and reproductive biology of this species likely have a primary role in this resilience. Fixation indices obtained in this study indicate that outcrossing rates in this species are likely high. Flowering in *E. sieberi* on the study site has been shown to be highly synchronised (Bassett 1995), promoting outcrossed pollination. Furthermore, large annual seed crops are produced in *E. sieberi* and seed production has been shown to be quite uniform across stands (Bassett 1995). These factors, combined with the fact that *E. sieberi* trees hold seed crops for up to 6 consecutive years (Basset 1995), make it likely that a large number of pollen parents contribute to the seed crop of any individual tree. As it is the earliest cohorts of *E. sieberi* saplings that make up the bulk of the regeneration after logging (Faunt et al. in prep.), coupes are primarily regenerated with seed derived from fertilisations that occurred prior to harvest. Hence, on the seed tree coupes, genes of any harvested tree have a very good chance of being represented in the sapling regeneration. Similarly, a very large number of pollen parents are likely to have contributed to the seed used for aerial re-sowing of the clear-cuts (collected from 12 scattered trees).

The abundance of polymorphic marker loci in *E. sieberi* make it a convenient model species for studies of the genetic effects of forest management practices. However, since DNA markers are most likely selectively neutral, reductions of the variation at particular loci due to the selection of seed trees for form or seed crop are not likely to be diagnosed. On the other hand, substantial disruptions of drift and/or the mating system by silvicultural practices (i.e., severe bottlenecks or drastic reductions in the outcrossing rate) can dramatically impact adaptive variation. Such disruptions are definitely detectable with DNA markers – indeed, they are more accurately quantified if the employed markers are neutral.

In terms of preserving genetic diversity, the seed tree system was slightly poorer than clear-felling with aerial re-sowing. Also cluster analysis indicated that genetic drift was more prominent with the seed tree system. It is intuitive that a higher degree of genetic control is attainable under clear-felling. As it would be possible to collect seed from many more than 12 trees for aerial re-sowing, enhancement of genetic diversity is quite feasible. Collection of seed from 50 or more trees from a wider area (e.g., within a 50 km radius) for re-sowing would likely result in a higher level of genetic diversity in the regeneration relative to the pre-harvest stand, particularly in terms of the number of alleles. Rare alleles that may prove useful for surmounting future adaptive challenges are more likely to be preserved under this scenario.

E. sieberi was the most common eucalypt species on the study site prior to harvest. Furthermore, the percentage composition of this species was even higher in the sapling regeneration (Lutze 1998). Given the local abundance of *E. sieberi*, coupled with a reproductive biology that makes it resilient to potential genetic erosion due to silviculture, it is not all that surprising that no significant impacts on genetic diversity were found in this study. As locally rare species are far more likely to be adversely impacted genetically by harvesting and regeneration practices, the next step in this research will be to examine such a species. *E. botryoides* Smith is an excellent candidate on the study site, as it was rare prior to harvest and is an even smaller component of the eucalypt regeneration (Lutze 1998). The results of such a study of a rare species should give an interesting contrast to the study of *E. sieberi* reported herein.

Acknowledgement

We gratefully acknowledge financial support received from the Forest and Wood Products Research and Development Corporation. We also thank John V. Owen, Josephine Morosin and Chris McElhinny for technical assistance. Special thanks go to the Victoria Dept. of Natural Resources and Environment (Mark Lutze and Peter Geary in particular) for access to the study site, information and advice.

Literature cited

Bassett, O.D. 1995. Development of seed crops in *Eucalyptus sieberi* and *E. globoidea* in a Lowland Sclerophyll Forest in East Gippsland. VSP Internal Report No. 26, Dept. Conservation and Natural Resources, Victoria.

Boland, D.J., Brooker, M.I.H., Chippendale, G.M., Hall, N., Hyland, B.P.M., Johnston, R.D., Kleinig, D.A. and J.D. Turner 1984. *Forest trees of Australia*. Nelson Wadsworth and CSIRO, Melbourne.

Buchert, G.P., Rajora, O.P., Hood, J.V. and B.P. Dancik 1997. Effects of harvesting on genetic diversity in old-growth eastern white pine in Ontario, Canada. *Conserv. Biol.* 11:747-758.

Byrne, M., Moran, G.F., Murrell, J.C. and W.N. Tibbits 1994. Detection and inheritance of RFLPs in *Eucalyptus nitens*. *Theor. Appl. Genet.* 89:397-402.

Byrne, M., Murrell, J.C., Allen, B. and G.F. Moran 1995. An integrated genetic linkage map for eucalypts using RFLP, RAPD and isozyme markers. *Theor. Appl. Genet.* 91:869-875.

Byrne, M., Parrish, T.L. and G.F. Moran 1998. Nuclear RFLP diversity in *Eucalyptus nitens*. *Heredity* 81:225-233.

Commonwealth of Australia 1992. National forest policy statement: a new focus for Australia's forests. Commonwealth of Australia, Canberra.

Cumming, F. and P. Black 1993. Cabbage Tree pre-harvest assessment. *VSP Internal Report No. 21*. Dept. Conservation and Natural Resources, Victoria.

Faunt, K. A., Geary, P.W., Cunningham, R. and P. Gibbons (in prep.) Silvicultural Systems Project: Germination and early survival of eucalypt regeneration following a range of harvesting and site preparation treatments in a lowland sclerophyll forest. *VSP Technical Report No. 31*, Dept. of Natural Resources and Environment, Victoria.

Frankel, O.H., Brown, A.H.D. and J.J. Burdon 1995. The conservation of plant biodiversity. Cambridge University Press, Cambridge.

Geary, P.W. (in prep.) Establishment of the Silvicultural Systems Project experiment at Cabbage Tree Creek, in East Gippsland, Victoria. *VSP Technical Report No. 25*, Dept. of Natural Resources and Environment, Victoria.

Gregorius, H-R. 1980. The probability of losing an allele when diploid genotypes are sampled. *Biometrics* 36:643-652.

Hardner, C.M. and B.M. Potts 1997. Postdispersal selection following mixed mating in *Eucalyptus regnans*. *Evolution* 51:103-111.

Hartl, D.L. and A.G. Clark 1989. Principles of population genetics. Sinauer Associates Inc., Sunderland, Massachusetts.

Leberg, P.L. 1992. Effects of population bottlenecks on genetic diversity as measured by allozyme electrophoresis. *Evolution* 46:477-494.

Lincoln, S.E., Daly, M.J. and E.S. Lander 1991. PRIMER: a computer program for automatically selecting PCR primers. Version 0.5, May 1991. Whitehead Institute for Biomedical Research, Cambridge, Massachusetts.

Lutze, M. 1998. Silvicultural Systems Project: species dynamics in eucalypt regeneration following a range of 38 harvesting and site preparation treatments in a lowland sclerophyll forest – progress report: species composition to age four. Internal Report, Dept. of Natural Resources and Environment, Victoria.

Marshall, D.R. and A.H.D. Brown 1975. Optimum sampling strategies in genetic conservation. *In* Frankel, O.H. and Hawkes, J.G. (eds.): *Crop genetic resources for today and tomorrow*. Cambridge University Press, Cambridge, 53-80.

Namkoong, G. 1988. Sampling for germplasm collections. *HortScience* 23:79-81.

Nei, M. 1972. Genetic distance between populations. *Am. Nat.* 106:283-292.

Nei M. 1973. Analysis of gene diversity in subdivided populations. *Proc. Nat. Acad. Sci., USA* 70:3321-3323.

Nei, M. 1987. Molecular evolutionary genetics. Columbia University Press, New York.

Nei, M., Maruyama, T. and Chakraborty, R. 1975. The bottleneck effect and genetic variability in populations. *Evolution* 29:1-10.

Reed, K.C. and D.A. Mann 1985. Rapid transfer of DNA from agarose gels to nylon membranes. *Nucleic Acids Res.* 13:7207-7221.

Sambrook, J., Fritsch, E.F. and T. Maniatis 1989. Molecular cloning: a laboratory manual. Cold Spring Harbor Laboratory Press, Cold Spring Harbor, New York.

Squire, R. 1990. Report on the progress of the Silvicultural Systems Project: July 1986-June 1989. Internal Report, Dept. Conservation and Environment, Victoria.

Wagner, D.B., Furnier, G.R., Saghai-Maroof, M.A., Williams, S.M., Dancik, B.P. and R.W. Allard 1987. Chloroplast DNA polymorphisms in lodgepole and jack pines and their hybrids. *Proc. Nat. Acad. Sci., USA* 84:2097-2100.

Yeh, F.C., Yang, R-C., Boyle, T.B.J., Ye, Z.-H. and J.X. Mao 1997. POPGENE, the user-friendly shareware for population genetic analysis. Molecular Biology and Biotechnology Centre, University of Alberta, Canada.

Influences on the international exchange of forest genetic resources - An Australian perspective

Stephen Midgley and Doug Boland[1]

Introduction

The international transfer of germplasm and cultivation of agricultural, tree crop and other exotic plants have long histories. Globally the introduction and domestication of the staple food crops such as wheat, rice, barley and maize have had enormous positive social impact. Apart from meeting basic needs of food and fibre, exotic plants have demonstrated a capacity to change history - in the late fifteenth century a European craving for pepper and spices influenced the Portuguese, Italian and Spanish exploration of other continents; sugar cane and the need for labour for its cultivation led to the infamous trans-Atlantic slave trade and shaped the modern Caribbean; the humble potato and its role in the great famine in Ireland of 1845-46 led directly to one of the great emigrations of that time (Hobhouse 1985).

Between 1960 and 1995, world population almost doubled and the world economy (as measured in GDP in real terms) increased three and a half times. Over the same period, world production of grains more than doubled, fuelwood production doubled and paper production more than tripled (FAO 1997). The area of forest plantations in tropical regions grew dramatically from 18 million ha in 1980 to 44 million ha in 1990 (FAO 1995a) and industrial wood production from farm forestry and from agroforestry systems has become increasingly important in several countries. For example, in addition to the current estimated plantation area of eucalypts in China of 670 000 ha, there are over 1.8 billion eucalypts (equivalent to approximately 1.2 million ha) in the "four-around" plantings (those around roads, canals, village and homes) and in Vietnam the estimated 300 - 400,000 ha of eucalypt plantations is supplemented by 700,000 ha equivalent as scattered trees around farms, homesteads, roads and villages (Tran Xuan Thiep 1995). The steadily increasing demand for forest products is likely to result in an expansion of current efforts in forest plantation development. Many exotic species will play a major role in this expansion as they offer promise of high productivity to meet the escalating need for wood products.

---

[1] This paper reflects the personal views of the authors and does not represent an official position of the Australian Government.

C. Mátyás (ed.), Forest Genetics and Sustainability, Volume 63, 197–210.

Australia is both a major user (e.g. *Pinus radiata, P. caribaea, P. pinaster* and other species) and a major source of forest genetic resources (e.g. eucalypts, acacias, casuarinas, grevilleas) and this provides an opportunity to offer a bipartisan perspective in addressing issues of access to forest genetic resources.

Australia's forest resources comprise 43 million ha of native forests (mostly eucalypts) and 1.1 million ha of plantation forests (mostly pines). Public and private owners have created a plantation resource for wood production which is comprised predominantly of *Pinus* species 87% although it is increasingly supplemented by plantings of eucalypts.

In mid-1996 the Federal and State Governments, in association with industry, jointly adopted a vision of trebling Australia's plantation resource to 3 million ha by 2020[2] (Blyth et al 1998). This vision recognises Australia's capacity for import substitution, export market opportunities and the opportunity to use plantations to help address environmental concerns. However, to realise the vision the annual planting rate will have to be increased from around 25 - 30 000 ha currently, to around 80 000 ha. As a consequence, forestry production is expected to expand into non-traditional areas, including agricultural land in low rainfall and irrigation areas. Exotic species, especially pines, will play a major role in this expansion and this will inevitably result in a need for further access to genetic resources from Central/North America (*Pinus radiata, P. caribaea, P. elliottii*) and southern Europe/northern Africa (*P. pinaster*).

Australia is a major supplier of forest genetic resources and there are now in excess of 10 million ha of eucalypt plantations in the world, more than 2 million ha of acacias and 1 million of casuarinas. Table 1 provides an estimate of the area of eucalypt, acacia and casuarina plantations in Asia. In addition, species of *Araucaria, Callistemon, Grevillea, Macadamia* and *Melaleuca* are widely planted internationally in plantations, on farms, in agroforestry systems and as ornamentals. FAO (1993) reported in their assessments of forest plantations in tropical countries that eucalypts accounted for 23 percent of the area and acacias for nearly 8 percent, and high demand in recent years for seed from the Australian Tree Seed Centre (ATSC) suggests that these proportions are growing.

In 1961, recognising the social, economic and environmental importance of Australian trees outside Australia, FAO requested the Australian Government to establish a "eucalypt clearing house" which has evolved to become the Australian Tree Seed Centre (ATSC)[3]. The ATSC's mission is to explore, domesticate and conserve Australian forest genetic resources. The ATSC's collection of tree seed has grown to include a wide range of multipurpose trees of Australian origin and now holds about 30,000

---

[2] For further details see: http://www.dpie.gov.au/agfor/forests/2020/vision.html
[3] For details and seed lists see: http://www.ffp.csiro.au/tigr/atscmain/

accessions comprising 1200 species from several thousand collection sites. This valuable research asset has provided base populations for international breeding programs of *E. globulus, E. nitens, E. grandis, E. pellita, Acacia mangium, A. auriculiformis, A. crassicarpa* and many other species. Eucalypts make up about half the species in the collection while multipurpose genera represented include *Acacia, Casuarina, Grevillea, Melaleuca, Sesbania* and *Terminalia*. Most accessions come from natural populations but the Centre is establishing and managing an expanding network of seed orchards. The Centre responds to over 2500 requests per year, about half of which come from other countries.

Table 1. Areas planted with Australian trees in Asia (hectares)[4]

Country	Eucalypts	Acacias	Casuarinas
Bangladesh	12 000		
China	670 000[5]	200 000[6]	300 000
India	2 500 000[7]	45 000	800 000
Indonesia	80 000	700 000[8]	
Laos	6 250		
Malaysia	8 000	200 000	
Myanmar	40 000		
Nepal	6 000		
Pakistan	10 000		
Philippines	10 000	45 000	
Sri Lanka	29 000	6 000	3 500
Thailand	195 000	20 000	
Vietnam	350 000[9]	80 000	120 000

There is a steady export trade in native Australian tree seed to some 130 countries and this totals some 25 - 30 tonnes annually valued at AUS$ 9 million (Doran and Gardiner 1993) and there are currently very few restrictions on the export of tree seed.

The importance of quality seed remains a constant in international exchange. At the 1989 meeting of the IUFRO Project Group P2.04-00 "Seed Problems" in Queensland, Australia, Professor Chin from UPM, Malaysia, reminded the meeting that Good Seed Does Not Cost - It Pays reviving an FAO slogan from the 1950s aimed at encouraging farmers to use the best

[4] Eucalypt figures derived primarily from papers presented in FAO (1995b). Blank spaces indicate no reliable information has been located.

[5] Does not include "four-around" community plantings equivalent to approx. 1.2 million ha of plantation.

[6] Turnbull *et al* (1997)

[7] Dewees and Saxena (1995)

[8] PIMA's Papermaker, August 1998, P.12.

[9] Does not include "scattered trees" equivalent to 700 000 ha of plantation.

seed possible. This slogan is as apt for forestry as it is for agriculture and the supply of high quality seed is central to the success of tree planting projects. Germplasm (including seed, cuttings, tissue cultures and pollen) supply is now being recognised by a number of governments and donor agencies as a development issue and this is reflected in the increasing number of tree seed centre and forest germplasm-related initiatives.

Although Troup (1932) observed "...reliance on supplies of seed from its original home may prove an obstacle to the formation of future plantations on a large scale.", the background search for this paper revealed that people did not regard access to genetic resources as an issue in the process of domesticating exotic species and subsequent plantation development. There is an underlying assumption that germplasm will be made available. It has been the nature of those involved in forest research to generously share forest genetic resources, especially those used for wood production and, as a result, access has traditionally been unencumbered. This has perpetuated some widely held misconceptions that seed is a cheap or free good.

If the expectations and research needs of plantation forestry are to be met, there will remain an important need to transfer of germplasm between countries. This germplasm will be used for plantation establishment, selection and breeding, and genetic conservation. Influences to effective international exchange of forest genetic resources occur at a number of biological and non-biological levels. In addition to technical issues which science can address, there are social, policy and legal issues which influence the effective exchange of germplasm.

1. Collection issues

Through its Project Group P2.04-00 "Seed Problems", IUFRO has recognised many of the technical constraints facing successful collection, storage and handling of tree seed, and the deliberations of this group can be accessed through its regular conference proceedings. One of the most straightforward of constraints is that of species identity:

> *I was staying in the country, about 50 miles south of Sydney, in the year 1896, when I came across a large sheet spread on the ground, with limbs of Eucalyptus piperita, laden with fruit, thrown on to it, and the seed had already begun to shed... [The collector] was quite honest, and perfectly frank as to his operations... Pointing to the E. piperita seed, and knowing it to be an almost useless species, at all events in this district, I said, "What is this, and why are you collecting it?" He replied, "E. longifolia, and its a fine timber". I said, "I am afraid what you are collecting might not be as valuable as you think it is," explained the harm he was doing, and gave the man my card... I pity*

the poor buyer and I wonder how many other collectors... are as ignorant as that one.

From J. H. Maiden (1929).

Australia, with over 700 species of *Eucalyptus* and 900 species of *Acacia* (Maslin and McDonald 1996) and a broad mega-diversity of other species, presents particular challenges with regard to taxonomy and species identity. Taxonomic changes are a regular feature of such a diverse flora and users of Australian species are advised to keep up to date with new nomenclature such as the changes in the taxonomy of the *E. microtheca* group which has been split into 8 taxa (Hill and Johnson 1994). Another example is the proposed change to the *Acacia aulacocarpa* complex where nine species, some new to science, will be described (McDonald and Maslin, in press).

The OECD Scheme for the Control of Forest Reproductive Material Moving in International Trade was established in 1974 to "encourage the production and use of seeds, parts of plants that have been collected, transported, processed, raised and distributed in a manner that ensures their trueness to name" (Bonnis 1993). A view has been expressed by Doran and Gardiner (1993) that the OECD Scheme is inappropriate for Australia and compliance would be extremely difficult. Most of the widely-used Australian species are easy to propagate and have the capacity to produce seed from an early age. Most users can develop their own breeding programs based upon a broad range of Australian foundation material and, in the case of the tropical acacias, these can produce significant quantities of seed within four years. The rationale for extended dependence upon natural stands for operational quantities of seed is not strong for many species. Under Australia's constitution, the States and Territories have responsibility for managing resources within their jurisdiction. Ownership regimes for biological resources vary between States. In many instances forests are owned and managed by the State. With 8 States and Territories, each with a number of agencies claiming ownership, nomination of a designated agency is administratively difficult. The diversity of the tree flora (over 2 500 tree species), remoteness of occurrence, active policies aimed at reducing Government interference, unclear ownership between private landowners, indigenous communities and Government agencies, and difficulty of effective policing, all contribute to the view that the OECD Scheme is not uniformly applicable.

There is no alternative tree seed certification scheme operating in Australia, and purchasers should ascertain details of the seed origins before entering into any purchase agreement. The minimum details which might be expected are the precise locality of collection including latitude and longitude co-ordinates, altitude of the collecting site, year of harvesting and number of trees sampled. The Tree Seed Suppliers Directory recently published by ICRAF (Kindt et al 1997) offers a thorough coverage of guidelines to be followed when ordering seed.

2. Plant health issues

The relative lack of pests and diseases has contributed to the success of exotic species when introduced into new environments. Seeds and other germplasm may be infected or contaminated with a number of pathogens such as viruses, bacteria, fungi and nematodes or seeds of weedy species. In addition to the risks of spreading pests or pathogens to new areas, many pathogens also reduce seed longevity. Attention to the quarantine risks associated with the international transfer of germplasm and adherence to phytosanitary regulations is imperative to the sustainable use of an exotic plantation species.

Australia, New Zealand, and Chile are the world's largest producers of radiata pine timber with plantations totalling close to 4 million ha. Radiata pine in these countries has enjoyed a relatively disease-free environment. Certainly the best method for control of plant diseases, if they are not already present, is to restrict their introduction. However, there are numerous examples where this approach has failed, e.g. the introduction of Dutch elm disease to North America and New Zealand.

Pitch canker, caused by *Fusarium subglutinans f. sp. pini*, appears to have been introduced to northern California around 1986. *Pinus radiata* is highly susceptible to the disease and thousands of trees have been killed. All three of the native mainland populations of *P. radiata* are severely threatened. There is also concern that other native pines in the Coast, Cascade, and Sierra Nevada Ranges may be at risk. Pitch canker also occurs in the southeastern United States, Haiti and Mexico, and has recently been introduced to South Africa, Japan and Spain. The occurrence of this disease in the native populations has given new importance to *ex situ* genetic conservation plantings, established before the disease became a problem. It also presents a problem to future use of germplasm from the native stands as the disease would present serious plant health problems to users of germplasm derived from affected areas.

Guava rust (*Puccinia psidii*), a native of South America, is regarded as the most significant plant health risk to the cultivation of *Eucalyptus* spp. and occurs as a pathogen on other species within Myrtaceae. Coutinho et al (1998) described P. psidii as "...one of the most serious threats to *Eucalyptus* spp. in the world today". It poses a potential threat to native forests in Australia where entire ecosystems could be damaged (Ciesla et al 1996). In addition, the pathogen presents a threat to Indonesian cloves which are the basis of that nation's kretek (Indonesian cigarettes) industry which was valued at more than US$ 5 billion in 1996. This risk remains one of the greatest concerns for Australian scientists wishing to bring genetically improved material of *Eucalyptus* from South America and South Africa to Australia for testing as part of Australia's expanded plantation program.

The Australian Tree Seed Centre routinely examines seed accessions for fungal contamination. The fungal pathogens recovered from seed and seedlings of Australian origin in a study by Yuan et al (1989) were widely distributed geographically and were not seen to present a plant health threat to countries receiving shipments of seed from the ATSC. Such findings do not however remove the responsibility to remain vigilant regarding transfer of pathogens with germplasm. Import of most eucalypt germplasm to Australia requires strict quarantine isolation. Australia is not alone in imposing strict controls; Singh and Rana (1989) provided an overview of the existing quarantine systems in India and listed some of the pests and diseases intercepted on imported tree seeds.

3. Self-imposed constraints and policing issues

Industry groups in some countries have adopted informal, self imposed constraints on seed movement aimed at protecting local industries. In the 1950s, the export of seed of *Eucalyptus polybractea* (an important species for the production of cineole-rich foliar oil) was prohibited to protect the local industry. The ban was lifted when it proved impractical to enforce (Fryer 1994). In deference to an informal industry request, the Australian Tree Seed Centre refers all international requests for seed of *Melaleuca alternifolia* (from which the foliar tea tree oil is produced) to the Australian Tea Tree Industry Association for attention. This does, however, not prevent private suppliers meeting international demands. Other countries have adopted similar informal restrictions - germplasm of *Pilocarpus* spp. (the sources of the alkaloid pilocarpine and the drug Jaborandi) is difficult to obtain from Brasil, access to germplasm of wild coffee from Ethiopia is restricted, Malaysia has prohibited the export of rattan germplasm (other than for small research quantities; Appanah, pers. comm.) and there are restrictions on access to *Santalum album* seed from India and Indonesia.

Effective policing of any regulations regarding transfer of germplasm is difficult. This is especially the case for species of *Eucalyptus* which have been transferred outside Australia for over 200 years. Where the seed is small, long-lived and easily transportable and there remains free public access to forests, the costs of effective policing are likely to outweigh the benefits.

4. Legal/policy issues

IUFRO recognised the emerging issues regarding germplasm exchange and in the early 1980s established the IUFRO Working Party on Legislation of Forest Reproductive Material (S2.02.21). For the purposes of legal and policy discussion, genetic resources are described in the Convention on Biological Diversity as genetic material of actual or potential value. Genetic material means any material of plant, animal, microbial or other origin containing functional units of heredity.

The most significant emerging influences to exchange of forestry germplasm are those relating to issues of determining and allocating ownership and use rights. Forestry, and the issues relating to forest genetic resources for wood production, are now part of a larger international debate regarding biological diversity conservation, use and benefit sharing, following the entry into force of the Convention on Biological Diversity. Biological resources, and in particular the genetic resources they contain, provide the building blocks for pharmaceuticals and agents to fight disease, provide control of pests in agriculture and help develop environmentally friendly industrial processes (such as bacteria to pre-treat wood chips). The future development of these industries will depend increasingly on continuing access to the world's biodiversity (Anon. 1994).

Rights to own and use biological resources are assigned in accordance with the sovereign rights of countries through policy and regulatory regimes. Increasingly, agreed policy approaches through a range of international treaties are influencing the development of genetic resources policy.

In the case of Australia, there are many different policies and regulatory approaches which relate to the ownership, management and use of biological resources. There is no nationally consistent approach on access to forest biological resources within Australia where the States have responsibility for management of forests and collection of seed from native forests is controlled by State and Territory land management agencies. For example, to gain access to a comprehensive genetic sample of *Eucalyptus camaldulensis* seed from throughout its natural range requires permits from over 15 agencies and organisations in 6 States and Territories, permission from several indigenous communities and many other private landowners. With such a diversity of ownership between State, community and individual landowners, it is difficult to frame policy which meets the needs of all stakeholders. There is, however, a prevailing attitude expressed at meetings of stakeholders that, as a matter of principle, Australia should benefit from providing access and this is particularly the case for bio-prospecting. Australia is in the process of developing a nationally consistent policy to cover access to biological resources (Anon. 1996).

5. International policy issues

International exchange of forest genetic resources has been without major debate or restriction, but may in the future be influenced by regimes on access which countries implement as a consequence of developments under the Convention on Biological Diversity. The following provides a brief overview of the Convention on Biological Diversity and some other agreements which may have a bearing on how countries implement policy on access to forest genetic resources. Many of the issues covering genetic resource access and use regimes, benefit sharing, technology transfer, biotechnology and intellectual property considerations arise in a number of fora.

5.1. Convention on Biological Diversity (CBD)

The CBD has three objectives:

- the conservation of biological diversity;
- the sustainable use of its components; and
- the fair and equitable sharing of the benefits arising out of the utilisation of genetic resources (including through technology transfer).

The CBD is a framework agreement, setting out generic policies and principles for signatories, individually and collectively, to achieve its objectives. It recognises the sovereign rights of the signatories over their natural resources. The Convention does not establish or assign any new ownership rights over either biological or genetic resources. It recognises the value of genetic resources, and that the authority to determine access to genetic resources rests with national governments.

By virtue of its general policy framework, covering terms of access to and use of genetic resources (which do not distinguish between access to material held *ex situ* or found *in situ*), the CBD covers many spheres of economic activity, including agriculture, forestry and pharmaceuticals. It provides an agreed policy framework through which Parties will facilitate access to genetic resources (as opposed to biological resources) and how they will use and transfer them. Benefit sharing, technology transfer, biotechnology, intellectual property and environmental objectives are covered.

Elaboration of the policy principles in the CBD is a major influence on policy approaches being developed by countries in exercising their sovereign rights over biological resources (and especially genetic resources). This encompasses activities on access to genetic resources, intellectual property and agricultural biological diversity under the CBD's work program.

5.2. World Trade Organisation (WTO)

A number of WTO Agreements are directly relevant to trade-related considerations in respect of genetic resources, including issues of genetic resources policy, technology transfer, biotechnology, intellectual property considerations, trade and sustainable resource use.

For example the General Agreement on Tariffs and Trade (GATT) 1994, and other multilateral agreements on trade in goods, cover trade in genetic resources and the products of biotechnology. WTO Agreements, which include the Agreement on the Application of Sanitary and Phytosanitary Measures, allow Members to take certain measures to protect human, animal or plant life or health, and to protect the environment. The application of these measures is subject to certain conditions to prevent unnecessary barriers to trade or unjustifiable discrimination in trade.

It is likely that part of the future international exchange of germplasm for forest genetic resources will involve genetically modified organisms (GMOs). Many countries are likely to adopt regimes to control the release and use of GMOs. How these will impact on access and transfer of forest genetic resources is not clear. The OECD Scheme (mentioned earlier) has suggested there may be specific provisions, by way of additional requirements or special dispensations, applicable to forestry or forest reproductive material. Factors such as the difficulty of testing, the peculiarities and long delays in genomic expression, the limited food impact, the long-term consequences of decisions and the lack of knowledge of environmental effects all require further discussion and consideration.

Negotiations are currently being held under the Convention on Biological Diversity for a Biosafety Protocol, aimed at the safe transfer, handling and use of Living Modified Organisms resulting from biotechnology which may have adverse effects on the conservation and sustainable use of biological diversity.

The role of intellectual property rights in respect of biological and genetic resource issues, including technology transfer, arises under the Agreement on Trade Related Aspects of Intellectual Property (TRIPs), especially Article 27.3 (b) which provides flexibility for Members on the patenting of plants and animals. The TRIPs Agreement is relevant to the discussions about whether there is a possible role for existing or new forms of intellectual property in providing incentives for the conservation and sustainable use of biological diversity and contributing to the fair and equitable sharing of benefits arising from the utilisation of genetic resources, including biotechnology. Matters relating to the protection of knowledge, innovations and practices of indigenous local communities embodying traditional lifestyles also arise in this context.

5.3. International Union for the Protection of New Varieties of Plants (UPOV)

UPOV is an intergovernmental forum under whose auspices countries may establish rights in relation to new varieties of plants which enable the holder of rights to capture benefits from the development of a new plant variety, for example under a licensing or marketing arrangement.

The acts of the International Convention for the Protection of New Varieties of Plants (1961, as amended 1972, 1978 and 1991) under the administration of UPOV enable countries to establish, within their jurisdiction, a registration scheme for new varieties of plants, often described as a plant breeder's rights (PBR) regime. Plants qualifying for registration under PBR need to fulfil certain criteria.

Some thirty-six countries, including Australia, have established legislation which enables registration of PBR in conformity with one of the UPOV Acts. In Australia , PBR rights for tree varieties have been applied in horticulture rather than the production forestry sector, except for a few cases for insect-resistant eucalypts. Cultivars of several species of *Eucalyptus*, *Grevillea* and *Callistemon* have been registered and a central register of all registered cultivars is available on the Internet at: www.dpie.gov.au/agfor/pbr/pbr.html. It can be anticipated that eucalypt cultivars for wood production will be registered in the future. Cultivars developed through controlled crossing (including hybridisation) and through genetic modification will increasingly be considered for PBR registration.

5.4. The FAO International Undertaking on Plant Genetic Resources

The FAO Undertaking is a non legally-binding technical cooperation agreement setting out the principles for exchange, conservation, scientific investigation and development of plant genetic resources for food and agriculture. The Undertaking was first agreed in 1983; Australia agreed to adhere to the Undertaking in 1992. It is broad in scope and covers forest genetic resources

The Undertaking forms the policy element of the multilateral approach on access to plant genetic resources for food and agriculture which was devised in 1983 as a global system. It is based on the premise that large numbers of plants have economic value as a consequence of the genetic characteristics they possess, including those in spheres such as agriculture and forestry. The system comprises functional and policy elements.

The functional elements revolve around the use of, and the maintenance of, material in a series of key biological resource collections through large-scale efforts in plant breeding, conservation, evaluation and utilisation of economically important species, along with national and international cooperative breeding programs.

The Undertaking is being revised to harmonise it with the Convention on Biological Diversity, and to find solutions to certain issues involving plants for food and agriculture which were unresolved during the Convention negotiations. These issues include access regimes to ex situ collections not acquired in accordance with the Convention and the issue of farmers' rights.

The revision negotiations have been moving slowly, and there are significant differences between countries. The extent to which the revised Undertaking will cover or have implications for forest genetic resources, or international cooperation is not clear.

6. Conclusions

There are no significant export regulations for the international transfer of forest genetic resources for either research or commercial purposes. Most current arrangements are based upon a history of institution-to-institution agreement, scientist-to-scientist agreement and more recently, seek to follow guidelines established through schemes such as that of OECD. All recognise the importance of appropriate quarantine procedures.

The influence of international policy developments means that it is likely that there will be changes to the way that nations deal with access to forest genetic resources. It is important that forestry researchers take note of developments in this area and establish strong links to, and avoid duplication with, activities underway in a range of international fora, including in particular, the scientific and technical body under the Convention on Biological Diversity (SBSTTA).

Review of the current arrangements is timely in view of changing national and international attitudes towards sovereignty over forest genetic resources and any proposed changes must be consistent with policies relating to wider access issues for agriculture, pharmaceutical plants and biological resources. Seed can no longer be regarded as a free good and future transfer will be on "mutually-agreed terms". In the case of the operations of the Australian Tree Seed Centre, these are likely to inevitably proceed towards a modified Material Transfer Agreement (MTA) arrangement originally developed within the framework of the Consultative Group on International Agricultural Research (CGIAR). MTAs will accompany each dispatch of seed and will require the user to abide by some minimum basic requirements which might include future access to information and/or germplasm.

Change is occurring and those in the forestry community involved in the international transfer of seed will become accustomed to a greater level of control and legal procedure. Australia, with its dual role as a supplier and a user of forest genetic resources, has maintained an active role in the negotiations regarding access to genetic resources. The issues are complex and it is likely to be some years before agreement is reached regarding the future shape of germplasm exchange. Forestry enjoys a tradition of exchange and good will among international collaborators, especially for germplasm for wood production, and this will underpin future arrangements as they evolve.

Acknowledgments

The contributions of our colleagues, Lex Thomson and Tim Vercoe are recorded with thanks.

Literature cited

Anon. 1994. Access to Australia's biological resources. A discussion paper. Office of the Chief Scientist, Department of the Prime Minister and Cabinet, Canberra.

Anon. 1996. Managing access to Australia's biological resources: Developing a nationally consistent approach. A discussion paper prepared by the Commonwealth-State Working Group on Access to Australia's Biological Resources. Environment Australia. Canberra. (http://www.biodiversity.environment.gov.au/biocon/cswg/)

Anon. 1977. Dictionary of Gardening. The Royal Horticultural Society. Chittenden, F. J., ed. Oxford: Clarendon Press.

Blyth, M. J., Midgley, S. J. and Kile G. A. 1998. The changing face of Australia's forest research. *In* Enters, T., Nair, C. T. S. and Kaosa-ard, A. (eds.): Emerging Institutional Arrangements for Forest Research. FORSPA Publication No. 20/1998. 403.

Boland, D. J. 1996. International exchange of forest genetic resources for wood production - towards a uniform Australian position. Institute of Foresters of Australia, Newsletter 37 (3), 14 - 19.

Bonnis, G. 1993. The OECD scheme for the control of forest reproductive material moving in international trade. *In* Wolf, H. (ed.): Seed Procurement and Legal Regulations for Forest Reproductive Material in Tropical and Subtropical Countries. Proc. of an International Symposium, Nairobi, Kenya. 4 - 10 October, 1992. 555.

Ciesla, W. M., Diekmann, M. and Putter, C. A. J. 1996. FAO/IPGRI Technical Guidelines for the Safe Movement of Germplasm. No. 17. *Eucalyptus* spp. Food and Agriculture Organisations of the United Nations, Rome/ International Plant Genetic Resources Institute, Rome.

Coutinho, T. A., Wingfield, M. J., Alfrenas, A. C. and Crous, P. W. 1998. Eucalyptus Rust: A disease with the potential for serious international implications. *Plant Disease*, July 1998. 819 - 825.

Dewees, P. A. and Saxena, N. C. 1995. Wood product markets as incentives for farmer tree growing. *In* Arnold, J. E. M. and Dewees, P. A. (eds.): Tree Management in Farmer Strategies: Responses to Agricultural Intensification, Oxford University Press. 198 - 241.

Doran, J. C. and Gardiner, C. A. 1993. Extent, control and documentation of tree seed trade at an international level - the Australian experience. IUFRO First International Symposium on Seed Procurement for Forest Reproductive Material in Tropical and Subtropical Countries, Nairobi, Kenya.

FAO 1993. Forest Resources Assessment 1990 - Tropical countries. FAO Forestry Paper 112, FAO, Rome.

FAO 1995a. Forest Resources Assessment 1990 - Tropical forest plantation resources. FAO Forestry Paper 128. FAO, Rome.

FAO 1995b. Proceedings of the Regional Expert Consultation in *Eucalyptus*. 4 - 8 October, 1993, FAO Regional Office for Asia and the Pacific, Bangkok. 196pp.

FAO 1997. State of the World's Forests, 1997. Food and Agricultural Organisation of the United Nations. Rome.

Fryer, J. 1994. Access to Australia's forest genetic resources. Working Paper. Bureau of Resource Sciences, Department of Primary Industries and Energy, Canberra. 23p.

Hill, K. D. and Johnson, L. A. S. 1994. Systematic studies in the eucalypts. 6. A revision of the coolibahs, *Eucalyptus* subgenus *Symphyomyrtus* section *Adnataria* series *Oliganthae* subseries *Microthecosae* (Myrtacae). *Telopea* 5(4), 743-771.

Hobhouse, Henry 1985. Seeds of Change - five plants that transformed mankind. Sidgwick and Jackson, London. 245pp.

Kindt, R. with Muasya, S., Kimotha, J. and Waruhui, A. 1997. Tree seed suppliers directory: sources of seeds and microsymbionts. Nairobi: International Centre for Research in Agroforestry.

Maiden, J. H. 1929. A Critical Revision of the Genus *Eucalyptus*. Vol. 7, p 86.

Maslin, B. R. and McDonald, M. W. 1996. A key to useful Australian acacias for the seasonally dry tropics. CSIRO, Australia.

McDonald, M.W. and Maslin, B.R. (in press). A taxonomic revision of the salwoods: *Acacia aulacocarpa* Cunn. ex Benth. and its relatives (Leguminosae: Mimosoideae: section Juliflorae). *Australian Systematic Botany* (submitted).

Perry, M. S. and Willan, R. L. 1957. Tanganyika - exotic forest trees in the British Commonwealth. Statement prepared for the seventh British Commonwealth Forestry Conference.

Singh, B. P. and Rana, R. S. 1989. Germplasm exchange of multipurpose trees: an Indian perspective. *In* J.W. Turnbull (ed.): Tropical Tree Seed Research : Proceedings of an international workshop held at the Forestry Training Centre, Gympie, Qld., Australia, 21-24 August 1989. ACIAR Proceedings No. 28. 156.

Tran Xuan Thiep 1995. *Eucalyptus* plantations in Vietnam (its history and developmental process). *In* FAO, Proceedings of the Regional Expert Consultation in *Eucalyptus*. 4 - 8 October, 1993, FAO Regional Office for Asia and the Pacific, Bangkok. 196.

Troup, R. S. 1932. Exotic Forest Trees of the British Empire. Clarendon Press, Oxford.

Turnbull, J. W., Midgley, S. J. and Cossalter, C. 1997. Tropical acacias planted in Asia - an overview. *In* Turnbull, J. W., Crompton, H. and Pinyopusarerk K. (eds.) (in prep): Proc. Third International Acacia Workshop, Hanoi, Vietnam, 26-30 October 1997. ACIAR, Canberra.

Vietmeyer, N. D. 1978. Incredible odyssey of a visionary Victorian pedlar. Smithsonian Magazine, 18. August 1978.

Yuan Z, Qing, Old, K.M. and Midgley, S.J. 1989. Mycoflora and pathology of fungi present on stored seeds of native Australian Trees. *In* J.W. Turnbull (ed.): Tropical Tree Seed Research : Proceedings of an international workshop held at the Forestry Training Centre, Gympie, Qld., Australia, 21-24 August 1989. ACIAR Proceedings No. 28. 156.

6. Conservation of Forest Genetic Resources

Genetic conservation strategies: Who needs data?

Gene Namkoong

Introduction

As geneticists, it seems obvious that the dynamics of allelic, genotypic, populational, and species level changes are basic to understanding genetics and to designing conservation programs. If we are to conserve such biological systems as forests, then we must conserve its constituent systems including the fundamentally important genetic systems. In the belief that „nothing in biology makes sense in the absence of genetics" (Dobzhansky, 1970), and in the practical implication that without the genetic resource, there are no species or ecosystems, genetic understanding is fundamental. As geneticists, we can take smug comfort in that all conservation and use programs have basic dependence on our science. Our studies on the genetic structure and levels of genetic variation of forest trees are necessary and the more information on heritable traits, isozymes, molecular markers, and even DNA sequences, the better we can achieve our conservation goals. A tree genome project like the human genome project will provide the information needed. It is so obvious!

Yet, in spite of that obvious necessity, the number of conservation programs that use the genetic data that we so happily gather is vanishingly small. At most, genetic indicators are a small section of ecological and other biological indicators of forest sustainability. Furthermore, when it is considered that genetic data should be used, there is little difference among the kinds of data that can be obtained, the sampling design to be used, or how the information on multiple gene loci is to be integrated. It is therefore not surprising that genetic data is little used for planning sustainable forest management. Particularly in tropical forests where many hundreds of species may be impacted and forest interventions are expected within a few months or years, the financial, time, and personnel costs of obtaining genetic data may rarely justify the research we would prefer to do. Therefore, in the absence of useful genetic data, forest plans are made that are based on overly conservative estimates of possible threats, or the plans may unnecessarily endanger resources that could otherwise be protected.

C. Mátyás (ed.), Forest Genetics and Sustainability, Volume 63, 211–216.

The mismatch between the obvious genetic expectations and reality is startling and forest managers may believe that geneticists are misguided by their sense of self-importance, while geneticists may believe that forest managers are misguided in ignoring significant genetic information. As a first step at reconciliation, attempts to bridge the gap have focused on building bridges for better communication between managers and geneticists (CIFOR, 1996). One result of these first steps has been a growing awareness that we geneticists must target the various tools that we have at our disposal for genetic analysis, to focus on the kinds of genetic systems and threats that require one or another form of genetic information. We must also understand the decision system that managers use, as it is within that decision framework genetic data is used or ignored.

1. Decision analysis

When making decisions about conserving biological systems, we encounter two kinds of uncertainties. One is that there are inherently uncontrollable and unknowable factors that a program will fail with a certain probability. Our science is not exact enough to know with certainty when a species will survive or will go to extinction even in the short run. We are always subject to error in both directions and the role of genetic management programs is to increase the chances for successful conservation – not to guarantee it. If we establish a reserve park that contains a rare species, we may still fail to save it due to random climatic events or simply due to reproductive failures that normally occur at some frequency. If we knew the probabilities of failure, we could then estimate the chances of the success of a management plan, and estimate its costs and benefits as an expected value. For example, if the chances of success for a reserve program that would cost 100 units is only 50%, and the value of success is 1,000 units, we can deduce that the expected value of the program is 400 units.

The other kind of uncertainty is that we are usually highly ignorant of the actual genetic status of the populations we deal with and have to make general guesses as to their level of endangerment. We often don't know what the probabilities of failure are, but regardless of our ignorance, we still have to make decisions about whether to invest in reserves or in another management system. In this case, genetic information can be useful to indicate more precisely, what is the extent of threat, and assist in our evaluation of the need for management actions. It wouldn't change the actual level of threat, but it would change the conditional probability of correctly assessing the chances of success or failure and the costs or benefits of management plans can then be more accurately evaluated. Information does not change the inherent probabilities of success or failure of the alternate management plans, but it does increase the chances that we are making the best decision. Whenever the first kind of uncertainty exists, there will always be failure that no amount of information can change, but if the probabilities are correctly assessed, then we can make reasoned judgments and choose among alternative management plans to balance expected costs and benefits. If, in addition to the first kind of uncertainty, the second kind exists, then our judgments should include the probability of an incorrect assessment of risks. Information would always decrease the second kind of error and would always be sought if it were cost-free.

However, information is rarely cost-free. As will be detailed later, direct information of the levels and distribution of genetic variations requires planned experimentation or expensive laboratory procedures, or both, and is carried out on relatively few species. Phenotypic traits require growth chamber or field tests, while molecular and isozyme analyses require costly laboratory techniques that must be adjusted for each new species analyzed. Many management decisions must be made within strict time and cost constraints, so detailed data on more than a few species in any one area is not feasible to expect.

In addition, the management decisions rarely require precise estimates of population size, migration rates, selection effects, or gene frequency changes. The questions usually involve the existence of threshold levels of population viability or levels of change in genetic variability, and not questions of whether the outcrossing rate is 85% or 95%. The genetic question may be whether the outcrossing rate has changed from 90% to 40%, and for this, exact estimators may not be important. Therefore, research to estimate outcrossing rate can tolerate high errors of estimate and may not require direct genetic estimates. Instead, surrogate measure of the abundance of pollinator insects or birds may be closely enough correlated that changes in their sizes by 50% may be sufficient to decide on the efficacy of a management plan. In this case, genetic guidance is needed on which of the genetic processes are significant and what ranges of parameters are critical, but direct genetic data is not needed. Similarly, if local population sizes are in the range of hundreds and all adults are sexually mature, there may be little need to estimate the effective population size with precision and a surrogate measure of numbers of adults with assurance of their successful mating could suffice. Again, genetic guidance is needed for what to consider, but direct genetic data may require so much time or effort that the costs and delays outweigh their utility. Therefore, in many cases, genetic data may be useful but not critical for deciding on such questions as the immediate regeneration or viability of a stand.

However, if the question for conservation is whether a particularly sensitive species is susceptible to even minor genetic erosion, then more exact data may be useful. If the question can be focussed to decisions on certain species and processes that are of critical significance then precision on specific process indicators can be more important to obtain. This requires that the decision analysis is focused on critical species and that decisions made on the basis of information on those species will be the basis of the management plan. For example, if certain species are chosen because of their sensitivity to mating system disruptions, they serve as indicator species for certain processes and if they go to extinction, then there is a certain probability of system failure. In a decision tree, this adds another step and the contribution of data to evaluating the management plan differs in this case. Thus, the need for genetic data depends on what management questions are asked and the decisions to be made within that management context.

Various forms of decision analytic tools exist and include our attitudes toward risk, assessing the costs of correct and incorrect decisions, and how we make evaluations of multiple factors such as when either the species or the management program has several different attributes of value. One of the complexities involved in genetic management is that several evolutionary factors act simultaneously and it is their joint effects on population evolution that determine the probabilities of success or failure. For processes such as selection where trait means may shift through dysgenic selection, population mean data may be required. For changes in mating system, outcrossing rate estimates may be useful, or for changes in the effective parental mating pool, individual tree DNA fingerprints may be useful. These and other processes can be estimated using different techniques, each with their own levels of precision and costs.

2. Uses of genetic data

As a rough guide, we can use a table of procedures (Table 1 adapted from Karp and Edwards, 1997) to indicate the kinds of molecular techniques that are useful for different levels of precision in identifying individual genotypes or individual genes. At one extreme, it is possible to examine the exact base pair sequence of a gene or chromosomal segment and to identify single base changes in individuals. This would provide information on mutation processes, and could be used to identify individual trees. However, this level of information may provide so many uniquely different elements that distinctions between populations may actually be less clear than with less detailed data. At the other extreme, individuals may not be genetically distinct if measured by allozyme alleles, but population differences may be very well characterized.

In addition to molecular and isozyme data, quantitatively inherited traits can also be used to estimate the rate of genetic change. For traits that have some level of genetic variability, experimental plantings can be used to retrospectively distinguish the level of heritable change that has occurred. In addition, if the heritability of a trait is known to some level of approximation, and the mean phenotypic difference between populations can be estimated, then the selection differential can be retrospectively estimated. Selection effects can therefore be estimated without molecular level data.

Table 1. Comparison of the different molecular screening techniques (adapted from Karp and Edwards, 1997)

Characteristics	RFLPs	RAPDs	Seq Tag SSRs	AFLPs	PCR-Seq
Development costs	Medium (100)	Low (none)	High (500)	Low (none)	High
Levels of Polymorphism	Low-Medium	Medium	High	Medium	Medium
Automation	No	Yes/No	Yes/No	Yes/No	Yes
Cost of automation	High	Medium	High	High	High
Reliability	High	Low	High	Medium	High
Level of skill required	Low	Low	Low-Medium	Medium	High
Cost ($ Per assay)	High (2.00)	Low (1.00)	Low (1.00)	Medium (1.50)	High (2.00)
Radioactivity	Yes/No	No	Yes/No	Yes/No	Yes/No
Samples/day (research)	20	50	50	50	20

3. Uses of genetic research

With these various tools available for obtaining genetic data, the gap between what geneticists are capable of measuring and how data can change management decisions remains large. One of the main problems in using genetic data for these purposes is that species differ in their susceptibilities so that we cannot just examine one of them and have any assurance that the total forest is well described. However if managers must examine multiple processes on hundreds of unique species, genetic information on just one of them is of little value. So there is a need to classify species into a small number of groups that face similar kinds of threats to one or another of the evolutionary processes. For this purpose, it would be useful to develop a series of model species that represent a range in their genetic architecture and susceptibilities to threats to their evolutionary system. If such a set could be developed, then species might be classified according to their reproductive system and geographic distribution and studied with respect to what threats are significant. A triage system might then be derived, sorting species by their amenability to management – either not salvable, amenable to management, or safe from extinction regardless of management.

One of the other problems in utilizing genetic data is that we don't know what critical levels the parameters must reach before threats of extinction are significant. A series of studies on critically sensitive species is also needed to define when critical decision points are passed. If this is known, then it would be possible to use surrogate measures that correlate well with genetic variables to guide management as long as the estimated parameters were not close to the critical points. For these species it is seldom true that only one threat exists to only one process, so research on the joint effects of threats on the joint effects of the processes on population viability and regeneration is needed. Without this, genetic data cannot be well focused on critical decisions and will be too expensive to use for non-critical cases.

4. Conclusions

If we believe that genetic information is indeed critically needed for conservation and management of forest resources, it is not clearly useful unless it is intelligently targeted. Merely providing excellent and interesting information about genetic processes is insufficient for decision making. The data that we gather must fit into the questions that decision makers need to ask. Without that targeting, genetic information is too expensive to obtain for very little marginal benefit and I would have to agree with those managers who honestly ask „Who needs genetic data?" The answer is that few managers need it if it is excessively detailed about trivial issues, and most managers need it, if it can inform and change critical decisions.

Literature cited

Dobzhansky, Th. 1970. Genetics of the Evolutionary Process. Columbia University Press. New York.

CIFOR (1996). Testing Criteria and Indicators for Assessing the Sustainability of Forest Management: Genetic Criteria and Indicators. (Compiled by Gene Namkoong, Timothy Boyle, Hans-Rolf Gregorius, Helene Joly, Outi Savolainen, Wickneswari Ratnam and Andrew Young). CIFOR Working Paper Series No. 10, Bogor, Indonesia.

Angela Karp and Keith J. Edwards, 1997. Techniques for the analysis, characterization and conservation of plant genetic resources. Molecular techniques in the analysis of the extent and distribution of genetic diversity. *In* Ayad, W. G., T. Hodgkin, A. Jaradat and V. R. Rao (eds.): Molecular genetic techniques for plant genetic resources. Report of an IPGRI Workshop 9-11 October, 1995, Rome, Italy.

The need for an international framework for the conservation and sustainable utilization of forest genetic resources[1]

Pierre Sigaud

1. The need for an holistic approach to the management of forest genetic resources

Forests and woodlands provide a wide range of goods and services in support of human sustenance and local and national economies. In addition to wood and wood products such as fuelwood and charcoal, posts and poles, these include food, fodder and medicines, as well as shade shelter, the protection of watersheds, and stabilization and amelioration of soil.

The well-being and productivity of forests are dependent on the structure and dynamics of their genetic foundation. Neglect or mismanagement will lead to a loss of genetic resources reduction in levels and quality of goods and services provided, increased ecological vulnerability, and foregone potential gains from breeding.

Threats to the integrity of forest genetic resources include deforestation and changes in land use, inappropriate harvesting practices for wood and non-wood products, over-grazing, and climate change. A further threat may be posed by the often uncontrolled and undocumented movement of exotic forest germplasm for tree planting and plantation establishment, which may lead to future loss of local adaptation due to hybridization of genetically diversified, local populations with non-local species or seed sources (provenances).

Overall solutions in genetic conservation and management of forest genetic resources will vary according to national environmental, social and economic circumstances, institutional and legal frameworks, and prevailing needs and priorities of countries concerned. Decisions on conservation strategies to be applied by countries will, in turn, vary depending on the amount of basic knowledge available about the genetics and variation of priority species and about their management and silviculture, the immediacy and type of resource use and threats or pressures present. They will, also,

1 This is an information paper based on note AFWC\98\INF.5 presented at the Eleventh Session of the FAO African Forestry and Wildlife Commission, held in Dakar, Senegal, 14-17 April 1998

C. Mátyás (ed.), Forest Genetics and Sustainability, Volume 63, 217–222.

largely be determined by availability of institutional capacities and funding to ensure continuity of action over time.

Many forest tree species are characterized by inherently high levels of diversity and extensive natural ranges which often cross national borders. Regional and international cooperation is therefore a prerequisite for effective evaluation, conservation, sustainable management and wise use of forest genetic resources.

2. A proposal for action

The Fourth International Technical Conference on Plant Genetic Resources, held in Leipzig, Germany, in June 1996, adopted a Global Plan of Action on the Conservation and Sustainable Utilization of Plant Genetic Resources for Food and Agriculture which stated that forest genetic resources would not be included in the Plan, but that the need for action in this field should be reviewed in the light of the outcome of the work of the Intergovernmental Panel on Forests (IPF[2]). The IPF, which held its fourth and final session in February 1997, did not address the issue of forest genetic resources.

In view of the urgency of the matter, FAO raised the issue of forest genetic resources for substantial discussion at the Thirteenth Session of the Committee of Forestry (COFO). In line with the recommendations of COFO, subsequently supported and further elaborated upon by the Tenth Session of the FAO Panel of Experts on Forest Gene Resources, FAO has initiated action to help plan and coordinate a series of regional forest genetic resources workshops, to be carried out in close consultation and collaboration with national and international partners.

3. Regional and sub-regional workshops: goals and aims

The overall goal of the planned regional and sub-regional workshops on the conservation, management, sustainable utilization and enhancement of forest genetic resources is the development of dynamic, country-driven and action-oriented regional and sub-regional plans, agreed upon and to be used by countries concerned to help ensure that forest genetic resources are conserved and sustainably utilized as a basis for local and national

[2] The IPF was established by the Economic and Social Council of the United Nations (ECOSOC) in June 1995, following a recommendation of the Third Session of the UN Commission on Sustainable Development in June the same year. The IPF held a total of four sessions. It has been recently succeeded by the Intergovernmental Forum on Forests.

development including food security, poverty alleviation, environmental conservation, economic and social advancement, and the maintenance of cultural and spiritual values. These action plans should, without fail, be compatible with national and regional strategies in other fields, contributing together with these to dynamic, multidisciplinary plans aimed at overall sustainable development.

National forest genetic conservation programmes will constitute the building blocks of the regional and sub-regional action plans. In this regard, it is acknowledged that national plans and programmes will vary according to local biological, social and economic environments and according to national needs and priorities. The purpose of the planned workshops is to help streamline concerted action at regional level; the aim is not the development of one, single model for conservation, but elaboration of a framework for coordinated action at sub-regional and regional levels.

While national plans are at the bases of regional sub-regional action plans, these latter ones can, in turn, provide a point of reference for national activities in the exploration, collection, evaluation, conservation in situ and ex situ and improvement and breeding of forest genetic resources. Common agreement on principles and mechanisms for the determination of priorities for species and specific, conservation-related activities, and on optional strategies for action, will help justify such work and will help strengthen the impact of national activities at regional level.

Coordination of action at regional level will, furthermore, help make best use of scarce resources by avoiding duplication and overlap of effort, and by facilitating the sharing of technologies, information, know-how and genetic materials, on mutually agreed terms.

The sub-regional and regional action plans on forest genetic resources which will be developed in the planned workshops and which will serve as dynamic tools underpinning action by countries concerned could later, if countries so wish, be placed within a larger context, contributing to a comprehensive, international framework. A coherent global framework for action on forest genetic resources could decisively help promote overall coordination of action and help further cooperation between and among geographical regions and, above all, between and among countries in ecological regions of the world in which environmental conditions and social and economic aims and aspirations are similar, and in which such collaboration could therefore bring tangible benefits to all concerned.

4. Regional and sub-regional workshops: issues for discussion

The contents and details of regional and sub-regional workshops and the resulting action plans in forest genetic resources are likely to differ by region according to biological, social and economic conditions and needs and priorities of countries concerned. However, the development of agreed, regional and sub-regional plans for action which are compatible between them, and which are consistent and mutually supportive, will facilitate international dialogue, and will provide a reliable mechanism for flagging important regional and global issues and thus help ensure that forest genetic resources are not lost due to lack of information or by neglect.

Action plans to be developed might conceivably include the following common, main elements:

- review of the status of forest genetic resources at regional level; their links with other sectors/activities/programmes;
- identification of needs and requirements common to the region or to a sub-set of countries, identification of problems amenable to a regional coordinated approach, definition of overall goals;
- identification of overall regional priorities by species and by activity, based on national priorities, programmes and needs;
- assignment of relative priorities;
- incorporation of top priority species/activities, into regional action programmes;
- incorporation into action programmes of supporting activities;
- dissemination of information;
- training and research;
- exchange of seed and germplasm for research purposes;
- exchange of know-how and technologies;
- identification of international, regional or national mechanisms or instruments for supporting and/or implementing action programmes;
- estimation of approximate costs of elements of the proposed plan;
- identification of national, regional and international institutions capable of and willing to assume operational responsibilities for given components and programmes or activities;
- identification of projects amenable to external finance in support of national and regional work, identification of possible donors or supporters in-kind;
- clarification of overall responsibility for programmes at regional/sub-eco-regional level.

5. Action taken to date

In line with priorities expressed by countries in international fora in which the issue has been discussed, FAO, the International Plant Genetic Resources Institute (IPGRI) and the International Centre for Research in Agroforestry (ICRAF) have jointly organized a Workshop on the Conservation, Management, Sustainable Utilization and Enhancement of Forest Genetic Resources in Sub-Saharan Dry-Zone (Sahelian) Africa in 1998, in close collaboration with the Secretariats of the Permanent Interstate Committee for Drought Control in the Sahel (CILSS) and Intergovernmental Authority on Development (IGAD), and with support from other concerned international and bilateral organizations with an interest and know-how in the subject.

Preliminary plans have also been made for holding of a sub-regional workshop covering Southern and Eastern Africa, in collaboration with the Southern African Development Community (SADC) Forestry Units; and the Pacific Islands, in collaboration with the South Pacific Regional Initiative on Forest Genetic Resources (SPRIG). Experiences and information from the Sahel workshop will help facilitate planning of these and other workshops preliminarily planned to be held in various parts of the world.

Technical and scientific support to the above, and other forthcoming forest genetic resources workshops, will be provided by the FAO Panel of Experts on Forest Gene Resources; and by a Task Force on Forest Genetic Resources established by the International Union of Forestry Research Organizations (IUFRO), following a recommendation of the Tenth Session of the FAO Panel of Experts on Forest Gene Resources, with the explicit purpose of providing scientific know-how and guidance in support of the workshops to be coordinated by FAO.

6. Conclusions

There is an acknowledged loss of habitats, tree species and intra-specific variation as well as widespread lack of implementation of sustainable management of forests and woodlands in all regions of the world. There is also widespread lack of understanding of the role that trees and shrubs play in local sustenance and in local, national and regional economies, as providers of a wealth of goods from wood and fuel to food, fodder and medicines; and in the protection of soil and water resources, and the maintenance of cultural and spiritual values. There is, further, a general lack of acknowledgement, at all levels, from policy makers to the public, of the fact that the conservation of forest genetic resources is compatible with managed and wise utilization of the resources being conserved. This has lead to insufficient action and attention to the conservation of forest genetic resources at national, regional and global levels.

As stressed by a number of countries at the Thirteenth Session of COFO and in a number of other international fora, including the XIth World Forestry Congress held in Turkey in October 1997, action to safeguard and sustainably utilize forest genetic resources is an urgent priority. Delays in the conservation of forest ecosystems, species and genetic resources of trees and shrubs will be costly, implying environmental, economic and social risks, needs for expensive and at times difficult remedial action, and foregone opportunities in management and sustainable resource utilization in support of overall national development.

The regional and sub-regional workshops on the conservation, management, sustained utilization and enhancement of forest genetic resources, which FAO plans to help coordinate and organize in collaboration with relevant Centres of the Consultative Group on International Agricultural Research (CGIAR), IUFRO and other international, regional and bilateral players, are a first step towards concerted action in response to these urgent needs. This country-driven and action-oriented process complements other initiatives such as the preparation of national Biodiversity Status and Action Plans, within the framework of the Convention on Biological Diversity.

Data gathered, analyzed and synthesized in the course of sub-regional and regional workshops will complement information made available and regularly up-dated by the Panel of Experts on Forest Gene Resources, and information stored in the FAO World-Wide Information System on Forest Genetic Resources (REFORGEN). Information will be disseminated through the FAO Internet Homepage on forest genetic resources, and through the Clearing House Mechanism of the Convention on Biological Diversity.

7. GENETICS AND TREE IMPROVEMENT RESEARCH: PROBLEMS AND PROSPECTS

Problems and prospects for forest tree improvement research in the United States

Scott E. Schlarbaum

Introduction

Tree improvement research has played an important role in production of wood and fiber in the United States. Research programs have been established in universities, state and federal agencies, and private industry since the early 1900's. In the last 15 years, however, traditional tree improvement research has been declining. This paper discusses some of the contributing factors to reduction of effort in this important research area and what could be done to avoid further decline.

Forest tree improvement has been defined differently by North American authors. Wright (1976) considered tree improvement to be "the application of genetic principles to the betterment of forest trees." Zobel and Talbert (1984) defined tree improvement as "when control of parentage is combined with other forest management activities, such as site preparation or fertilization, to improve the overall yields and quality of products from forestlands." Both definitions have a connotation of field-oriented research, e.g., testing, selection and breeding, although Zobel and Talbert proposed a separate definition for forest tree breeding. In this paper, tree improvement research is considered to encompass traditional selection, breeding, and testing activities alone or in combination with cultural practices.

1. Early period (1916-1940's)

Genetic testing of forest tree species was initiated shortly after the turn of the century in this country (Van Haverbeak 1976). The first provenance test [*Pseudotsuga menziesii* (Mirb.) Franco] was established in 1916. Other tests of coniferous and hardwood (angiospermous) species were established during this period at various locations across the country. An early center for tree

C. Mátyás (ed.), Forest Genetics and Sustainability, Volume 63, 223–233.

improvement was the Eddy Tree Breeding Institute near Placerville, California. Initially, the Institute was a privately funded venture, but was donated to the USDA Forest Service in 1936 and became the Forest Service's first Institute of Forest Genetics. Provenance and progeny (common garden) tests were established on the Institute grounds. Over the years, the Institute became a center for pine hybridization under the direction of N. T. Mirov.

Other notable programs of this era were poplar breeding and testing programs in northeastern States lead by E. J. Schreiner and A. B. Stout (Syrach Larsen 1956). At the close of this era, the existence of genetic variation in tree species was established and potential to mirror gains in agronomic and horticultural crops was becoming generally recognized.

2. Rapid Program Expansion (1950's - mid 1960's)

Widespread proliferation of tree improvement research programs began to occur in the early 1950's. Research programs involving selection of "plus" trees from natural stands, establishment of provenance and progeny tests and breeding were established in different parts of country. In the early years of this period, the first forest genetics course was taught at Yale University by Professor F. Mergen, and forest geneticists began to be hired in professorial public and private positions.

During this time frame, private industry, the federal government, and many universities became involved in tree improvement research. University-based programs were formed often in cooperation with the wood products industry (University-Industrial cooperatives) in order to pool resources. These cooperatives used expertise in wide variety of disciplines available at state universities, to aid industry and in some instances state and federal governmental agencies in improvement of different species. Other University-based tree improvement programs were developed in the 1950's and early 1960's and were not formally affiliated with forest industry. The resident forest geneticist often worked closely with State Divisions of Forestry in construction of seed orchards and progeny testing. The Federal Government, primarily through the United States Department of Agriculture (USDA) Forest Service and the Tennessee Valley Authority, established testing and eventually breeding programs. Generally, research by private industry, university-industrial cooperatives, and federal government programs was directed toward increased yields of commercially important timber species, while university and State programs often encompassed a broader range of species for multiple uses.

Tree improvement research in this period was field oriented. Phenotypic selections were made from natural stands and grafted to form orchards for eventual breeding and/or seed production. Characterization of variability in commercial species was approached through provenance and progeny testing and evaluation of wood properties. The commercial potential of native, and occasionally exotic, species was evaluated in species comparison tests. As the number of scientists participating in tree improvement research increased, conferences began to organize in the eastern States throughout the 1950's to facilitate communication and cooperation.

3. Stability in the mid 1960's-1970's

After the rapid expansion of tree improvement research in the 1950's and early 1960's, a period of stability and progress occurred. Initial and advanced generation breeding activities, particularly in coniferous species, began to generate results allowed for construction and refinement of seed orchards that produced genetically improved seed.

In this period, private citizen groups concerned with different aspects of forest management and practices began to increase in number and size and become more politically influential. These groups are generally referred to as "environmentalist groups" in the United States. The country experienced an "ecology" movement in the early 1970's, as the effects of industrial waste, widespread pesticide spraying, and air and water pollution on the environment became more apparent. Concern also grew over loss or diminishment of forest-dwelling and forest-dependent species. The environmentalist groups generally targeted harvesting practices, particularly clear cutting in the northwestern forests of the Pacific coast. While not specifically targeting tree improvement research and development, the growing opposition to certain harvest practices posed an indirect threat to artificial regeneration of species that had been improved.

4. Changing emphasis in the late 1970's-early 1990's

Tree improvement research and development was impacted by five developments in this era: budgetary constraints in State and Federal agencies, retirement of forest geneticists in universities and the USDA Forest Service, the direction of Federal programs that provided extramural funding for research, a change of administrative philosophy, and the emergence of new technologies that had the potential to augment tradition tree improvement efforts. These factors combined to change the direction and emphasis of tree improvement research, primarily in universities and the USDA Forest Service.

Budgetary constraints began to negatively affect tree improvement program activities in this time frame. The long-term, relatively low profile of tree improvement research involves personnel, a lengthy commitment of land and resources to establish and maintain plantations, and a permanent operating budget. A significant loss to the general tree improvement effort was the suspension of Tennessee Valley Authority (TVA) tree improvement activities in 1982. The TVA had an extensive commitment in upland hardwood species, in particular oaks Progeny testing of phenotypic selections of white oak (*Quercus alba* L.) and chestnut oak (*Quercus prinus* L.) were established in 1968. In addition, the TVA established the largest North American genetic study of northern red oak (*Quercus rubra* L.) in 1972 with multiple locations throughout the Tennessee River Valley. Moreover, the TVA effort was foresighted by being diversified in addressing both timber and wildlife needs.

In the 1980's, a number of forest geneticists and breeders, who initiated tree improvement programs in the 1950's and 1960's, began to retire. The vacant positions often were not filled with scientists trained to continue tree improvement research due to a combination of factors. The cost of operating a tree improvement program in an era of reduced budgets and/or the lack of ability to secure extramural funding, e.g., National Science Foundation or USDA competitive grants, weakened the commitment to maintain tree improvement research. The advent of "biotechnology" in plant sciences, e.g., molecular biology and *in vitro* culture, and the projections for use in improving plant species caused administrations to examine the feasibility of continuing traditional tree improvement breeding and testing. Molecular biology and *in vitro* propagation on the other hand, research were eligible for extramural funding. In some instances, scientists with tissue culture or molecular backgrounds were hired and the resulting research program shifted from field to laboratory-based. Correspondingly, field testing and breeding programs were often reduced, eliminated or suspended. This trend was particularly apparent at universities in northeastern, Midwestern and Great Plains States that did not have formal University-Industrial cooperatives.

Within the USDA Forest Service, forest genetic and tree improvement research was centered in Experiment Stations and implemented by the Regional Tree Improvement Programs (now called Genetic Resources Programs) in the National Forest System. The research was conducted at three Institutes of Forest Genetics and various projects throughout the United States. Regional Genetic Resources Programs (RGRP) existed in all Regions except for Region 10 (Alaska). In 1985, a change in administrative philosophy, coupled with budget reductions, directed USDA Forest Service Research to switch from long-term projects to relatively short-term studies. This change significantly affected long-term tree improvement research, and many projects were closed over the next decade. Scientists replacing the retiring forest geneticists were more oriented toward basic research, as at the universities. This change in philosophy began to remove research support for the RGRP. Other changes in USDA Forest Service policy, discussed below, had more devastating effects on the RGRP.

The reduction of university and USDA Forest Service tree improvement efforts were evident in the gradual decline of Tree Improvement Associations in different regions throughout the United States. These Associations would sponsor regional tree improvement conferences. In the eastern states, the Southern Forest Tree Improvement Conference (SFTIC) was initiated in 1951. The first Lake States Forest Tree Improvement Conference and Northeastern Forest Tree Improvement Conference were held in 1953. The Central States Forest Tree Improvement Conference series began in 1959 in Wooster, Ohio. Tree improvement personnel in western states attended the Western Forest Genetics Association meeting and North American Forest Biology Workshops, which generally published abstracts of presented papers.

The eastern conferences generally met every 2 years until Lake and Central States groups merged in 1979, forming the North Central Tree Improvement Conference, following the closure of the USDA Forest Service's Lake States Forest Experiment Station. This Conference met every two years until merged with the Northeastern Conference in 1991 to form the Northern Tree Improvement Conference. The resulting conference series was short-lived, however, meeting in Vermont (1991) and Minnesota (1992) before discontinuing due to lack of participation. Another eastern-based conference was a multi-State Division of Forestry initiative, the North Central Fine Hardwoods Tree Improvement Cooperative. This group met annually from 1988 to 1995, but was disbanded due to declining interest. Only the SFTIC, Western Forest Genetics Association and North American Forest Biology Workshops currently continue to meet.

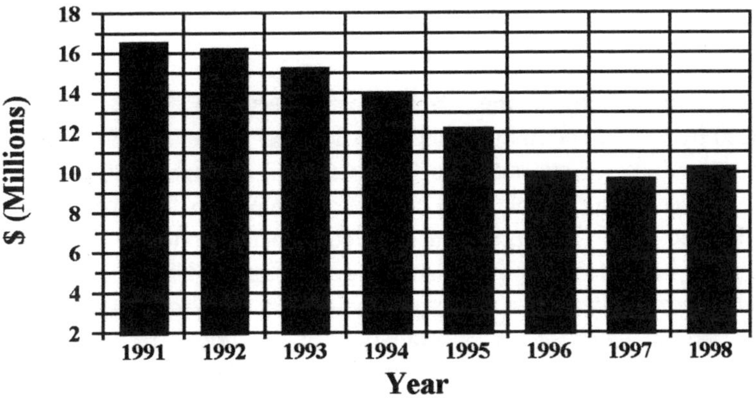

Figure 1. Annual budget (1991-97, projected 1998) for existing USDA Forest Service Regional Genetic Resources Program Regions 1, 3, 5, 6, 8, 9).

A number of State Divisions of Forestry (or equivalent agency) programs were diminished or eliminated in this period or the tree improvement position has been combined with a nursery-oriented position. Although seed can be still collected from existing orchards, further development, e.g., advanced generation breeding, of selected materials generally does not occur. The reduction of Forest Service programs caused a lack of support for remaining State tree improvement programs. While this reduction of effort did not immediately affect the wood product industries in these States, the stagnation of tree improvement research will have a future effect on forest production.

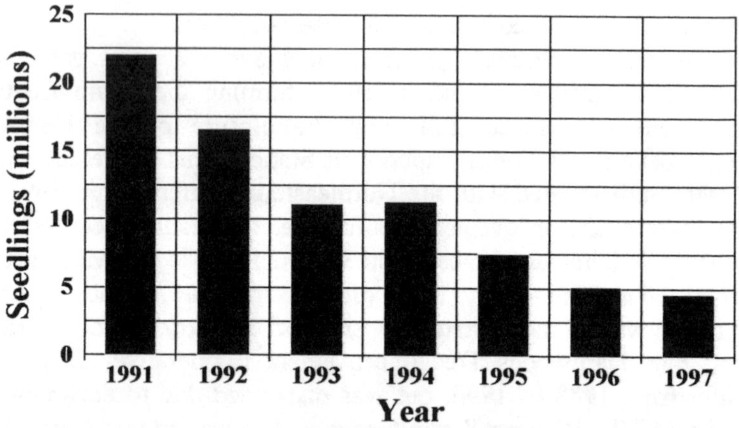

Figure 2. Pine seedling production (1991-1997) at the Ashe Nursery, USDA Forest Service,Southern Region.

Changes philosophy in the USDA Forest Service Research were greatly influenced by environmentalist groups that lobbied for equal recognition of the non-timber values forests provided to citizens. In the late 1980's, the USDA Forest Service publicly declared a change of philosophy in a program called "New Perspectives." This program evolved into the concept of "Ecosystem Management," a philosophy that currently permeates the USDA Forest Service. The concept of ecosystem management holds that knowledge and technology can be used in actions to encourage desired conditions of ecosystems for current and future environmental, economic, and social benefits (cf. Salwasser 1994). Although the concept promotes multiple use of National Forests, in effect this philosophical change severely impacted timber harvesting, which in turn impacted USDA Forest Service budgets. With the reduction in harvest, the need for improved seedlings correspondingly was reduced and RGRP budgets generally have declined since 1991 (Figure 1). In some instances, the RGRP were essentially eliminated and the responsibilities were divided among adjacent regions. USDA Forest Service nurseries also were impacted, and some have closed due to lack of need. For example, pine seedling production figures at the federal Ashe Nursery (Southern Region) for southern National Forests have steadily declined from 22.1 million in 1991 to 4.6 million in 1997 (Figure 2).

Not all developments in this period were negative for tree improvement research. The creation of a number of non-genetic cooperatives at different universities, e.g., nursery and fertilizer cooperatives, augmented the genetic research through important contributions to the silviculture component of tree improvement.

5. Integration of basic and applied research (mid 1990's to present)

The integration of basic research scientists into the forest genetics community generally has been successful. Teams of scientists have been formed at universities, the USDA Forest Service and in industry, with expertise ranging from molecular biology to tree breeding. Research directed toward early identification of important wood properties and disease resistance has been successful and breeding programs can use molecular markers to identify desirable genotypes. *In vitro* clonal propagation has advanced from the experimental stage to operational planting in some species. Although the number of tree improvement conferences remains small, both the SFTIC and the Western Forest Genetics Association Meeting are still strongly attended. Papers presented range from basic molecular research to provenance/progeny test results.

The University-Industrial tree improvement cooperatives are continuing to significantly contribute to improving tree species, with some programs operating for over four decades (Li *et al.* 1997). Some University-Industrial cooperatives, however, have been weakened through the mergers of member corporations, which reduce budgets based on membership number.

Universities without industrial cooperatives have begun hiring forest geneticists with a combination of basic and applied training. These scientists will be able to capitalize on materials developed by their predecessors. However, it is questionable whether they can resume directing a tree improvement program of previous magnitude. A philosophical change is needed in universities that traditionally have allocated only one position to forest genetics/tree improvement. University administrators should recognize the increasing sophistication and diversification of basic research and the difficulty for one scientist to manage laboratory and field-based research. In addition, university administration should recognize the long-term nature and service aspects of a university-based traditional tree improvement program and be prepared to provide permanent funding. Continuity of effort over time is not best served by relying upon grants and contracts.

Despite the optimistic outlook and apparent stability of tree improvement programs, most USDA Forest Service RGRPs and some state Division programs continue a tenuous existence. The 1998 budget (FY98) for the RGRPs was projected at just over $10,000,000 for the entire National Forest System (77,646,000 ha.). However, this figure is not evenly distributed across Regions. The FY98-99 projected budget for Regions 1, 2, 3, 4, 8, 9, 10 (all states except California, Hawaii, Oregon and Washington), which contain 76.7 percent of all National Forest land, was under approximately $3,500,000 for fiscal year 1998.

The reduced RGRP budgets will limit response to a number of problems National Forests have or will encounter. Resident exotic pests are impacting or destroying a number of native arborescent species (Campbell and Schlarbaum 1994, Schlarbaum *et al.* 1999). Incorporating genetic resistance is a viable option in many affected species including the once dominant hardwood species in eastern States, American chestnut [*Castanea dentata* (Marsh) Brok]. Additionally, an increasing number of exotic pests are being intercepted at ports-of-entry (Haacke and Carvey 1997). The opening of Siberian forest to harvest and export to the United States and increased importation from eastern Asian countries will probably result in more exotic pest problems. The RGRPs could be important in future gene conservation efforts by participating in activities ranging from establishing *ex situ* plantations of representative genotypes to collecting seed for storage. Species in some ecosystems, e.g., *Quercus* on good upland sites (site index > 65), are not naturally regenerating in eastern forests. Oak seed orchards could be a source of seed for use in reforestation via enrichment plantings on these problem sites. Changes in global conditions, i.e., global warming, and increased air pollution problems could cause forest decline. Massive forest fires, wind and ice storms and other natural disasters can destroy or damage large areas. The RGRPs could respond to these challenges by selecting suitable genotypes for reforestation and constructing seed orchards to provide adapted seed. In addition, the emphasis on National Forest use could change in the future, due to changes in local, national, and international demand and supply of wood. If the American public deemed it desirable for cutting levels to increase in the future, seedlings, preferably genetically improved, will be needed to reforest harvested areas.

National Forests are valued from many perspectives (Oliver *et al.* 1997). Trees originating from the seed orchards of RGRPs replace not only the timber in the forest, but provide habitat and mast for wildlife, scenic beauty, and other values. Although the USDA Forest Service has adopted an "Ecosystem Management" policy, changes have not occurred in the budgeting process to recognize the multiple values of planted trees. The RGRPs are funded through the National Forest - Forest Vegetation line item budget and are limited to working with species grown for timber production. A policy change needed to recognize the multiple values provided by the RGRP, and revision of functional budget lines should be made to allow other Regional programs, such as Recreation and Wildlife, to share in funding RGRPs.

Tree improvement programs in State Forestry Divisions currently exist in 32 of 50 States (Table 1). Some of these programs are large in size and broad in scope, while others have been abandoned or reduced to only maintaining seed orchards. Unfortunately, most State tree improvement programs are placed in a Reforestation Division, which must compete with landowner assistance and wildfire control for internal funding. Program alterations due to budget reductions in State Divisions are often made to emphasize short-term goals, and tree improvement program budgets can be reduced or eliminated without immediate impact. A policy change to raise tree improvement to a department level, e.g., equal with wildfire control, would increase the difficulty of diminishing these long-term programs. State Divisions of Forestry with limited or without tree improvement programs should recognize the need for delineation of seed movement, if creating or maintaining a tree improvement program is not feasible. Strong ties should be made with a state-based university or cooperative program, if present to ensure the planting of locally adapted stock.

Table 1. Survey of State Division of Forestry (or equivalent) for existence of a tree improvement program.

States with active tree improvement programs:
Alabama, Arkansas, Florida, Georgia, Hawaii, Idaho, Indiana, Maryland, Michigan, Minnesota, Mississippi, Missouri, Montana, Nebraska, Nevada, New Hampshire, New Mexico, North Carolina, North Dakota, Oklahoma, Oregon, Pennsylvania, South Carolina, Tennessee, Texas, Washington, Wisconsin
States combining tree improvement with nursery operations:
Kentucky, New York, Virginia
State without tree improvement programs:
Alaska, Arizona, California, Colorado, Connecticut, Delaware, Illinois, Iowa, Kansas, Maine, Massachusetts, New Jersey, Ohio, Rhode Island, South Dakota, Utah, Vermont, Wyoming

– The growing lack of tree improvement infrastructure for federal land bases extends beyond the 77,646,000 ha, managed by the USDA Forest Service. Other federal lands will be affected by exotic pests, pollution etc., regardless of their management policies. The National Park Service contains 33,705,000 ha, of properties that are managed as preserves. However, no permanent funding is available for gene conservation or species restoration in case of a natural disaster, exotic pest decimation etc. The Bureau of Land Management has 18,219,000 ha, of forested land

and allocates approximately $2,000,000 in funds (FY98) for tree improvement-related activities.

In total, 129,600,000 ha of federal land are being supported through tree improvement activities by only $12,500,000 (approximate); an average of $0.104 per hectare. This amount is dangerously low, given the values most Americans place on forested lands and the current and future challenges to forest/species integrity. The role of Federal Genetic Resources/Tree Improvement Programs on Federal land bases needs to be examined and appropriate policies adopted to stabilize, protect, and/or improve forests on Federal lands in the future.

Although generally against timber-oriented forestry activities, environmentalist groups have the potential to be strong advocates for tree improvement research. Genetically improved trees produce more fiber on less acres, leaving more land for alternative purposes. Additionally, tree improvement programs support genetic conservation and preservation of diversity in natural areas. A narrow genetic base can be dangerous to a forest's future, and tree improvement programs strive to keep maintain an appropriate level of diversity in improved seedlings. Changes in wood product demands and manufacturing processes can alter selection criteria, causing new selections from natural populations to be made. Therefore, it is important to have reservoirs of natural diversity available to meet the changing needs of society and technological advances.

6. Closing comment

A myriad of challenges to forested lands in the United States can be addressed by tree improvement research. An increasing population will use more fiber-based products, yet harvesting restrictions on governmental and private lands have become more prevalent. Increased fiber yield per hectare will be needed to sustain demand. Tree improvement research can contribute to resolving public concerns about supply and demand, while advocating genetic diversity and gene conservation. Although the integration of biotechnology techniques with traditional breeding and testing approaches is promising, the tenuous or declining infrastructure of tree improvement research programs at some universities and within certain State Divisions of Forestry and federal agencies is a critical threat to the integrity of United States' forests. Philosophical and policy changes are needed at educational and governmental agencies to ensure the wise, multiple use of forested lands in the United States.

Acknowledgement

The author would like to acknowledge USDA Forest Service employees: Dr. R. A. Cecich, Dr. T. La Farge, and Mr. T. Tibbs and Professor Howard B. Kriebel for helpful suggestions and comments. The author also would like to acknowledge the assistance of USDA Forest Service Regional Geneticists (or equivalent): Mr. J. Dick, Dr. J. Kitzmiller, Dr. M. F. Mahalovich, Dr. S. Martinson, Mr. R. Meier, and Mr. T. Tibbs and Mr. D. Stewart, Bureau of Land Management, in obtaining budget figures.

Literature cited

Campbell, F. T. and S. E. Schlarbaum 1994. Fading forests. North American trees and the threat of exotic pests. Natural Resources Defense Council report, 47 pp.

Haack, R. A., and J. F. Cavey 1997. Insects intercepted on wood articles at United States ports-of-entry and two recent introductions: *Anoplophora glabripennis* and *Tomicus piniperda*. *In* International Forest Insect Workshop Proceedings, 18-21 August, Pucon, Chile. Corporacion Nacional Forestal, Santiago, Chile.

Li, B., S. McKeand, A. Hatcher and R. Weir 1997. Genetic gains of second generation selections from the NSCU-industry cooperative tree improvement program. Proc. 24th Southern Tree Improve. Conf., p. 234- 238.

Oliver, C., D. Adams, T. Bonnicksen, J. Bowyer, F. Cubbage, N. Sampson, S. Schlarbaum, R. Whaley, H. Wiant, and J. Sebelius 1997. Report on forest health on the United States by the Forest Health Science Panel. Center for International Trade in Forest Products, University of Washington, RE43. 300 pp.

Salwasser, H. 1994. Ecosystem management. Can it sustain diversity and productivity? *J. For.* 92: 6-7; 8-10.

Schlarbaum, S. E., R. L. Anderson, and F. T. Campbell 1999. Exotic pests: insects, pathogens and plants. *In* Peine, J. (ed.): Ecosystem Management for Sustainability: Principles and Practices Illustrated by a Regional Biosphere Cooperative. St. Lucie Press, St Louis. 291-306.

Syrach Larsen, C. 1956. Genetics in Silviculture. Oliver and Boyd Publishers, Edinburgh. (Translated by M. L. Anderson). 224 pp.

Van Haverbeak, D. F. 1976. Forest genetics and tree improvement. *In* Shelterbelts on the Great Plains. Proc. Symp. Great Plains Agric. Council Publ. 78: 154-159.

Wright, J. W. 1976. Introduction to forest genetics. Academic Press, New York. 463 pp.

Zobel, B. and J. Talbert 1991. Applied forest tree improvement. Waveland Press, Inc. 505 pp.

Conservation and management of genetic resources of commercial forests in New Zealand: Challenges of institutional changes and new technology

Rowland D. Burdon and Michael J. Carson

Introduction

Forestry in New Zealand has recently arrived at a clear division into intensive commercial plantation forestry based on exotics and indigenous wildland management. Alongside this has come almost complete privatisation of plantations from around 50% of state ownership, an almost 50% withdrawal of the state from research funding, and rapid development of laboratory-based biotechnology. Plantation forestry is around 90% based on the strongly preferred *Pinus radiata.*

While seemingly simple, the conservation and management of genetic resources face technical problems arising from the very success of *P. radiata,* and institutional challenges. Privatisation and 'user-pays' have brought greater focus on current breeding objectives, but generally much less on the long-term management of genetic resources. While genetic resources in privatised forests may be subject to covenant protection they can remain vulnerable to accidental loss, and the *de facto* policing agency is dependent on the forest owners for funding.

The new technology has brought changes in attitudes, which are often aimed at more proprietary control; this, along with tight funding, tends to undermine traditional collaboration in tree breeding. Classical field-based breeding and gene resource management operations come under funding pressure, yet the successful application of new technology may depend on sustained or increased inputs in these areas. More directly, new genetic technologies can serve either to erode genetic diversity or as powerful tools in conserving it. Use of laboratory techniques to measure functional diversity and to monitor its changes is very attractive, but hard to develop.

As a treaty, the International Convention on Biodiversity imposes limited obligations on commercial forestry in New Zealand. Environmental certification is seen as open to abuse, in erecting non-tariff trade barriers and hampering the contribution that highly productive plantations can make to global conservation. New Zealand's Resource Management Act, which emphasises sustainability, is a promising but still untested instrument for

C. Mátyás (ed.), Forest Genetics and Sustainability, Volume 63, 235–246.

fostering genetic conservation. Public opinion can be a major influence, but is easily misinformed.

Recent developments, therefore, have brought both benefits and problems for forest genetic conservation in New Zealand, with important challenges for the future.

1. Background and present situation

When large-scale European settlement of New Zealand began in the mid-nineteenth century there was abundant high-quality native timber. However, while the species were excellent as virgin timbers they have not been successfully domesticated as commercially renewable resources. By contrast, a number of exotic species performed excellently, and proved eminently suited to commercial plantation forestry. The preferred conifers almost all come from the coast of California and Oregon, USA, while the preferred hardwoods come mainly from Australia.

By 1913, with native timber supplies dwindling fast, timber supplies for the future clearly had to come from plantations of exotic species. This was addressed vigorously by both the state and the private sectors in a planting boom during 1925-1935. While several exotic species grew very well, *Pinus radiata*, an obscure conifer from the Californian coast, was pre-eminent, because of its very fast growth, wide tolerance of New Zealand sites, and its ease of establishment. Meanwhile, research was begun on the utilisation of *P. radiata*, since the species had been chosen with almost no track record as a commercial timber. Ultimately, its utilisation proved to be a huge success, its wood being suited to a remarkable range of processes and end-uses. That there have been problems with the utilisation has reflected a combination of several factors, notably early neglect of tending, then having to learn how best to tend the species and, more recently, the results of pressures to reduce growing costs by shortening rotations. This experience, however, basically represents a new paradigm of adopting fast-grown plantations as a major source of wood.

Tending and exploitation of the plantations from the planting boom received little attention during World War II. By 1960, however, the potential of plantation-based industries to broaden a narrow and vulnerable export base was becoming clear. A second planting boom began, building up until the early 1970s, since when it has continued with some fluctuations. Because species diversification had proved unsatisfactory, on biological as well as economic grounds, *P. radiata* became more than ever the preferred species, such that it now occupies about 1.5 million ha in New Zealand, or around 90% of the plantation forest area.

Pinus radiata, despite its natural advantages, was an obvious candidate for genetic improvement, because its huge and seemingly heritable variation among trees in health, growth rate, branching habit, and general quality of tree form, indicated great scope for genetic gains. Accordingly, an intensive breeding programme was begun in the early 1950s at the New Zealand Forest Service Forest Research Institute, and it continues unabated. It started with selections within local 'land-race' material, followed by provenance testing and estimation of genetic parameters. This approach proved basically correct, but while the local material was well-adapted and variable, it proved to have originated from a very incomplete and unbalanced sample of the species' natural range (Burdon 1992; Burdon, Broekhuizen and Zabkiewicz 1997, Burdon *et al.* 1997). Indeed, it appeared that there were probable long-term advantages in making fresh seed importations from native stands. That was done in 1978-80 in collaboration with CSIRO, Australia (Eldridge 1978), leading to the establishment of numerous follow-up provenance trials and some gene-resource plantings (Burdon 1988, Eldridge 1997, Burdon *et al.* 1997, 1998).

With some secondary species, provenance choice was clearly crucial, so a large network of provenance trials of many different conifers was established during the mid-1950s. Some further trials, largely with hardwoods, have since been planted. Our provenance trials include some internationally important collections. Several secondary species, notably Douglas-fir, *Cupressus macrocarpa*, *C. lusitanica*, *Eucalyptus nitens* and *E. regnans*, have warranted their own breeding programmes which are now running, but all on a lesser scale than with *P. radiata*.

In the mid-1980s New Zealand began major economic reforms and institutional changes. The Forest Service was disestablished on 31 March 1987, largely through the combined objections of economists and environmentalists to its broad mix of responsibilities, in a dramatic repudiation of multiple-use forestry. This change separated the Forest Research Institute (FRI), which ran the genetic improvement programme, from the commercially-oriented Forestry Corporation which ran the state-owned plantation forests. Also, the state plantation forests were separated from the indigenous wildland management, which was placed under a new Department of Conservation, thus formalising a final phase-out of native forests as a significant timber source. Meanwhile, a shift began from 100% government funding of FRI to substantial dependence on outside funding, which is now over 40% of its total funding and around 50% of funding for research on genetic improvement and related biotechnology. In the case of *P. radiata* breeding, substantial external funding now comes from the New Zealand Radiata Pine Breeding Cooperative (NZRPBC), which was established out of necessity in 1987 and now has 14 industry members. In addition, various other external funding arrangements have developed, including other cooperatives, some multi-client ventures, one-to-one deals with individual companies and overseas consultancies. In 1992 a further change was made, FRI becoming the New Zealand Forest Research Institute Ltd, one of 11 Crown Research Institutes, each owned by the state, but with a company structure and a politically appointed Board of Directors.

The initial corporatisation of state exotic forests turned into outright privatisation during the 1990s, which is very nearly complete. The new owners of these forests include both pre-existing multinational corporations in New Zealand and several newcomers, so the plantation estate is now over 60% foreign-owned. On top of this second phase of change has come further changes of forest ownership: some basically rationalised land bases, and some also served to reduce high debt levels resulting from major acquisitions. These developments have come on top of a 'boom and bust' of 1987, and the inherently cyclical behaviour of forest industries. The economic views driving these institutional and ownership changes have also brought a high degree of deregulation but a heavy monetarist emphasis in economic policy, which has pushed real interest rates to high levels by world standards. Finally, the east Asian recession placed New Zealand's forestry sector under severe pressure. All these factors have contributed to what is currently a tight financial situation. In addition there have been frequent and drastic corporate restructurings, often on a two-year cycle, which have accentuated an alarming loss of 'institutional memory' in forest management.

Thus New Zealand, while hardly unique in any one respect, epitomises a set of global trends: rapid, large-scale privatisation of forests; changes in research funding; institutional changes in both the research and production areas; deregulation and the associated reliance on the market; and pervasive emphasis on monetarism.

Despite these problems new planting has continued apace. This reflects some taxation advantages, long-term confidence, and conversion of large areas of marginally economic pastoral land. Importantly, over half of all planting, including restocking, is now being done by 'small players' which include small companies, syndicates and even individuals.

2. Technical issues

In the area of gene conservation *P. radiata* has been a priority, for several reasons:

- The species' sheer importance to New Zealand's economy;
- New Zealand's major role as steward of the species' genetic management;

- The fact that the species' pre-eminence, its greatly preferred status, and it being an exotic, all give greater importance to the various imponderable biotic threats;
- The success of the intensive genetic improvement and a strong acceptance of its benefits are leading to a rapid replacement of the huge secondary gene resource represented in the land-race stocks.

Serious concerns about genetic resources of *P. radiata* in New Zealand were first voiced by Bannister (1959), prompted largely by the writings of Vavilov. This led to systematic testing of material from the species' full natural range (Bannister 1963, Burdon 1988, Burdon *et al.* 1992). In parallel came a major expansion of the breeding population based on the land-race material (Shelbourne *et al.* 1986, 1989). With input from Libby (1973) a scheme of genetic management evolved (Burdon *et al.* 1977, Shelbourne *et al.* 1986, Burdon 1988). Briefly, it entailed a three-level hierarchy of populations, with the **production population** at the top, underpinned by the **breeding population** which would serve as the main 'engine room' for achieving cumulative genetic gain, and which in turn would be underpinned by the **gene resources**. This structure reflects the trade-off between the level of genetic improvement and the breadth of the genetic base, and it was envisaged that any movement of genetic material (at least between these main strata of the hierarchy) would be upwards. Population sizes envisaged, in terms of parents for each generation, were in line with a wide consensus that has since emerged (Burdon 1997), for a few tens, a few hundred and a few thousand in the production, breeding and gene-resource populations respectively.

Within this scheme it was envisaged that native-population material could be maintained *ex situ* under dynamic management that was designed to keep sizeable samples of the native populations in an essentially pure state, while capturing appreciable genetic improvement for basic silvicultural traits without materially eroding the genetic base (Shelbourne *et al.* 1986, Burdon 1986, 1988). The land-race stocks, however, were addressed on a different basis; during 1985-88 nearly 1000 new plus trees were selected therefrom, before these stocks were replaced and/or subject to heavy pollen contamination from stands of seed-orchard origin.

Since then, concerns have been expressed over how much pollen contamination will occur without prohibitively expensive controlled pollination in *ex-situ* gene resources, because *P. radiata* is wind-pollinated and abundant almost throughout the country, while any pollen contamination will come increasingly from narrowly-based seed-orchard stocks (Burdon 1995, 1997). Such contamination could compound strongly over successive generations, and could severely compromise the gene resources as potential sources of new and unrelated genetic material. The contamination is not quantified, but needs to be. To preclude it could entail very heavy direct or indirect costs, which would not be welcome in the industry. A notable exception to this situation is the critically endangered Guadalupe Island population of *P. radiata*, where use of controlled pollination to maintain a pure population *ex situ* is commercially attractive as well as ethically laudable. We will return to this population later.

Even with the native populations of *P. radiata* on the Californian mainland, which might be gene resources of last resort, there is now a major complication, created by pitch canker (Storer *et al.* 1997). This disease is a real threat to the populations themselves, while it poses such quarantine risks that further importations of seed from those stands to New Zealand are out of the question.

Two other planks of the scheme are now in question, regarding how to provide for capture of spontaneous new genetic variation that might be needed (Burdon 1997). The first involves the upwards-only movement of genetic material within the population hierarchy. Having the commercial stands as a genetic dead-end might close off important opportunities for capturing the favourable new genetic variation that has always arisen in the past in natural stands (albeit at very low frequencies) through various types of mutation. An associated concern actually arises from the other plank in question, namely maintaining pedigree in breeding populations. While this is aimed at controlling both the rate of inbreeding and losses of existing alleles each generation, it entails a cost of deploying very finite numbers of candidates at the start of each new generation, which could also limit the scope for capitalising on the occasional favourable mutations. If a biotic crisis arose, e.g. a new and serious disease with a very low incidence of host resistance, there might be a real need to deploy candidate material in extremely large numbers to make it possible to reinstate a robust breeding population with the required resistance. To manage large areas of commercial plantation to provide for active capture of favourable mutants may not be impossible, but it would present major technical and logistical challenges for breeders and forest management.

Yet the role of mutation (in the broad sense of the word) can be viewed otherwise. It may allow continued long-term genetic gains from quite small breeding population units (30-50 individuals) (Eriksson *et al.* 1994, 1995, Namkoong and Khoshy 1995, Khoshy and Namkoong 1996), while total genetic diversity can be boosted by addressing different breeding objectives in different subdivisions of the breeding population under the "multiple populations" scheme (Namkoong 1976).

Herein we have an unresolved technical issue, namely whether one needs the really large numbers, at least early in each generation, in order to achieve the desired protection against a biotic crisis, or whether relatively modest numbers will suffice. The answer may depend strongly on the situation. For example, in a species like *P. radiata,* which is grown massively as an exotic in the absence of various known and potential diseases and pests, the 'big numbers' option may be preferable. Yet with *P. taeda* within its native range, where it should already have a much more complete biota of pests and pathogens to all of which it already has significant resistance, there may not be the same need for sheer numbers.

With other exotic tree species, which often have essentially clinal patterns of native provenance variation, the technical issues tend not to be pressing. This is because New Zealand is only a minor grower, and they have not been intensively improved, thus pollen contamination problems do not exist.

3. Impacts of privatisation and new funding system

When the Forest Service was broken up, all potentially important genetic material came under covenant protection, to be felled only with permission from FRI, and these covenants remained, after some review, upon privatisation. Such protection, however, has not been complete, because some material has been felled, often inadvertently, by hard-pressed management in organisations that have been largely stripped of institutional memory by ownership changes, reorganisations and staff layoffs. Serious losses have been few, but remain a disturbing possibility. Even where oversights do not occur, requests to fell often arise at very short notice. Often the individual plantings are not demonstrably crucial, but their loss may erode safety margins against the effects of climatic damage or fires. In any event, it can be difficult to deny requests from managers who are hard-pressed, and when we depend on their organisations for both funding and in-kind contributions.

When the NZRPBC was formed in 1987, the emphasis was placed on operational breeding (Carson *et al.* 1990). This reflected the perceptions of the members in relation to short- and medium-term benefits. Work on gene resources was either deferred or covered by government funding. What has been done was mainly on evaluation of material in existing plantings, with the notable exceptions of new Douglas-fir collections from the coastal fog belt of California and Oregon and the selection of some plus trees within *ex-situ* plantings of native provenances of *P. radiata*. Deferring planned operations, however, can have insidious effects. For example, if selected trees of *P. radiata* are not climbed promptly for genetic material, their increased size can make the task prohibitive. And, as Eldridge (1997) forcefully stated, gene resource plantings are subject to progressive loss if not actively managed and renewed. In fact, active management of the gene resources of *P. radiata* is now being considered by the NZRPBC.

The NZRPBC has not only supported the general advance of the breeding population. With increasing emphasis on wood properties it is now supporting a strong differentiation of 'elite' elements of the breeding population into a number of 'breeds'. This addresses the opportunities that arise from the wide range of processes and end-uses for the species combined with its great phenotypic plasticity across environments (Jayawickrama *et al.* 1997, Burdon in press). Such differentiation will augment and utilise the genetic variation that is expressed within the breeding population.

The state research funding agency, FRST, has continued to fund research on gene resources of *P. radiata*, and has actively supported research on secondary species out of a concern over the dependence on a single species. There has also been active support from industry of breeding work with Douglas-fir, eucalypts and cypresses, although that support is vulnerable to periodic downturns and changes of corporate strategies. But whatever the funding, there is still a great dependence on the forest owners to host and tend genetic material in the field.

4. New biotechnology

The new biotechnology includes: *in-vitro* propagation; use of DNA markers for 'genetic fingerprinting', measuring genetic diversity and gene mapping; and genetic engineering and the underpinning studies of gene function. The advent of this technology has attracted major resource inputs and brought about important changes (Burdon and Moore 1997). This expenditure has occurred despite a tight funding situation, reflecting in part a perception of the promise of that technology, and in part defensive moves to capture market share of what is seen as an expanding market. While this has undoubtedly increased the total funding for genetic improvement and associated research, it does appear to have diverted significant funds and management focus from the more traditional, field-based genetic management, even though much of the application of the new technology will depend very heavily on field-based work. Indeed, most such technology must serve to enhance traditional genetic management rather than supplant it (Burdon 1992, 1994).

The developments in biotechnology, along with consolidation of sector organisations into larger companies and a brief boom in wood prices, have been accompanied by much more proprietary attitudes towards technology. There is now a joint venture for the development of new gene technologies between New Zealand Forest Research Institute Ltd and GEENZ Ltd which represents various private companies. In addition, however, the main forest-owning companies have mounted their own biotechnology ventures, sometimes with large budgets and considerable secrecy. Such work, running in parallel with traditional breeding under the auspices of the NZRPBC, but without the open exchange of information that has been a cornerstone of traditional tree breeding cooperatives, is placing additional pressure on the organisational structures for research/industry collaboration.

The focus on proprietary biotechnology may well have further diverted attention from the gene resources that underpin the breeding populations. This may partly reflect a perception that genetic transformation can largely obviate any need for future gene conservation. But, while transformation may bring in genes that are not currently available, it seems dangerously optimistic to see it as an answer to any new biotic challenges. However, we now have controlled pollination and vegetative propagation technology that lets us readily apply the "multiple populations" principle in managing breeding populations of *P. radiata* (Burdon 1986, Carson *et al.* 1990, Jayawickrama *et al.* 1997), which can amplify the genetic variation that is expressed within the species. Also, new technology gives additional options for preserving and propagating rare and valuable gene-resource material.

The modern laboratory-based genetic markers in principle afford very powerful tools for measuring and monitoring genetic diversity. However, if marker diversity is to be used well it must be cross-calibrated against 'functional' genetic diversity (cf Libby 1995), which can be a huge task. It appears that the informativeness of the various types of markers can depend greatly on the species and even on the taxonomic subdivision of a species. Furthermore, we need markers that give good resolution of the early stages of a decline in genetic diversity.

In fact, a spectacular success for marker technology has been in the area of 'genetic fingerprinting', specifically with DNA microsatellites, which has been used to verify clonal identities and pedigrees. The application so far has been in the area of operational breeding and associated research. However, the potential is great for developing further applications when the right types of markers are identified for specific situations.

Using the new gene technologies to best effect, and avoiding dangers of misuse, will require care and skill. Success with their use will depend on how well they are integrated with the genetic management throughout the population hierarchy.

5. Treaty obligations

New Zealand's obligations under the International Convention on Biodiversity are notably limited with respect to commercial forest trees. Our own native tree species, while they include attractive ornamentals, command no interest for commercial production forestry anywhere else, let alone in developing countries. Concerning their own genetic conservation, the large woody plants now enjoy considerable protection at the level of the remaining native forest, while the often scattered remnants of these forests should afford very efficient conservation of the species' total genetic diversity.

Of the commercial exotic forest species, few come from developing countries. A very important exception, though, is the Guadalupe Island population of *P. radiata* (*P. radiata* var. *binata*), which in the wild is fast becoming extinct (Libby *et al.* 1968, Ledig *et al.* 1996) because of an uncontrolled population of introduced goats. This population is now of significant commercial interest for use as F1 hybrids with select material from our land-race stocks (Low and Smith 1997). The hybrids show excellent vigour and health, with few of the adaptive problems shown by pure, somewhat inbred Guadalupe material, and the Guadalupe parentage contributes superior stem straightness, higher wood density and, hopefully, better resistance to exposure and even winter cold.

In maintaining a pure Guadalupe Island population, we are actively observing the Convention. However this is being done because of commercial interest in the population, rather than to avert further risks of accidental loss. Also the University of California Genetic Resources Conservation Program is interested in helping protect the native mainland stands of *P. radiata* which are threatened by a combination of urbanisation and the advent of pitch canker, and it is seen fit to cite the species' commercial importance (D.L. Rogers, pers. comm.).

Certification of forest products for the environmental practices observed in their production is also a potentially important outside influence. It is seen as open to being abused by parties aiming to erect non-tariff barriers to trade by exploiting unfounded environmental fears of plantation forestry, and is being actively addressed within the New Zealand forestry sector. Yet highly productive plantations have a major potential to relieve pressures on tropical rainforests which may be capable of producing barely one-tenth of the wood volume per ha per annum.

6. Legislative and policy mechanisms within New Zealand

New Zealand has both legislative and policy mechanisms to promote the conservation of genetic resources. The key legislation is the Resource Management Act (RMA) of 1991. The RMA is very wide-ranging in its provisions, addressing wildlife conservation, wildland management, management of farmlands and plantation forest, prevention of pollution, and town and country planning. A pervasive theme is sustainability of management and exploitation. In general, the RMA strongly endorses the provisions of the International Convention on Biodiversity. Policy mechanisms exist in the area of research funding. "Sustainability" and "Biodiversity" are among the themes selected for emphasis in FRST funding. This has translated into increased government funding for research on forest plantation species that are alternatives to *P. radiata*. These are mainly species for the relatively low-volume special uses to which *P. radiata* wood is not well suited, rather than being 'contingency' species that would become commercial only in the event of major biotic problems with *P. radiata*.

Public opinion can strongly influence forest management, and concern about it has varied widely among forestry companies. Most parties are probably wary of public opinion, however, fearing that it could it could press for uneconomic practices based on major ecological misconceptions. Non-profit organisations can make significant financial contributions and exert major political influence. But while several in New Zealand have engaged strongly in forestry issues, they may have little influence on gene conservation of plantation species. Historically, they opposed the replacement of native species by exotics. While it is now accepted that exotics are needed, and that their availability has probably saved much of the remaining native forest, it will still require a very big change of attitude for those organisations to champion gene conservation in exotics.

Acknowledgement

We thank Dr C.J.A. Shelbourne for comments on a draft.

Literature cited

Bannister, M.H. 1959. Artificial selection and *Pinus radiata. New Zeal. J. For. 8*: 69-90.

Bannister, M.H. 1963. Planning a genetic survey of *Pinus radiata. Proc. World Consultation on Forest Genetics and Tree Improvement, Stockholm, Aug. 1963. FAO, Rome. Paper 4/1.*

Burdon, R.D. 1986. Clonal forestry and breeding strategies - a perspective. *In Breeding Theory, Progeny Testing and Seed Orchards.* Proc. IUFRO Meeting, Williamsburg, VA, USA. 645-659.

Burdon, R.D. 1988. Recruitment for breeding populations: objectives, genetics and implementation. *In* Weir, B.S. *et al.* (eds.): *Proc. 2nd Internat. Conf. on Quantitative Genetics* Sinauer, Sunderland, MA, USA. 555-572.

Burdon, R.D. 1992. Tree breeding and the new biotechnology - in damaging conflict or constructive synergism? *In Proc. IUFRO Meeting "Breeding Tropical Trees - Resolving tropical forest concerns through tree improvement, gene conservation and domestication of new species"*, Cartagena and Cali, Colombia. 1-7.

Burdon, R.D. 1994. The place of biotechnology in forest tree improvement. *In Forest Genetic Resources 25*, FAO, Rome. 2-5.

Burdon, R.D. 1995. Future directions in tree breeding: Some questions of what to seek and how to manage the genetic resource. *In* Lavereau, J. (ed.): *Evolution and Tree Breeding, Part 2,* Proc. Canad. Tree Impr. Ass./Western For. Genet. Ass. Conf. 28 Aug. - 1 Sept., Victoria, BC, Canada, 24-39.

Burdon, R.D. 1997. Genetic diversity for the future: Conservation or creation and capture? *In* Burdon and Moore (1997). 237-246.

Burdon, R.D. (in press). *Pinus radiata. In* Last, F.T. (ed.): *Ecosystems ot the World.* Vol. 19: Tree Crops. Elsevier, Amsterdam, The Netherlands. Chapter 4.

Burdon, R.D. and J.M Moore (ed.) 1997. *IUFRO '97 Genetics of Radiata Pine. Proc. NZ FRI - IUFRO Conf. 1-4 Dec. and Workshop 5 Dec. 1997,* Rotorua, New Zealand. FRI Bull. 203.

Burdon, R.D., P. Broekhuizen, and J.A. Zabkiewicz 1997. Comparison of native-population and New Zealand land-race samples of *Pinus radiata* using cortical oleoresin monoterpenes. *In* Burdon and Moore (1997). 50-56.

Burdon, R.D., C.J.A. Shelbourne and M.D.Wilcox. 1998. Advanced selection strategies. *In Proc. 3rd World Consultation on Forest Tree Breeding,* Vol, 2. CSIRO, Canberra, Australia. Paper 6/2.

Burdon, R.D., A. Firth, C.B. Low and M.A. Miller, M.A. 1997: Native provenances of *Pinus radiata* in New Zealand: performance and potential. *New Zeal. Forestry* 41 *(4)*: 32-36.

Burdon, R.D., A. Firth, C.B. Low and M.A. Miller (1998). Multi-site provenance trials of *Pinus radiata* in New Zealand. *Forest Genetic Resources 26 (1998)*. FAO, Rome. 3-8.

Carson, M.J., R.D. Burdon, S.D. Carson, A. Firth, C.J.A. Shelbourne and T.G. Vincent 1990. Realising genetic gains in production forests. *Proc. Joint Meeting of Western For. Genet. Ass. and IUFRO Working Parties S2.02-05, -06, -12 and -14 on Douglas-fir, contorta pine, Sitka spruce and* Abies *spp.* Breeding and Genetic Resources Session. Olympia, WA, USA, Aug. 1990, Paper 3.1.

Eldridge, K.G. 1978: Seed collections in California in 1978. *CSIRO Div. For. Res. (Canberra) Ann. Rep. 1977-1978*, 6-17.

Eldridge, K.G. 1997. Genetic resources of radiata pine in Australia and New Zealand. *In* Burdon and Moore (1997). 26-41.

Eriksson, G., G. Namkoong, and J.H. Roberds 1994. Dynamic gene conservation for uncertain furures. *For. Ecol. and Manag.* 62: 15-37.

Eriksson, G., G . Namkoong, and J.H.Roberds 1995. Dynamic conservation of forest tree gene resources. FAO, Rome. *Forest Genetic Resources* 23: 2-8.

Jayawickrama, K.J.S., M.J. Carson, P.A. Jefferson and A. Firth 1997. Deployment of the New Zealand radiata pine breeding population. *In* Burdon and Moore (1997). 217-225.

Khoshy, M. and G. Namkoong 1996. Futuristic breeding - some plausible options. *In* Dieters, M.J. *et al.* (eds.): Tree Improvement and Sustainable Tropical Forestry, Proc. QFRI-IUFRO Conf., Caloundra, Qld, Australia, 27 Oct. - 1 Nov. 1996. 333-337.

Ledig, T. and D.L. Rogers, (Ed.) 1996: The status of temperate North American forest genetic resources. *University of California Gene Resources Conservation Program*, Davis, CA, USA. Report No. 16.

Ledig, F.T., J.J. Vargas-Hernandez and K.H. Johnson 1998. The conservation of forest: Case histories from Canada, Mexico and the United States. *J. Forestry 96*: 32-41.

Libby, W.J. 1973. Domestication strategies for forest trees. *Canad. J. For. Res. 3*: 265-276.

Libby, W.J. 1995. General synthesis: How knowledge of population structure and mating patterns of forest-tree populations can help improve genetic conservation procedures. *In* Baradat, Ph. *et al.* (eds.): *Population Genetics and Genetic Conservation of Forest Trees.* SPB Academic Publishing, Amsterdam, The Netherlands: 1-2.

Libby, W.J., M.H. Bannister and Y.B. Linhart 1968. The pines of Cedros and Guadalupe Islands. *J. Forestry 66*: 846-853.

Low, C. and T. Smith 1997. Use of the Guadalupe provenance of *Pinus radiata* in New Zealand. *In* Burdon and Moore (1997). 57-61.

Namkoong, G. 1976: A multiple index selection strategy. *Silvae Genet. 25*: 199-201.

Namkoong, G., and M. Khoshy 1995. Maintaining the genetic variance. *In* Lavereau, J. (ed.): Evolution and Tree Breeding, Part 2, Proc. Canad. Tree Impr. Ass./Western For. Genet. Ass. Conf. 28 Aug. - 1 Sept., Victoria, BC, Canada. 17-23.

Shelbourne, C.J.A., M.J. Carson and M.D. Wilcox 1989. New techniques in genetic improvement of radiata pine. *Comm. For. Rev. 68*: 191-201.

Shelbourne, C.J.A., R.D. Burdon, S.D. Carson, A. Firth and T.G. Vincent 1986. Development Plan for Radiata Pine Breeding. *New Zeal. For. Serv., For. Res. Inst.*, Rotorua, New Zealand.

Storer, A.J., T.R. Gordon, D.L. Wood and P. Bonello 1997. Pitch canker disease of pines: Current and future impacts. *J. Forestry 57*: 21-26.

Cooperation with private woodland owners in funding and establishing programs and priorities for tree breeding research

Yves Lesgourgues

Introduction

With 15 million hectares of productive woodland, the French forest is characterised by a wide diversity of species (Table 1). It is composed of 60 % broadleaved species (oak, beech) and 40 % coniferous species (fir, spruce, maritime pine, Douglas fir, Scotch pine).

Table 1. Forests in France

Surface	14 million hectares, 26 % of the country's surface	
Ownership	69 % private (1/3 farmers) 19 % public cooperatives 12 % state	
Species	64 % broadleaved species 40 % *Quercus* spp. 10 % *Fagus sylvatica* 14 % others	36 % conifers 22 % *Pinus* spp. 10 % *Abies* spp. and *Picea* spp. 4 % others

For administrative purposes, France is divided into 22 regions. Among these, Aquitaine is the leading forestry region with 1.7 million hectares of forests, including 1 million hectares on podzolic acid soil set aside solely for pure plantations of Maritime pine (*Pinus pinaster* Ait.). This area was deliberately created in the 19th century from a native species, and is essentially private woodland, belonging to thousands of owners.

Maritime pine is native in both the Atlantic (France, Spain, Portugal) and the Mediterranean coastal regions (Spain, Italy, Morocco, France, including Corsica). The economic impact of the species is considerable along the Atlantic coast, where it occupies approximately 3 million hectares. In France it has long been stated that the Maritime pine is the "leading coniferous tree in France". In Aquitaine it represents 90 % of the timber economy, which explains the substantial amount of research dedicated to this particular species.

C. Mátyás (ed.), Forest Genetics and Sustainability, Volume 63, 247–251.

1. Genetic improvement of Maritime pine in France

Maritime pine silviculture in Aquitaine is a remarkably dynamic activity. Reforestation every year involves some 15,000 hectares planted either by direct sowing or by planting. The genetic improvement programme of this species is now one of the oldest and most advanced in Europe, and probably in the world; the third generation of selection is currently underway.

A public research body (INRA, the National Institute for Agronomic Research) initiated the programme in the 1960s, with the aim of producing seed selected for better wood production in terms of both quality and quantity. In the selection programme, 380 „plus trees" were initially chosen, with a further 255 trees added at a later date, bringing the total number of „plus tree" selections in the tree improvement programme to 635.

These trees were grafted, control-crossbred and progeny tested, and the first family seed orchards were then planted at Cabanac, Sore and Lavercantière on approximately 250 hectares of land. Since 1981, seeds from the first generation orchards have been used for the production of plants in nurseries intended for reforestation. A 24 % increase in volume and a 20 % improvement in straightness have been achieved as a result.

In the last ten years, selection cycles have been carried out both by INRA and AFOCEL (the Cellulose Forest Association, financed by the paper industry). This work has resulted in the planting of a second generation of orchards. These orchards are either designed for polycrosses by free pollination or for controlled crossbreeding, 180 ha and 8 ha in size respectively. The theoretical improvements expected are 50 % in volume and 40 % in straightness.

Finally, a new way of improving the species, namely interracial hybridisation using Landes and Corsican seed sources, needs to be examined. By doing so it is hoped that trees with a straighter trunk and finer branches will be obtained.

2. A cooperative programme

In the 1990s, after having started a joint project involving a "cooperative of data on the growth of the Maritime pine", the main organisations in the public and private forest sectors became concerned about the continuation of two parallel programmes related to the genetic improvement of Maritime pine. There was a risk of insufficient financing for the two research organisations (INRA, AFOCEL) to cover the costs necessary for completion of the projects.

After several consultative meetings, 1994 saw the creation of a GIS structure (Scientific Interest Group) called "The Cooperative Programme for the Maritime Pine of the future", the aims of which are:

- to define together the characteristics of new varieties and to develop a short- and medium-term improvement programme adapted to users' needs;
- to inform partners in the wood sector of the characteristics and performance of the improved Maritime pine varieties.

Public (INRA, ONF1) and private (CPFA2, AFOCEL) organisations are equally represented as the four founding members.

2.1. Operating principles

The structure of GIS allows it to operate in a very flexible way but requires constant unanimosity. The system is based on the pooling of all basic equipment held by INRA and AFOCEL from the start, as well as complete accountability for all operations. Principles on which GIS operates follow.

- Principle 1. Pooling of all basic equipment held by INRA and AFOCEL within GIS.
- Principle 2. New organizations resulting from GIS's activities will be jointly owned by those operating them (INRA, AFOCEL).
- Principle 3. One single member represents GIS in searching for financing.
- Principle 4. Unanimosity must be sought. Contributions must be balanced.
- Principle 5. Every year GIS consults representatives from the entire sector with regard to the direction to be taken and provides progress reports of its work.

2.2. Organisation and expected benefits

The structure of GIS (see Table 2) is very different from the traditional structure of organisations. The advantage of GIS is that it allows representatives from the private forest sector to play a concrete role in improvement programmes. Private forest owners, who represent 90 % of Maritime pine forests, now feel far more closely involved in genetic improvement research. This system also has the advantage of allowing private land to be used for experimentation at reduced cost.

---

[1] The National Office of Forest (ONF) manages the 4.5 million hectares of public forest in France.

[2] The Aquitaine Centre for Forestry Productivity and Action (CPFA) is the organization representing private woodland owners in Aquitaine and is dedicated to forestry extension and applied research.

The expected benefits include:
- accountability;
- close relationship between researchers and users;
- adapting of new varieties to the sector's needs;
- economy of means.

Table 2. Structure of G.I.S.

Executive board	**Strategy board**
INRA	Authorities
AFOCEL	Forest owners
CPFA	Forest cooperatives
ONF	Paper industries
	Sawing industries
	Nurseries
	Silvicultural companies
	etc...
Defines technical goals and action	Examines new varieties
Determines annual program	Defines varieties, „pine of the future"
Seeks financial resources	Examines reports on work and finances
Consults and reports to the Strategy Board	Carries out consulting tasks
Meets twice a year	

2.3. Action currently in progress

The basic outline of the GIS programme for the next ten years will include the following tasks.

Selection of the improvement population.This involves taking an inventory of plant material (families, clones) from two programmes independently led by INRA and AFOCEL from the beginnings. More than 3,000 clones were selected from the Landes and Corsica from 3 successive generations (G_0, G_1, G_2). About 5,000 half- or full-sib families are currently undergoing progeny testing on an area of over 500 hectares in Aquitaine. As a result of the pooling of 1,095 G_1 INRA clones and 1,527 G_1 AFOCEL clones, a database of 2,622 clones was created and 800 G_1 clones were selected to act as an improvement population for the GIS.

Testing of selected progenies. The aim is to select the most successful G_1 progenies in terms of growth and quality through progeny tests. The multi-location progeny tests are set up by each of the organisations on public or private land. In the long term this program will encompass nearly 80 hectares of tests, with three tests set up per year.

Diversification of varieties. This is the central task of GIS's work, and the area that most directly concerns the steering committee. It involves defining and creating new varieties that are adapted to growth conditions in the forest. The decision was initially made to work towards the creation of three new varieties. These are:

1. a well-adapted, straight-growing variety suitable for the reforestation of the most fertile parts of the region (moist, well-drained land), where Maritime pine often has shape defects;
2. a "sturdy" variety adapted to the very dry environment of the dune areas;
3. a "Landes-Corsica" hybrid variety in the region's border areas that is straighter than the local populations.

Financing for all of the above operations comes to approximately 2 million francs a year (US $350,000), to be financed by the partners and by various grants from the State, the region and Europe.

3. Conclusions

GIS was formed out of the desire for cooperation between the partners and bringing together research organisations and representatives from both the private and public forestry sectors. It has the advantage of complete accountability for the improved varieties of Maritime pine that will be used in the future.

After five years of operation, it can be said that the GIS "Pines of the Future" has fully achieved its aim. It is cited as an example throughout southern Europe. Nevertheless, its impact will be judged in the long term, in particular when the time comes to set up contracts for the multiplication and marketing of the newly-created varieties.

Selected literature

Alazard, P. 1993. Les vergers à graine de pin maritime. *Information forêt 1 No. 450:*1-14.

Arbez, M. 1988. Résultats et perspectives de l'amélioration génétique du Pin Maritime. *CR Académie d'Agriculture de France*, 3 pp.

Baradat, P., Durel, C.E., Pastuszka, P. 1992. The polycross seed orchard: an original concept. *Symposium AFOCEL IUFRO Bordeaux* Vol. 2 (Mass production technology for genetically improved fast-growing forest trees), 53-62.

Chaperon, H. 1991. Amélioration génétique et sylviculture. Application au Pin Maritime . *Revue Forestière Française XLIII* (1):17-26.

Durel, C.E. 1994. Amélioration multi-génération : exemple du Pin Maritime. *Forêt Entreprise 96*:24-25.

Durel, C.E. 1992. Gains génétiques attendus après sélection sur index en seconde génération d'amélioration du Pin Maritime. *Revue Forestière Française XLIV* (4):341-355

Research advances in forest tree genetics and breeding in China

Jusheng Hong

Introduction

China is short of forest resources. The area of forest land is 1.32 million sq.km, which is 13.94% of the total land area in China. The total growing stock, including forest land, open forest and solitary trees is 11,558 million cubic meters. Each year the shortage of timber is about 90 million cubic meters. The Chinese Government pays close attention to establish man-made forests, and has decided in 1978 that 20 million hectares of fast-growing and high-yielding plantations should be set up in 30 years. By the end of year 2000, 6.6 million hectares of plantations will be established. It was also decided to establish 10 great forest environment projects. This requires a large quantity of improved planting material. Therefore tree improvement programs became an important field of forestry reconstruction in China, and developed very fast during the past 20 years.

Research programs on tree improvement and tree breeding were first established in the early 1950s. During the period 1950-60 great attention was paid to the investigation and classification of the genetic resources of tree species for wood production and economic purposes, as well as to the introduction of cross breeding of poplar. Research work concerning the selection of plus trees and establishing of seed orchards intensified in the 1970s. In the 1980s, a research program entitled "Breeding and Improvement of Main Species Used in Afforestation in China" was given priority. This program covered provenance trials, selection of outstanding stands, seed crop stimulation techniques, species selection for planting on farmland, seed orchard management of conifers, breeding to improve commercial traits and gene conservation of rare and endangered species. In the early 1990s, during the eighth "Five-Year Plan" period, a national key research program on "Selection and Breeding of Superior Planting Material for Short Rotation Industrial Forest" was conceived and implemented. This program included genetic improvement and tree breeding of 17 main timber species, seven of which were coniferous and ten broadleaved.

C. Mátyás (ed.), Forest Genetics and Sustainability, Volume 63, 253–266.

Generally it can be said that based on the research work carried out over the past 15 years, when national key research programs were conducted during the sixth, seventh and eighth "Five-Year Plan" period, many achievements have been reached in the field of forest genetic improvement and tree breeding, but also in the development of improved propagation material.

Table 1. Provenance trials of main species in China.

Species	Range-wide provenance test				Limited provenance test			
	Exp. series	Year	Number of sites	Number of test provenances	Exp. series	Year	Number of sites	Number of test provenances
Cunninghamia lanceolata	2	1978	25	20	1	1985	37	5-15
		1981	60	207				
Pinus massoniana	2	1981	30	22	1	1986	5	21
		1984	28	35				
Pinus yunnanensis	1	1988	13	64				
Pinus taiwanensis		1982	5	15				
	3	1983	5	16				
		1984	5	23				
Pinus tabulaeformis	2	1979	10	19	1	1989	13	5-10
		1983	19	26	1	1986	4	8
Pinus sylvestris var. mongolica		1980		10				
	3	1981	24	8				
		1983	9					
Pinus armandii	2	1980	9	30				
		1982	3	12				
Pinus koraiensis	2	1983	11	26				
		1988	22	17				
Pinus taeda	2	1981	7	12				
		1983	18	43				
Pinus elliottii	2	1981	7	11				
		1983	20	25				
Pinus caribaea	2	1973	2	8				
		1983	18	18				
Larix olgensis		1980	13	10				
	3	1986	6	13				
		1988	19	9				
Larix gmelinii	1	1983	13	16				
Platycladus orientalis	2	1983	24	20-30	1	1987	6	24-28
		1988	20	51				
Ulmus pumila	2	1979	5	15	2	1980	3	5
		1980	41			1984	1	5
Eucalyptus spp.	2	1986	16	258				
		1983	7	20				
Tectona grandis	1	1976	7	72				

1. Provenance trials

Provenance trials in China started in the mid 1950s with a few species. A small number of provenances was included for each species but the trials were probably not very well designed. Until the mid-1970s provenance selection was considered to be the basic element for tree improvement, and therefore provenance tests were systematically run with some important tree species. Provenance tests have so far been established with 41 species: 22 of them softwoods, 18 hardwoods and shrubs. The 17 species of national importance for afforestation projects are listed in Table 1. The total area of tests amounts to 1761 ha.

Most species, for instance *Cunninghamia lanceolata, Pinus massoniana, P. yunnanensis, P. sylvestris var. mongolica, P. koraiensis, P. taiwanensis, Larix gmelinii, L. olgensis, Platycladus orientalis, Ulmus pumila, Melia azaderach* and *Tectona chinensis* are by nature continuously distributed and show clinal variation by latitude, some also by longitude and latitude. Temperature and rainfall are affecting the variation.

For some species, for example, *C. lanceolata, P. massoniana, U. pumila, L. gmelini, P. elliottii, P. taeda, Eucalyptus spp.* and *Tectona grandis*, provenance trials were established early. Some outstanding provenances with a genetic gain of more than 15% in volume growth were screened out and are used in operational forestry in different ecological zones (Table 2).

Table 2. Superior provenance regions of main species used in afforestations.

Species	Superior provenance regions	Mean volume gain (%)
Cunninghamia lanceolata	Southern part of Nanling Mountains. Reaches of Qingyi River in west Sichuan, mountainous areas in South Guangxi and Yunnan	30.5
Pinus massoniana	Areas around Yunkaida Mountain. Dalou Mountain and Wuyi Mountain	50.1
Larix olgensis	Areas around Xiaobeihu Lake in Heilongjiang	79.2
Larix gmelini	South part of Xiaoxinganling Mountain	15-25
Ulmus pumila	Reaches of Yellow and Huaihe River	15-20
Pinus elliottii	North part of Florida state and south part of Georgia in USA	15-20
Pinus taeda	Gulf of Mexico coast in the USA South Carolina and Florida states	20-25
Pinus caribaea	Mountain Pine Ridge of Belize, Alamicamba, Poptun in Cuba	15-20
Acacia mangium	Papua New Guinea, Queensland in Australia.	>20%

Following studies of the geographical variation pattern, provenance regions have been established. The "National Standards of Tree Seed Zones" were set up for 13 important timber species (Table 3).

2. Seed orchards

Since the 1960s and in order to establish seed orchards, much attention has been paid to the survey, collection, testing and utilization of plus tree resources in China. During the 1970s, a large scale plus tree selection was carried out for the main tree species in all provinces, autonomous regions and municipalities except for Xizang. Qinghai and Taiwan. In the last twenty years a total of about 50,000 plus trees of 36 tree species had been selected and plus tree collections and archives of 614 hectares had been set up.

Table 3. Division of provenance and seed zones.

Species	Provenance zone		Seed zone	
	Prov. zone	Sub-zone	Seed Zone	Sub-zone
Cunninghamia lanceolata	9		10	8
Pinus yunnanensis	4	4	6	6
P. massoniana	6		9	17
P. tabulaeformis	7		9	22
P. sylvestris var. mongolica	3		4	6
P. armandii	5		3	
P. koraiensis	3		2	
Larix rupprechtii			3	
L. olgensis	4		2	
L. gmelinii	4		3	
Platycladus orientalis	7		4	7
Picea asperata			3	4
Ulmus pumila	4	3	3	

Clonal orchards were first established in China in 1964. First generation seed orchards were established for 45 species, with 30 coniferous, 12 hardwood and 3 fruit species, comprising a total area of 6,710 hectares, financed by the Ministry of Forestry, with an annual seed yield of 400,000 kg (Table 4). This amounts to about 10% of the seed used for national afforestation, in case of *C. lanceolata* and *P. elliottii* 20-25% and 35-40%, respectively.

Concerning seed orchard technology, extensive research was conducted on the establishment and management techniques of orchards in order to increase seed yield.

Table 4. Improved propagation material producing units in the forestry administration (Unit: mu, 15 mu=1ha)

Species	Seed orchard	Seed stand	Scion garden stool bed	Propagation area
Aleurites fordii	1000		30	2
Camellia oleifera	250	800	100	
Castanea mollissima			166	57
Catalpa bungei	200		380	37
Catalpa fargesii	1050	200		
Cunninghamia lanceolata	27257	3856	770	112
Cupressus funebris	600			
Eucalyptus globulus	150			
Eucalyptus sp.	200		400	
Fraxinus mandshurica		700		
Juglans regia	1400	520	400	220
Juniperus virginiana		1200		
Larix gmelinii	1325	9906		
Larix leptolepis	5467	3328		
Larix olgensis	4033	6808		
Larix principis rupprechtii	5648	13290	6	
Liriodendron chinensis	300	200		
Metasequoia glyptostroboides	700	5426		
Paulownia sp.	250	500		
Pimus sylvestris var. mongolica	5840	12566		
Pinus armandii	773			
Pinus caribaea	600			50
Pinus elliottii	6784	3515		
Pinus khasya	350			
Pinus koraiensis	1000	9472		
Pinus massoniana	14600	6216	180	50
Pinus rigida	100	400		
Pinus tabulaeformis	8288	11882		
Pinus taeda	2484	2215		
Pinus taiwanensis	350	2000		
Pinus yunnanensis		150		
Platycladus orientalis	330		30	20
Populus sp.	545	820	210	480
Populus tomentosa			1150	220
Pseudolarix amabilis	850	750		
Robinia pseudoacacia	4260	674	440	215
Salix sp.			2	2
Taxodium ascendens	1400		52	
Ulmus pumila	1079		150	100
Ulmus sp.	545	220	180	300
Ziziphus jujuba			230	90
Other broadleaved species	88	770	486	612
Total	100197	98384	5362	2567

In order to increase genetic gains and to establish advanced seed orchards, progeny testing has been conducted with 22 species since the late 1970s. 4510 ha of progeny trials have been established so far in the country, in most cases using half-sib family seeds. Progeny trials of full-sib families have only been established on an area of some 150 ha with *C. lanceolata, P. tabulaeformis, L. olgensis, L. gmelinii, P. massoniana* and *P. armandii*.

Based on studies carried out over the past 10 years, 2500 outstanding families have been identified with an average volume gain of 10-15%. For *C. lanceolata* (Chinese fir), 40 second-generation parents have been selected with an average volume gain of 32-85% in comparison to the first generation parents. Since the late 1980s tandem selection techniques, considering growth per plant, growth per unit area, timber production rate from volume, fiber characters, wood density, adaptability and stress resistance etc., have been applied *for P. massoniana*, larch, *C. lanceolata* and eucalypts.

Early prediction methods resulted also from progeny tests. Early selection can be done at the age of 8 years, final screening at 12 years for *L. leptolepis* while *P. massoniana* final selection can be set at 9 years. As far as *C. lanceolata* is concerned, the first selection is usually made at 3 years after planting, the second at 6 or 7 years, with a precision of about 80 %.

3. Crossbreeding

Since the 1950s, crossbreeding has been carried out in China with poplars, pines, eucalypts, larches,. Chinese fir, *Cryptomeria* sp., *Camellia* sp., *Carya* sp., *Castanea mollissima, Paulownia* sp., *Ulmus* and bamboos etc. A number of artificial hybrids have been applied in afforestation and have brought great economic gains. For example, *P. cathayana* cv. Beijingensis, *P. simonii* cv. *Populus 35-44, P. pyramidalis* cv. opera 8277, *Populus pyramidalis* and *P. deltoides* crosses show great hybrid vigor with a 5-23% increase in volume growth, better rooting ability and disease resistance. The cultivars 'I-214' and 'I-72' are used as standards, *Larix leptolepis* was crossed with *L. gmelini, L. olgensis, L sibirica*, respectively. The hybrids performed significantly better than their parents. Further research was conducted on the taxonomy, geography, ecology and flowering biology of the above mentioned species. Cross affinity and compatibility have been established for intersectional, inter-species or intra-species crosses of poplar, larch, arboreal and shrub species of willow, pines, *Camellia*, and walnut, as well as hybrids of *Cunninghamia lanceolata* and *Crytomeria fortunei, Taxodium ascendens, T. distichum* and *Juniperus* sp.

In recent years, scientists of the Forestry Research Institute of the Chinese Academy of Forestry, have carried out studies on reproductive biology of crossbreeding. The cause of crossing incompatibility between Section Turanga (female) and Section Leuce (male) poplars has been clarified through this research program and immature hybrid embryos were cultured in vitro to obtain plants that were finally transplanted into nursery. Similar research was also carried out with *Salix matsudana*.

In terms of new breeding techniques, the raising of haploid plants from pollen grain has been applied on more than 20 species and hybrids of poplars, on rubber tree and *Lycium chinensis*. The ovule was successfully cultured in vitro in order to achieve haploid plants of *L. chinensis* and *Actinidia chinesis*. In the Poplar Research Institute, in Liaoning Province, a salt resistant cultivar, 'Liao River No.9', was obtained by selecting salt resistant cells. Some progress was also made in DNA transfer in the cytological genetic research with a disease resistant hybrid of *Paulownia* sp. At Chinese Academy of Forestry. The scientists from the Beijing Forestry University have developed trisomic poplar hybrids. Nanjing Forestry University researchers have created new poplar plants by cell fusion. The scientists from the Chinese Academy of Forestry have been successful in isolating the living protoplasm from the callus produced by the cells of *Populus x Xiaozhuannica, P. x xiaohei* and *P. x euromericana* cv. Robusta or the cambium cells of *Populus x euramericana* cv. I-214. The Research Institute of Subtropical Forestry, Chinese Academy of Forestry, has developed early crossing techniques for *Carya* sp.; already the fourth generation orchard has been set up and the fifth generation orchard is under establishment.

4. Clonal breeding and vegetative propagation

Besides poplar, willow, chestnut, walnut and other species, clonal breeding has also been employed in recent years with *Metasequoia glyptostroboides, Taxodium ascendens, Cunninghamia lanceolata, Casuarina* spp., *Eucalyptus* spp., *Tectona grandis, Robinia pseudoacacia, Ulmus* spp., *Larix* spp., *Pinus massoniana, P. elliottii, P. taeda, Acacia* spp., *Betula* spp. and other timber species, as well as with *Camellia oleifera, Aleurites fordii, Sapium sebiferum* and other economic species. The strategy of clonal breeding in China is to focus clonal selection and multiplication on tree species with an easy vegetative propagation combined with a longer sexual breeding cycle and low seed production, so that the principle of „sexual breeding and application of vegetative propagation" can be followed.

In recent years clonal breeding of Chinese fir (*Cunninghamia lanceolata*) and eucalypts has gained impetus. The clonal breeding material of Chinese fir derives from the outstanding individuals from superior families in progeny test stands (full-sib or half-sib) chosen by means of "8-tree selection". Plus trees from the superior provenance stands are also used as clonal breeding material. Up to now, a total of 20,379 outstanding individuals or plus trees have been selected and 135 hectares of clonal test stands have been set up for testing 4,300 clones. 382 superior clones have been preliminarily screened out and about 120 superior clones have been used for establishing production hedging orchards and are being applied in plantations. In addition about 5,000 outstanding individuals of *E. urophylla, E. grandis x urophylla, E. urophylla x tereticornis* and *E. leizhou* have been selected and used for clonal breeding, and 336 superior clones have been preliminarily screened out: about 63 superior clones have been applied in large-scale eucalypt plantations.

The vegetative propagation techniques have been greatly improved for many species. The sprouts of Chinese fir, for instance, can be used as cuttings for root cutting-propagation, one plant being able to produce up to 10,000 cutting-propagated plants in the course of three years. This means that one hectare of Chinese fir hedging orchard can, in three years, produce cuttings for establishing 500 hectares of plantations. For the eucalypts, the main species successfully propagated vegetatively are: *E. urophylla, E. grandis x urophylla, E. tereticornis,* E.12ABL and *E. globulus*. The eucalypt ortets are planted in pots or, sometimes, in the field into a hedging orchard. Some eucalypts, for instance *E. urophylla, E. grandis x urophylla and E. Maidenii* can get rooted to 40-80%.

Cuttings taken from one to three years old seedlings of *Larix leptolepis and L. rupprechtii* could be rooted to more than 90%. *P. massoniana* can reach a rooting percentage of 80-90% and *P. eliottii, P. taeda* one of 80-94%. Species of the white poplar section for instance, *Populus tomentosa, P. hopeiensis* are often regarded as difficult to get rooted but now the rooting percentage can reach 95-100%. For aspen, a rooting percentage of more than 80% is reached. Survival of *Tectona grandis* can reach 95-100% using short cuttings. Short cuttings were taken from fine roots of *Robinia psedoacacia*. Up to 80 cuttings can be obtained from one tree in the first year, 6,000 in the second year and half a million in the third year. Improved methods of grafting have been developed for *Ulmus pumila, Camellia* sp., *Aleurites* sp. and *Sapinum chinensis.*

Since the 1970s, the tissue culture of woody plants has developed very quickly. Tissue culture has so far been used in the propagation of more than 70 species. Poplars, *Eucalyptus*, Chinese fir, teak, *Paulownia, Taiwania, Magnolia, Sassafras tsumu, Davidia involucrata, Phellodendron amurense, Canarium pimela, Cathaya argyronhylla* and other species have been mass propagated for operational forestry.

5. Tree introduction and domestication

China is a country with a history of tree introduction of almost 2000 years, but systematic and planned tree introduction was not initiated until the 1950s. During the 1950s the Chinese Academy of Forestry made detailed investigations on exotic species and created lists of species to be introduced into China based on ecogeographic conditions. Some species of *Populus, Eucalyptus, Casuarina* and also rubber tree were introduced. American southern pines, olive, black wattle, cypresses and some valuable tropical species have been grown since the 1960s. The seventies saw many exotic species widely extended into plantations. Research projects on growing exotic trees were given priority since 1983, which accelerated research on exotic forestry in China.

More than 1,000 exotic species have been introduced into China so far, about 100 species of which are worth of being extended into forestry practice. More than 30 species have become very important in afforestations. Hybrid poplar clones 'I-214' and 'Sacrau 79', for example, are more suited for the plains of central China; poplar clones I-69' and I-72' are mainly planted south of latitude 37° N and in the Changjiang River valley with a total area of about 1.93 million hectare. Eucalypts have become very important in plantation forestry in tropical and subtropical zones of China with a total area of more than half a million ha. *E. exserta* and *E. citriodara* were widely used before the 1980s; however, *E. urophylla, E. tereticornis, E. grandis, E. camaldulensis (N. Queensland)*, *E.* ABL 12, *E. Maidenii* and *E. nitens* tend to be very common nowadays.

It has been demonstrated that *Pinus elliottii* is a species suited for altitude below 300 m in the south and central subtropical regions, while *P. taeda* can be grown further north until the Changjiang River valley in the northern subtropical area. It is estimated that the total area of plantations with these two pines reaches 2.55 million ha, 25% of which consisting of *P. taeda*. Other species, for example *Larix leptolepis, Robinia pseudoacacia, Casuarina equisetifolia, Pinus nigra* and *P. densiflora* have also become important for afforestation. In addition, the planting of Pinus caribaea, Acacia mangium and *Acacia auriculiformis* has recently been developed rapidly in tropical and subtropical regions, where they have become the main afforestation species.

In the course of more than 40 years of introduction practices, great experience has been gathered. A monograph entitled "Introduction to Exotic Tree Species" was published. Its editor is Prof. Wu Zhonglun, who died in 1995. This monograph gives a summary of the introduction theories and practices for exotic tree species in China, as well as of exotic tree species that were successfully planted in China. Based on natural geographic conditions, 10 tree introduction regions for the whole country are mentioned with descriptions of their ranges, introduction conditions, introduction purpose, suitable introduction species and their sites as well as the related technical details, including the experience accumulated in the course of introduction of 570 tree species belonging to 85 families.

Besides the introduction of exotic trees, the expansion of the cultivation of some native tree species was also emphasized. The main native species planted extensively since the 1970s are: *Cunninghamia lanceolata, Metasequoia glyptostroboides, Pinus sylvestris* var. *mongolica, Larix principis rupprechtii, L. olgensis* and *Phyllostachis bambusa* etc. There are some other species that have been successfully domesticated beyond their natural distribution areas, such as *Taiwania flousiana, Michelia macdunei* var. *sublanea, Phellodendron amurense, Cinnamomum camphora, Sassafras tzumu, Phorebe manmu, Cupressus forkenia, Castanopsis kawakamii, C. hystrix, Schema supenta, S. wallichi, Liriodendron chinensis, Erythophloeum fordii, Mytfilaria laosensis, Alnus cremastogyne, Gmelina chinensis, Keteleeria fortunei* etc.To enhance exotic forestry, the Chinese Academy of Forestry has established a research network over the whole nation, and more than 40 institutions are carrying out trial work with exotic species. 400 ha of various experimental plantings have been established so far.

6. Forest genetic resources conservation

Systematic collection, conservation, investigation and utilization of forest genetic resources was started in the early 1980s and considered an important component of the forest genetic improvement programs.

Aiming at the in situ conservation of vegetation types or rare plant and animal species, China began to establish national reserves in 1956. Up to now, a total of 926 natural reserves has been established covering an area of 76.98 million hectares, which constitute 8% of the national land area. Among them, 640 forest reserves have been created, out of which 14 have been accepted by the "Man and the Biosphere" program.

A comprehensive resource inventory on the natural reserves was set up and management regulations were established for the natural reserves. Based on the studies and investigation carried out in the National Key Research Programs, a total of 3900 hectares of superior autochthonous stands of *Picea* sp., *Pinus yunanensis, Pinus massoniana, Pinus tabulaeformis, Pinus sylvestris* var. *mongolica, Pinus koraiensis and Larix gmelini* have been selected.

In China the conservation of rare and threatened tree species was particularly emphasized. A red book entitled "Rare and Threatened Tree Species in China", edited by the Institute of Botany, Chinese Academy of Science, has been issued, which lists 300 protected tree species. A series of research programs were conducted on the *in situ* conservation and artificial propagation strategies and methods for some very important rare tree species, such as *Taiwania flousiana, Davidia involucrata, Parashorea chinensis, Ostrya rehderiana, Carpinus putopensis, Alsophila spinulosa*.

Since 1990, a specific research program on the "Technologies for Conservation of Germ Plasm Resources of Main Coniferous and Broadleaved Species in China" financed as one of the National Key Research Programs, was conducted over a period of 8 years, which resulted in the establishment of 8 germplasm conservation banks, where 10,906 entries, populations, provenances (or stands), families, plus trees or clones of 41 industrial timber species are conserved. In addition, 170 hectares of ex situ conservation forests and 3.6 hectares of conservation gardens were set up. A database system was established for processing information on the country's forest genetic resources.

7. The application of biotechnology in tree breeding

Though biotechnology was applied late and developed slowly in forest tree breeding in China, since 1990s it develops fast. Some examples of research in this field are briefly described.

Transfer of *Agrobacterium tumefaciens* strain C58 into *Populus simonii* and *P. euramericana* have been done in 1990 (Wang Y. *et al.* 1993). Root hair induced by *A. rhizogenes* 9402 was obtained in transgenic poplar, which can also be regenerated from root hair. Mannopine and agropine were detected by paper electrophoresis in the transgenic poplar .

Populus nigra, P. euramericana, P. deltoides were transformed with *Bacillus thuringiensis* toxin gene and the plants showed high insect resistance. The PCR analysis and western blot technique demonstrated that the BT toxin gene was not only inserted into the chromosome of poplar, but also expressed into protein (Cheng Y., *et al.*. 1996).

Protoplasts have been isolated from many trees such as *Populus tomentosa, P. simonii, Actinidia arguta, Paulownia* etc. since 1990 (Wang Y. *et al.* 1993). Many species are regenerated from protoplast culture in vitro, providing a solid foundation for foreign DNA transformation and somatic hybridization. The isolation of poplar protoplast and regeneration from protoplast opened a new way of cell engineering for poplars. The cell suspension cultures of poplar have been established in the Chinese Academy of Forestry. Somatic mutants resistant to salt were selected by adding NaCl into the culture medium. After four generations of subculture in meda containing NaCl, twelve cell clones were got and eight hundred plantlets were transplanted into the field. The plantlets are being tested by DNA markers.

The research of embryogenesis and artificial seed was started in 1988. *Abies chinensis* is the first tree which regenerated through embryogenesis in China (Shu and Zhang. 1998). Embryoids in embryogenic callus from juvenile embryo in 1993. Embryoids of *Eucalyptus urophylla* were induced from hypocotyl in 1996 (Wang 1996), the whole plantlets with root and shoot have been regenerated through somatic embryoid germination.

The research of secondary metabolic compounds has also made great progress. The cell culture of *Taxus chinensis* and the metabolism regulation of taxol have been researched since 1993 (Gang 1996). In 1997, a research group of the Chinese Academy of Forestry studied the callus induction from leaves of *Taxus chinensis* and got many high yielding cell clones. Through suspension culture, a cell clone with biomass of 0.49 Gfw/Gw.d was obtained. The taxol yield is 0.25 mg per liter. This research offers a great opportunity for the application and exploitation of taxol.

8. Molecular genetic markers

The linkage map of Masson pine was constructed using RAPD markers (Wei 1996). In order to screen the primers for amplification, eighty primers were used in RAPD reaction, sixteen tested primers yielded polymorphic PCR products. Through analysis of forty-nine linkage markers, twenty-nine marker were found distributed over twelve linkage groups.

The genetic variation of numerous poplar species, among others *Populus ussuriensis, P. suaveolens, P. maximowiczii* was studied by RAPD markers. All primers could generate polymorphic PCR products. High genetic diversity was observed within species. The molecular taxonomy confirmed the conclusions of traditional taxonomy (Li 1997).

The genetic diversity of clones of P. deltoides was studied using RAPD markers (Wang 1995). DNA fingerprinting was applied to indentify clones for the first time.

The genetic diversity of seven Chinese fir provenances was studied by Professor Ju Sheng Hong's research group. For the first time RAPD and RFLP markers were used. High genetic diversity was found within all provenances.

9. Outlook

Although great progress and achievements have been made in forest genetics and tree breeding over the past 40 years, which has contributed much to the development of forest seed management and afforestation in China, some shortcomings remain in this field: such as a lack of breeding strategy or breeding plan for most tree species. Wood quality or stress resistance characters have not been considered among breeding objectives; little emphasis has been given to basic research and bio-tech applications. The following aspects should, therefore, be emphasized in the future research on forest genetics and tree breeding in China:

1. Primary aim of Chinese forest tree improvement is to supply superior planting material for timber plantations and environmental protection.
2. In forest tree improvement practice traditional tree breeding should be the backbone, while molecular genetics and of biotechnology has to be promoted.
3. Breeding strategies and breeding plans ought to be elaborated for the main tree species. More emphasis should be placed on stress resistance breeding, selection and breeding of insect resistant planting material, disease and drought resistance or salt tolerance of some tree species. The same emphasis ought to be placed on hybridization and the vegetative propagation of results of multiple generation breeding. The progress of clonal breeding ought to be accelerated by specific propagation techniques. The application of biotechnological methods is to be encouraged for wood quality and stress resistance breeding and conservation of forest genetic resources..
4. Basic research in forest genetics and tree breeding ought to be strengthened, such as in reproduction, natural regeneration, genetic structure and variation at different levels, hereditary parameters of main characters, compatibility, cyclo- and topophysis effects in vegetative propagation of different tree species, maintenance of the juvenility of ortets and mechanisms of vegetative propagation etc., as well as research in molecular genetics, gene mapping, marker-assisted selection and genetic engineering.
5. The collection, conservation and utilization of forest genetic resources ought to be strengthened and a uniform management system ought to be established for genetic resource conservation. The natural reserves, forest reserves, seed stands and clone gardens should be utilized to realize a reasonable *in situ* and *ex situ* conservation system of forest genetic resources. A national information network on forest genetic resources has to be established.

Selection of pertinent literature from China

Cheng Y. *et al.* 1996. Research on a transgenic poplar with LCE gene. *Forestry Research*, 9(6):646-649

Chinese Academy of Forestry, 1991. *Genetic improvement of poplars*. Tree Breeding Division of the Research. Institute of Forestry. Beijing Agricultural University Press

Gang F .Y, 1996. The metabolic regulation of taxol in cell culture of *Taxus yunnanensis* . *Acta Botanica Yunanica*. 18(4):451-453

Hong Jusheng 1987. *Research on tree genetics and breeding*. China Forestry Yearbook, China Forestry Press

Hong Jusheng *et al.* 1993. Research on broadleaved species in Australia. China Forestry Press

Hong Jusheng *et al.* 1994. Research on variation of *Cunninghamia lanceolata* provenances. *Forestry Research* (special edition)

Hong Jusheng *et al.* 1994. Selection of superior provenances of *Cunninghamia lanceolata* for afforestation. *Forestry Research* (special edition)

Li K. L., 1997. Study on origin of *Populus tomemtosa Carr. Acta Phytotaxanomica Sinica*. 35(1):24-31

Ma Changgeng *et al.* 1992. Research on selection of *Pinus armandii* provenances. Beijing Agriculture University Press

Pan Zhigang, 1991. *Research on introduction of exotic trees*. China Forestry Press

Shen Xihuan 1997. Facing *21* century's forest breeding tasks in China. Procedings of the fourth symposium on forest tree breeding. *China Forestry Society Publ*. No.11

Shen Xihuan *et al.* 1992. Techniques for seed orchard management. Beijing Science and Technology Press.

Shu X. H. and Zhang. Q. W. 1998. The application and future of biotechnology in forest tree breeding in China. Forestry Science and Technology

Tu Zhongyu *et. al.* 1991. *Genetic improvement on broadleaved species*. Scientific and Technological Document Press

Wang M. L. 1996. The embroygenesis and amplification of *Eucalyptus urophylla. Shi Cuan Forestry Science*. 17(2):9-13

Wang S. J. 1995. Advances in poplar research. China Forestry Press.

Wang Y. *et al.* 1993. The research progress of forestry protoplast culture. *Journal of Nanjing Forestry University*, 17(2):91-96

Wang Zhangrong *et al.* 1990. Selected papers on establishment techniques of *Pinus massoniana* seed orchards. Science Press

Wei L. B., 1996. Segregation analysis of random amplified polymorphic DNA marker in *Pinus massoniana. Scientia Silvae Sinicae*. 32(5);476-450

Weng S. Q. 1993. The genetic transformation system of poplar induced by *A. tumefaciens* with hormone synthesis gene. *Biotechnology*, 7(2):11-14

Wu Zhonglun 1983. *Introduction of exotic tree species*. Science Press

Xu Huacheng *et al.* 1984. A preliminary study on geographical variation of Pinus tabulaeformis. *Bulletin of Beijing Forestry College*, 84(2):57-72.

Xu Huacheng *et al.* 1990, *Forest seed regionalization*. China Forestry Press

Ying Y., 1993. The embrogenesis and plant regeneration of unfertilized ovule. *Plant research*. 13(3):262-267

You. Y. and Hong. J. S. 1998. Study of genetic variation between the provenances of Chinese fir using RAPD. *Scientia Silvae Sinicae* (in press)

Zhang Songyun 1990. Selected papers on applied genetics and breeding of main coniferous species. China Forestry Press

International cooperation in forest genetics research in Southeast Asia: is it working and how can it be improved ?

Boonchoob Boontawee

Introduction

Forestry research in Southeast Asia started early in 1920 when Malaysia created the Forest Research Institute at Kepong (Razak *et al.* 1998), followed by the creation of the Silvicultural Research Section under the Silviculture Division of the Royal Forest Department of Thailand in 1942 (Kaosa-ard 1998). Similarly, the Philippines created the Forestry Research Institute (FRI) within the campus of the University of the Philippines at Los Baños to conduct research and development in forestry. In Indonesia, the Forestry Research Agency was established under the Ministry of Forestry with responsibility for forestry research and development to support the private and public sectors. Brunei Darussalam started forestry research in the 1970s with the creation of the Sungai Liang Forest Research Centre. Vietnam also became very active in forestry research after the internal war in the late 1960s, while other Southeast Asian countries such as Laos, Cambodia, and Myanmar are in an early stage of forestry research and development. Myanmar began active forestry research in the late 1970s after the development of the Forestry Research Institute in Yezin and received technical and financial support from overseas, from FAO/UNDP, JICA, and other organizations.

Currently there are more than 100 institutions in 20 countries in Asia conducting research and development in forestry (Razak *et al.* 1998), but the range of their capabilities is quite diverse. Major research problems are carried on by a handful of qualified researchers, often incorporated into government forestry departments, and typically research staff is managed in a bureaucratic manner. Apart from this, financial support, equipment, and technical assistance are still very limited in many countries. Collaborative research is thus an alternative in supporting national forestry research in some countries. Forest genetics research in Southeast Asia typically consists of collaborative programmes that have faced many problems; however, efforts have been made to solve these problems and to improve international cooperation and collaboration in this subject area.

C. Mátyás (ed.), Forest Genetics and Sustainability, Volume 63, 267–274.

1. Forest genetics research in Southeast Asia

Forest genetics research in Thailand began in 1906 when teak plantations were established in the north of the Kingdom. Teak seeds from various sources were used to study differences in growth habits and yield variation. Subsequently, a number of species trials and pilot plantations were established in the following decades. To support the teak planting programme, the Royal Forest Department (RFD) created the Silvicultural Research Section in 1942, which gradually developed into the Silvicultural Research Division with more than 50 research stations under its supervision by the 1990s. Thai-Danish cooperation in forestry began in the late 1960s; it has strengthened activities in forestry research in Thailand, especially in tree improvement and breeding. In 1965, the Teak Improvement Centre (TIC) was created under this collaborative programme, followed by the creation of the Pine Improvement Centre (PIC) in 1969. Both tree improvement centres are still active; their main activities include:

- seed procurement and handling;
- plus tree selection and propagation;
- nursery technology development;
- plantation establishment and management;
- breeding programmes;
- seed orchard and seed production area establishment and management.

International collaboration in forest genetics research in Thailand was initiated with (1) the establishment of the ASEAN Forest Tree Seed Centre in early 1981 and (2) the collaborative project of the Japan International Cooperation Agency (JICA) and the RFD Reforestation and Training Centre soon thereafter. These two projects became important cooperative programs in forest genetics research with the aid of overseas consultants, who also developed cooperative relationships with other international and national forestry research institutes, including technical expert exchange. The ASEAN Forest Tree Seed Centre Project, for example, has worked closely with other ASEAN member countries, including Brunei Darussalam, Indonesia, Malaysia, Philippines, Singapore, Thailand and more recent members Vietnam, Laos and Myanmar in a number of fields of forest genetics. These fields include seed procurement and handling, seed production and seed health, forest genetic resources, seed origin, nursery technology and planting stock production.

A strong research relationship has developed among ASEAN member countries. This has been extended to include cooperation with other agencies such as FAO, the International Plant Genetic Resources Institute (IPGRI), the Centre for International Forestry Research (CIFOR), the International Network on Bamboo and Rattan (INBAR), and the Australian Centre for International Agricultural Research (ACIAR). Many collaborative forest genetics research projects have been drawn up and carried out with the cooperation of these institutions. In 1991 FAO, with the cooperation of the United Nations Development Program (UNDP), established a regional project on "Improved Productivity of Man-made Forests Through Application of Technological Advances in Tree Breeding and Propagation", called "FORTIP". Its headquarters office at Los Baños in the Philippines provides coordination and cooperation among many Asian-Pacific countries by promoting the establishment of networks in each country.

Forest tree improvement and breeding have become increasingly active aspects of forest research in Southeast Asia. Research results from various countries have been published and distributed to other member countries, and many trial plots created earlier have been used as classrooms for other members. Thailand, for example, played a major role in improving teak (*Tectona grandis*) in the FORTIP species networks. Experience from the early Thai-Danish Teak Improvement Centre has been distributed world-wide through this network, and international cooperation has supplied interested members with both technical information and planting materials. Experience gained from the ASEAN Forest Tree Seed Centre Project led to the establishment of the Forest Genetic Research Section under the Silvicultural Research Division of the Royal Forest Department.

In Malaysia, forestry research started initially with the establishment of the Forest Research Institute (FRI) in 1929 (Appanah and Weinland 1993) and many experimental plantations were established in the succeeding years. After Japanese occupation during World War II, Malaysia re-established its FRI. In 1985, FRI was renamed Forest Research Institute of Malaysia (FRIM) and become autonomous with an area of 1,528 ha located at Kepong, 16 km northwest of Kuala Lumpur. Currently, FRIM is the most active forestry research agency in Malaysia and has a major role in international cooperation and collaboration in forest genetics research. The establishment of its tree breeding and tree improvement programme is making FRIM very active in forest genetic studies and research with financial support from Malaysia as well as from overseas organizations including ODA, OFI, GTZ, JICA and ITTO. Other institutes also work in close cooperation with FRIM, including the National Institute of Environmental Studies (NIES) of Japan, Wageningen University of the Netherlands, and the Ministry of Forestry of Indonesia.

During the last 3 decades, Malaysia has been able to financially assist its researchers in obtaining higher degrees from the U.K, the USA, and Australia to support its forestry research programmes. In addition, the very extensive forestry library at FRIM, as well as that at the University Putra Malaysia (UPM), helped to build a more active forestry research program than that of neighbouring countries. The private sector in Malaysia was recently encouraged to invest more in forestry projects, especially in industrial forestation. This private support has helped to make FRIM and UPM more actively involved in forestry research, particularly in forest genetics.

Although Indonesia has a Forestry Research Agency under the Ministry of Forestry, it has played a less active role in international forest genetics research than do Thailand and Malaysia. After the establishment of the Centre for International Forestry Research (CIFOR), Indonesia changed its role from national supporter of forestry research to an international contributor. In addition to aid from CIFOR, Indonesia received assistance from other developed countries in the development of forest genetics research. With support from FINNDA, a large-scale nursery project was launched in Kalimantan, Java, and Sumatra. Financial and technical support from JICA made possible the recently-established Forest Tree Improvement and Development Institute in Jogjakarta. This institute is also acting as a country coordinating agency for the UNDP/FAO regional project (FORTIP), and is involved directly with some species improvement programmes, e.g. those with *Paraserianthes falcataria* and dipterocarps.

In the Philippines, the Forest Research Institute (FRI) was once very active in forestry research, and was the only agency dealing with forest genetic studies in the 1960-1980s. With the reorganization of official agencies in the Philippines, the FRI became the Ecosystems Research and Development Bureau (ERDB) directly under the Department of Environment and Natural Resources (DENR). Subsequently, ERDB played a lesser role in forestry research. Because of the economic crisis in the 1980s and 1990s, ERDB has terminated some of its activities, especially in forest genetics research. Through the ASEAN Forest Tree Seed Centre Project, ERDB and the University of the Philippines at Los Baños (UPLB) also contributed their experience internationally in terms of forest tree seed technology and forest genetic resource management. After the establishment of the UNDP/FAO Regional Project on FORTIP, ERDB and UPLB actively resumed their roles is forest genetics research. Some private companies have also contributed greatly to forest genetic studies. The Manila Seedling Bank Foundation focused its activities on nursery technique development, planting stock production, and plantation establishment and management. The Pulp and Paper Industry Corporation of the Philippines (PICOP) also contributes greatly in the fields of forest tree breeding, improvement and propagation. The main species they are working with are *Paraserianthes falcataria, Eucalyptus deglupta, E.pellita, Gmelina arborea* and many dipterocarp species.

Vietnam has been an ASEAN member country since 1995. Vietnam lost time and effort during the internal war, which hindered advances in forestry research. However, after the war, Vietnam became actively involved. The establishment of the Research Centre for Forest Tree Improvement at Chem-

Tu Liem, Hanoi under the Ministry of made Vietnam known for its forest genetics research. Vietnam received assistance from various overseas countries, including Australia (ACIAR, CSIRO), Japan, Sweden, Germany and Yugoslavia to promote forestry activities of the country. Many trials for species screening and provenance selection of promising species such as *Eucalyptus camaldulensis, Acacia mangium, an Acacia* hybrid *(A. mangium* x *A. auriculiformis*) and other local species are under way (Kha 1996,1997).

Forest genetics research in Myanmar is rather limited due to the shortage of well-trained manpower and research equipment. The principal interest is in economic timber species like teak (*Tectona grandis*), padouk (*Pterocarpus macrocarpus*) and *Xylia xylocarpa*. There are trial plots of these species in the vicinity of the Forest Research Institute at Yezin. Myanmar also participated actively in the International Neem Network under the sponsorship of FAO and Danida, and it also plays a major role in the IPGRI-funded project concerned with the conservation of some endangered species within Myanmar, Thailand, Laos and Vietnam. JICA has assisted Myanmar in the establishment of some projects relevant to reforestation and seedling production. Forest genetics research is a part of this programme, with particular attention given to plus tree selection and propagation.

Laos is the latest ASEAN member country; its forest genetics research is also at an early stage of development because of limited manpower and budget. Australia helped Laos to create many trial plots of fast-growing species such as *Eucalypts* and *Acacias* for the selection of appropriate species adapted to conditions in the country. Technology for forestry research has mainly been available from Thailand, whose research results are well recognised. Laos also participated in both the International Neem Network and the conservation programme for some endangered species, as described above.

2. Major constraints on forest genetics research in Southeast Asia

Although some ASEAN member countries are very active in forest genetics research, their activities are still limited because of various constraints. These constraints can be categorized as follows:

2.1. Limited personnel

Forestry research was previously or is still of little interest. For example, most forest research institutes are part of a forestry department or ministry of forestry. They depend mostly on forestry activities including logging, exploitation, and wood industries. In general, the important role of forestry research has been ignored. Research staff is usually only a small part of the staff, receiving less attention and support because research is not a tangible benefit producer. During the last 3 decades, support for advanced education of research staff has increased and more funds have been made available for research and development. However, in some countries research manpower is still limited and further personnel upgrading is needed.

2.2. Limited budget

With inadequate attention paid to research, most countries in Southeast Asia allocate a very small budget for research and development (R&D). For example, the annual budget of the Forestry Research and Development Office in Thailand is around 300 million Baht per year, which is about 0.01% of the national Gross Domestic Product (GDP). This amount includes various activities in addition to research. The research budget for silvicultural and forest genetics research is only 56.8 million Baht or about 0.0016% of the national GDP (Kaosa-ard 1998). In Malaysia, total R&D expenditure in 1994 was about RM 611.2 million, which was 0.34% of the country's GDP. (Razak *et al*, 1998). Compared with other emerging economies, this figure is very low. Singapore, Taiwan, and South Korea, for example, are spending 1.12%, 1.82% and 2.33% of their GDP, respectively, on research.

2.3. Technology transfer

Within Southeast Asian countries, many research results have been published in local languages, such as Bahasa, Thai, Tagalog, or Vietnamese. This has been the major problem in technology transfer throughout the region. However, after the establishment of various regional and multilateral projects, many research papers and reports have been published in English, thus enabling wide readership throughout the region. Nevertheless, there is still a problem, since countries vary in ability to use English for communication. For example, Thailand, Vietnam and Indonesia are rather weak in English, whereas in Malaysia and the Philippines research workers can use English fluently.

2.4. Differences in priority

There are many forest tree species in each country, and species of economic interest vary from country to country. As examples, in Malaysia, *Hevea brasiliensis, Tectona grandis, Azadirachta excelsa, Acacia mangium* and dipterocarps are of particular interest. In Indonesia, the focus is on *Eucalyptus urophylla, Paraserianthes falcataria, Pinus merkusii, Acacia mangium,* and dipterocarps. Thailand's major species of interest are *Tectona grandis, Pinus kesia, P. caribaea, P. oocarpa* and *Eucalyptus camaldulensis.*

Species priority differences among the countries of Southeast Asia present problems in cooperation within and outside of the region.

3. International cooperation in forest genetics research

International cooperation in this field began with the national cooperation and contracted research programmes in some countries such as in Malaysia; cooperation was further extended to cover the regional level and eventually became global. For example, the ASEAN Forest Tree Seed Centre Project, initiated at the regional level, now also has strong international links to the region. AFTSC, in cooperation with IPGRI, Danida and FAO, tried to initiate some conservation projects for endangered species. This led to wider cooperation at the international level, including, for example, the International Neem Network. Some years after the establishment of AFTSC, FAO and UNDP initiated the FORTIP project at Los Baños to work within 12 Asian-Pacific countries, focusing on tree improvement and breeding, and many of its activities are similar to those of AFTSC. This development enabled the establishment of collaborative forest genetics research projects, especially the Species Improvement Network (SPIN), although with less support from UNDP/FAO. Similarly, CIFOR, in cooperation with AFTSC, initiated a study program on the effects of forest fragmentation on genetic diversity of some common species in the ASEAN region and other regions, including India and Central America. The research results have been published by CIFOR.

Malaysia's close cooperation with various agencies both locally and internationally enables FRIM to generate funds for their infrastructure development as well as for personnel upgrading. FRIM also produces forestry journals and other publications distributed to readers on the basis of payment. For these reasons, FRIM has become well-known and engaged in many cooperative activities. Its library is highly regarded. FRIM also provides information services on a fee basis. In Indonesia, the establishment of CIFOR at Bogor and the Forest Tree Improvement and Development Institute at Jogjakarta with financial support from JICA have increased the involvement of that country in cooperative forest genetics research. In Thailand, forest genetics research and tree improvement continues to make progress due to the upgrading of its research personnel in forest genetics under the AFTSC Project. International projects with outside support, such as those from CIFOR, ACIAR, IPGRI, FAO JICA, and DANCED, have made Thailand a centre for international cooperation. Moreover, its well known tree improvement programmes have been used as demonstration and training centres for forest genetics and tree improvement in the region.

These examples show that regional and international cooperation in Southeast Asia has been carried on successfully. Cooperation in information and materials exchange has actually been even more successful than collaborative research projects. For example, annually AFTSC publishes 7,000-12,000 copies of technical papers (research results), manuals, handbooks, and review publications, and distributes them to 72 countries worldwide. In addition, AFTSC supplies up to 800 kg of tree seed of various species each year to many tropical countries for testing, studying or multiplication.

4. Conclusions

Although regional and international linkages have been developed in some countries in Southeast Asia, in some cases with considerable success, there are a number of other countries that do not have forestry research as a mandated activity. Even where such activities do exist, it is very possible that cooperation in forest genetics research may decline after termination of joint regional or international projects such as AFTSC or FORTIP. Program gaps are still common among the research institutes endeavoring to conduct forest genetics research. The greatest constraint is the limitation in research funds and staff. A system of collaborative research programmes with outside funds supporting each participating country may be a good way to strengthen the research network. But without this support, such programmes will have great difficulty meeting their goals. Experience with networks, based mainly on national support, such as SPIN under the FORTIP project, has clearly shown that collaboration cannot work unless outside supporting funds can be generated and a sustainable project can become the core of collaboration among member countries. Such a structure can achieve very significant gains from forest genetic and tree improvement research.

Literature cited

Appanah, S. and Weinlaud, G. 1993. Planting quality timber trees in peninsular Malaysia: a review. Forest Research Institute of Malaysia, Kepong. Selangor, D.E., Malaysia.

Kaosa-ard, A. 1998. Private sector forestry research in Thailand. *In* Emerging institutional arrangements for forestry research. *FORSPA Publication* 20/1998.

Kha, L.D. 1996. Study on natural hybrids of *Acacia mangium* and *Acacia auriculiformis* in Vietnam. *In* Proceedings of the QFRI-IUFRO Conference on Tree Improvement for Sustainable Tropical Forestry, Caloundra, Queensland, Australia. 27 Oct.-1 Nov. 1996.

Kha, L.D.; Hai, N.D. and Vinh, H.Q. 1997. Clonal tests and propagation options for natural hybrids between *Acacia mangium* and *Acacia auriculiformis*. *In* The Third International Acacia Workshop. Hanoi, Vietnam. 26-30 Oct. 1997.

Razak, A.M.A.; Hong, L.T. and Appanah, S. 1998. Developing and managing collaborative research in Malaysia. Emerging institutional arrangements for forestry research. *FORSPA Publication* 20/1998.

Cooperation as a support to promote forest genetics research in developing countries of Latin America

Rodolfo Salazar

Introduction

Accelerated population growth has led to the rapid destruction of biodiversity around the world. This problem has reached a point at which, presently, there are many communities facing serious problems in satisfying some of their basic subsistence needs. The process of destruction of the forest and conversion to agricultural and animal husbandry use, without taking into consideration its utilization capacity and without knowing the significance of its biodiversity, has had a very severe impact on the human population, and the American tropics are not an exception.

Much has been reported about the floristic richness of the natural forests of the American tropics, but little effort has been made to evaluate their real composition and to protect them. The negative impacts of this accelerated process of destruction are leading mankind to face a series of problems that are not easy to solve and that have not been properly addressed by the responsible authorities in the countries of this region. The importance of wisely protecting and managing the floristic diversity of the remnant forests and the need to promote the development of silviculture to satisfy the growing demand for forest products are much discussed in international forums. However, the achievements of these international initiatives have not been as expected. One reason is the region's great ecological diversity; another is that most of the countries do not have the technical basis to develop plantation silviculture. The governments do not allocate the necessary financial resources to support research or to train professionals in the field of forest genetics and tree improvement; indeed, there are very few specialized professionals.

Studies carried out by the Inter-American Development Bank (BID) indicate that 36% of the land in Latin America is covered by natural broadleaf forests and 5% by coniferous forests. It is estimated that annually 4 million hectares are deforested and converted into agriculture and extensive livestock grazing. In terms of wood volume, approximately 350 million cu.m per year are extracted from the region, of which only 53 million cu.m per year are used by the timber industry; the remainder is used mainly as fuel.

C. Mátyás (ed.), Forest Genetics and Sustainability, Volume 63, 275–279.

By the year 2000, demand for timber from the natural forest of 84 million cu.m/year is expected (McGaughey and Gregersen 1983).

In the early nineties approximately 8.6 million ha was covered by plantations in tropical Latin America (excluding Chile and Argentina) as compared to 5 million ha in 1980. Less than half of this area is occupied by pine species, the rest by eucalypts and some other broadleaved species, such as teak and mahogany. Out of this plantations, Brazil alone has a share of 7 million ha, or 81% (FAO 1995). These figures clearly indicate the need of promoting sound forest management throughout the region, giving importance to native species due to their wide diversity and for their conservation. To reach this objective, it is necessary to strengthen forest research.

This document describes in a general way some of the achievements in forest genetics and tree improvement in the region and the main constraints on its application. It also discusses the need to make available to the countries involved the necessary tools to promote the development of silviculture.

1. Some important achievements in forest tree improvement

In the 1970s, some governments initiated reforestation projects as a reaction to the accelerated deforestation process. Most of these projects were funded by international organizations and implemented mainly at the national level. Later, also with international support, some countries started projects concerned with genetic improvement. The motivation was an absence of the technical information needed to assure the availability of quality germ plasm for establishing more productive plantations. Usually, genetic improvement work started with the identification and selection of promising species, followed by provenance testing. These projects were primarily undertaken by governmental institutions, research institutes and universities.

In Central America, forest genetic improvement projects have been implemented over a relatively long period (Mesén *et al.* 1992) as a joint effort of international organisms such as the Overseas Development Administration (ODA, now DFID) from the United Kingdom, the Department for Development Cooperation (DDC/Norway), the Swiss Office of Development Cooperation (COSUDE), the US Agency for International Development (USAID), and the Danish International Development Agency (DANIDA),

From 1977 to 1997 the Tropical Agricultural Research and Higher Education Center (CATIE) carried out intensive research on species selection and establishment of provenance and progeny trials in Costa Rica with the technical and economic support of ODA, COSUDE and DDC. As a result, about 150 genetic field tests were established, including 21 seed orchards of six native species and two introduced species. Most of them were on private farms (Mesen *et al.* 1992). Between 1980 and 1997, CATIE also conducted species evaluations and established genetic improvement trials with about 150 native and introduced species in Guatemala, Honduras, El Salvador, Nicaragua, Costa Rica and Panama with USAID financial support. These activities allowed the countries to identify appropriate species for the subsequent establishment of more than 70 provenance and progeny trials.

A number of forest genetics research programs were instituted within countries of the region during this period. In Honduras, the Corporation for Forestry Development and the National School of Forestry Sciences (CONSEFOR) jointly carried out a research program between 1987 and 1997 with twelve priority native species for the evaluation of provenances and progenies and for the establishment of seed orchards. The project had the technical and financial support of ODA. In Nicaragua, the Ministry of Natural Resources started provenance trials in 1993 with native and introduced species on the Pacific coast with the technical and financial support of DANIDA. In Colombia, the National Resources and the Environment Institute began species evaluations in the early 1970s, mainly with conifers of Central America and Mexico and with *Eucalyptus*. Later, as a follow-up to the species selection, several provenance and progeny trials were established with high-yield species in various regions of the country (Diaz *et al.* 1993). In Mexico, the Forestry Research Institute began a forest tree improvement program in 1967. After 34 years of research, it had established seed sources, provenance trials and small seed orchards, mainly with coniferous species. The Autonomous University of Chapingo, the Postgraduate College, the Autonomous University of Mexico and the Forest Genetics Center also participated In this program (Vargas *et al.* 1994).

The Central America and Mexico Coniferous Resources Cooperative (CAMCORE) has been implementing important work since 1980 in Mexico, Venezuela, Colombia, Brazil, Chile and South Africa. Research has been undertaken through private companies with 14 conifers, eucalypts and about a dozen native broadleaf species at the provenance, progeny and seed orchard levels (Dvorak and Donahue 1992).

2. Constraints on the implementation of results

Most of the countries in the region that have undertaken research programs in forest genetics and tree improvement have had the technical and financial support of international organizations. Although the results obtained have provided important support for forestry activities, their application has not been as successful as anticipated.

A major problem is that forest tree improvement research is a long and expensive process but the cooperative projects are usually short-term. Sometimes it is possible to carry the project to provenance selection and even to the establishment of first-generation seed orchards. However, once the project ends, national institutions are usually not prepared to allocate sufficient economic resources to provide follow-up to the results obtained or to promote the use of the results. Experimental areas and seed orchards are often located on private farms and the owner may decide to harvest the timber, due to lack of financial incentives. Furthermore, trained personnel are not supported to continue exploring new research options.

In most countries, reforestation plans do not have clear objectives, priority areas to be reforested are not established and species of interest have not been defined. Thus, when genetic improvement programs make improved material available, commonly the species in question are no longer of interest for reforestation. Consequently, research must begin again. In addition, there is not a clear awareness of the potential for the use of the genetically improved material in reforestation projects.

3. Cooperation needs

Biological, social and economic reasons for promoting reforestation are well known. This is a productive activity in which the private sector as well as the governments must participate. To successfully accomplish this task, authorities must change their attitude and show a greater interest in reforestation by providing the necessary support for training of professionals, regionalizing reforestation programs, defining objectives and priorities (including priority species) and assigning the necessary resources to carry out research. In the field of forest genetics and tree improvement, long-term international cooperation, technical as well as financial, is crucial for the development of silviculture at the commercial level.

Mistakes made during the implementation of some cooperation projects show that it is necessary to pay attention to the following points:

- reforestation programs must have production objectives clearly defined and governments committed to provide the necessary support and follow-up;
- there must be a clear justification for the species selected as priority for reforestation;
- cooperative programs must be long-term so that production of genetically improved material can be guaranteed to the countries;
- research projects in forest genetics and tree improvement must be implemented as a joint effort with non-governmental institutions to guarantee the successful implementation of the project as well as its follow-up once international cooperation finishes;

– governmental institutions must regulate and, if possible, make compulsory the use of genetically improved seed in reforestation projects using public funds.

4. Priorities

Results obtained to date and the constraints identified show that it is important to define priorities in a program of genetic improvement in order to promote the appropriate forestry development in the countries of the region. Among these priority needs, which include support needs, the following should be listed:

- to carry out well-founded analyses of species of interest at mid- and long-term, in order to guarantee the use of the research results obtained; preferably species with a wide adaptation range should be identified;
- to give preference to native species, especially endangered ones;
- to conduct the provenance and progeny trials necessary to establish seed orchards and assure provision of improved seeds; the trials must be located in the regions where reforestation projects are planned to be implemented and on sites that assure their permanence;
- to promote the application of techniques necessary to ensure *in situ* and *ex situ* conservation of species and populations of interest;
- to provide the necessary financial support to train the professionals, thus ensuring the implementation of genetic improvement research in their countries;
- to strengthen selected non-governmental institutions in order to ensure continuity, once cooperation projects are finished.

Literature cited

Diaz, J., L. Jara and J. Venegas 1993. 20 años de investigación forestal. Resultados INDERENA; Bogota, Colombia. 1105 pp.

Dvorak, W. and J. Donahue 1992. Reseña de investigacion de la Cooperativa CAMCORE 1980-1992. North Carolina State University, Raleigh, NC, U.S.A. 94 pp.

FAO 1995. Forest resources assessment 1990. Tropical forest plantation resources. FAO Forestry Paper No. 128, Rome

McGaughey, S. and H. Gregersen 1983. El desarrollo forestal en América Latina. BID. Washington. 234 pp.

Mesén, J.F., D.H. Boshier and J.P. Cornelius 1992. Genetic improvement of trees in Central America with particular reference to Costa Rica. *In* Tropical Trees: The Potential for Domestication and the Rebuilding of Forest Resources. Institute of Terrestrial Ecology, Edinburgh, Scotland, 249-255.

Vargas, J., J. Jasso and B. Bermejo 1994. El mejoramiento genético forestal como base para el establecimiento de plantaciones forestales. *In* Memoria: IV Reunión Nacional de Plantaciones Forestales, México, pp. 49-80.

8. CONTRIBUTION OF GENETICS TO THE SUSTAINED MANAGEMENT OF GLOBAL FOREST RESOURCES

Conclusions and recommendations of the 4th International Consultation on Forest Genetics and Tree Breeding, held in Beijing, China, August 22-28, 1998 organized by IUFRO in cooperation with FAO[1]

Eric Teissier du Cros, Csaba Mátyás and Howard Kriebel

More than one hundred forest tree geneticists and breeders of 33 countries contributed by their invited papers and their discussions to defining the role of forest genetics for the enhancement and sustainability of forest production at the threshold of the XXIst century. During the conference, participants proposed a series of statements and research priorities which are here summarized.

1. New challenges, differentiated priorities

Selection and improvement of forest trees is ongoing in about a hundred countries in the world. While increasing wood and fiber production to meet the ever growing demand and to alleviate the global pressure on natural forests has to remain among the top priorities in forest policies, it is stated with concern that traditional genetic improvement has lost its priority status and has been considerably reduced or even stopped in certain developed and developing countries. This leaves expensive basic research without the necessary applied and extension research background. Reasons nourishing this trend are not based on scientific considerations.

---

1 Compiled by the authors from conclusions and recommendations made by session moderators of the consultation: Jeffery Burley, Gerhard Müller-Starck, David Karnosky, Gösta Eriksson, Huoran Wang, Francis Yeh, Pierre Sigaud and Howard Kriebel.

C. Mátyás (ed.), Forest Genetics and Sustainability, Volume 63, 281–287.

- Decision makers and scientists have to deal with the challenge to accommodate long-term interests of tree improvement programs and short-term genetics studies within the context of social and political streams of society. Strategies and tools of tree improvement must be adapted to social needs, to the scientific and technical background and to financial possibilities. In developing countries, priority should be given to traditional tree improvement. In the more developed countries integrated programs applying traditional breeding and biotechnology will be critical to the success of forest tree improvement.

Advanced breeding should focus on a limited number of species used in intensively managed plantations. When appropriate, it should include biotechnology such as genomic mapping, marker-assisted selection, genetic engineering and short generation model species. Biotechnology should be combined with proper field testing to shorten breeding cycles and increase tree improvement efficiency. This should be supported by long-term multi-generation breeding using sound quantitative genetic tools.

While classical quantitative genetics appear to be relatively under-represented in current research activities, genomic science offers new opportunities to link it to molecular genetics in improvement programs. Improved computing techniques and statistical methods developed in other fields (*e.g.* animal breeding) should be applied when appropriate. At present, research priorities are somewhat distorted in many instances by the short-term characteristics of research projects.

Regardless of advances in modeling and genetic engineering, provenance research is still needed in many regions and traditional breeding will continue to be important to practical objectives such as improving adaptability, resistance and yield. New genetic improvement approaches such as small-population intensive breeding and the use of model species should be considered. Species priorities will have to be established for each region. Existing valuable field experiments and selected seed orchards must be preserved for future research.

2. Adaptive potential of forest trees

Genetic improvement requires testing of the adaptation potential of species, populations and individuals. New concepts are in progress in order to quantify adaptive potentials and adaptive capacities respectively. Testing trees in diverse environments, with inclusion of rare biotic or abiotic events, needs to be assisted by gene makers. Particularly, the following research topics should be considered :

- Studies on phenotypic plasticity in tree species with respect to both individuals and populations. Forest tree species deviate substantially in their adaptive strategies. Verification of specific adaptive strategies are needed instead of generalizing results obtained on "model" species.

- Utilization of proper markers adapted to specific purpose. Knowledge gaps are evident in the use of markers to study population dynamics and particularly reproduction processes.
- Differential modeling with respect to species and processes. Avoidance of over-simplification is particularly indicated with respect to disruptive, stabilizing and directional selection effects.

Generally, it is concluded that the existing knowledge gap between genetic markers and the expression of adaptive traits among individuals, families and populations needs to be bridged in close cooperation with physiologists. Genetic mapping of quantitative trait loci (QTLs) and of loci which are involved in response to biotic and abiotic stresses should be developed rapidly.

3. Adaptation to a changing environment

The global environment is changing. Although forest geneticists have at their disposal the methods and capabilities to study the genetic implications of global environmental change on forest sustainability, there is still insufficient research to date in this field. Genetic diversity and improved adaptability are important components of forest sustainability under global environmental change.

The following recommendations are made :

- Genetic diversity should be among the high priority research areas in biodiversity studies related to global environmental change.
- A better integration of genetics in forest ecosystem studies related to global environmental change is needed.
- Understanding genetic control of tolerance to environmental change will help determine how best to maintain adaptability and variability.

The following research avenues are recommended:

- Impacts of environmental change on forest productivity, genetic diversity, community dynamics and reproductive fitness.
- Reliability of genetic markers and bio-indication as verifiers of genetic change in populations due to environmental change.
- Conservation of genetic resources under the changing environment.
- Impact of global environmental change on reproductive biology in connection with reproductive fitness.

4. Maximizing productivity while maintaining sustainability

To protect natural forests, particularly in the tropics, the world's demand for wood will increasingly be met from plantation forestry with fast growing, often exotic, species. International exchange of germplasm, which has historically proceeded with little restrictions, should continue thanks to bilateral technical agreements between governmental and non-governmental organizations.

There is a widely held assumption that naturally regenerated forests are more stable than planted ones, because they display maximum diversity. Natural populations are seldom, perhaps never, in an optimum adaptation state, rather in an adaptable condition. Properly managed breeding can maintain or even increase genetic variation in plantations relative to natural populations.

Intensive forestry should include research on risk assessment particularly when high genetic gain varieties are used. Special attention should be paid to the deployment of clones in plantations. To maintain sufficient variation, 10-30 unrelated clones may be adequate in many circumstances.

The *Pinus taeda* breeding program in southeast USA is a good example to demonstrate how forest genetics and tree improvement can contribute significantly to sustainable forestry development. After more than 40 years of research, it has been shown that on 15% of the area, commercial forest plantations produce 50% of the required wood supply. Thus, improved wood production on limited commercial lands can reduce logging pressure on natural forests.

5. Conserving forest genetic resources

5.1. New tools

It is essential to raise awareness, at all levels, on the importance of conservation and sustainable utilization of forest tree gene resources to fulfil the needs of current and future generations. These needs may include genetic improvement, or new uses for which available genetic resources may become of the greatest importance.

The prerequisite to any conservation scheme includes the examination of the amount and distribution of genetic diversity. Molecular and biochemical markers have become a useful tool for this purpose. However adaptive, quantitative characters should also be considered in genetic conservation.

The amount and distribution of genetic diversity of a species can often be inferred from knowledge of its genetic system and life-history parameters.

Research in genetic diversity should continue to develop on the effect of mode of reproduction, breeding system, gene flow mechanism, geographic range, life cycle and population size, at scales varying from stand size to range size.

5.2. Networking

There was full agreement of an urgent need for concerted effort in genetic conservation and utilization across state boundaries, since forest tree species and typical ecosystems often extend over large areas. Country-driven actions should be linked and streamlined, and if needed, supported by regional/ecoregional and international initiatives when they exist. In certain regions capacity building and creation of conservation frameworks and mechanisms are still needed.

FAO's role in organizing regional workshops on conservation, management and utilization of forest gene resources was acknowledged and supported.

5.3. Priorities

In view of the large number of tree species of current or potential use it is necessary to carry out in-depth priority setting according to environmental, social and economic factors. All stakeholders should be involved in the dialogue and decisions.

Management of forest gene resources may be started with simple methodologies, even in absence of precise knowledge of genetic variation patterns of targeted species. Regarding management options, most forest tree species can be classified into three groups :

- species of current high socio-economic value, likely to be already included in improvement and conservation schemes;
- species with clear potential or future value, for which a main research objective could be to understand the extent and patterns of genetic variation;
- species of currently unknown value, into which a large majority of tree species falls, for which the continued existence of populations of adequate size could be the only management objective.

5.4. Access to and exchange of genetic resources

Emerging social, policy and legal issues identified among the objectives of the FAO Convention on Biological Diversity will influence the terms under which national authorities will exercise their sovereign rights over their genetic resources. Administratively robust, simple and fair systems have to be developed for germplasm access and exchange. A recently emerging issue is the deployment of genetically modified organisms. IUFRO members are encouraged to play a guiding role in shaping appropriate national policies and to maintain accessibility of genetic resources under mutually fair conditions for both donor and recipient countries.

6. Managing genetic resources under traditional silviculture

There is an urgent need to demonstrate to policy-makers and to the general public, that genetic conservation is fully compatible with a wise management of forest resources and that it is an integral part to the overall maintenance and enhancement of biological diversity.

Maintaining genetic diversity must be considered throughout all steps of tree domestication : selection, breeding, seed handling, nursery operations, planting, thinning, and also in natural regeneration. It should also be considered when planning protected areas and conservation forests and stands.

Gene conservation must be part of silviculture in managed forests since they dominate most landscapes. Management of forests must be carried out without genetic erosion or in such a way that genetic erosion is kept at an acceptable level. However, current scientific knowledge on the genetics of managed forests is mainly based on studying gene markers, the relevance of which is still disputed. Consequently, this gives cause for concern about the adaptive genetic potential of managed forests. Scientists should provide suitable indicators of loss of adaptive capacity, *e.g.* demographic parameters, markers and adaptive traits.

Accordingly, studies on the effects of different forest interventions on genetic diversity in sampled forest ecosystems – from natural reserves to fiber farms – should provide background data on genetic diversity in key species. Further suitable indicators of disruption of genetic processes should be looked for. IUFRO should play a key role in the coordination of international research, particularly in proposing standard methods to monitor changes and in establishing reference trials.

7. The right messages through the right messengers

Breeding and improvement of forest trees is a long-term process. Improvement strategies should be explained to all stakeholders to convince them of their benefits. The value and importance of genetic improvement must be conveyed to decision and policy makers through proper and credible mediators such as :

- seed merchants, nursery managers and forest owners making use of genetic gain;
- current opponents, such as enviromentalists, who have a strong ecological background and who, once they are convinced, would be better conveyors of the right message because of their public acceptance.

8. Funding policy and cooperative research

In North America and Europe, public sector support for tree improvement research has declined, often significantly, including federal, state and university programs. In New Zealand and other countries, these programs have been privatized to a large extent. In tropical developing countries, there is inadequate state funding and a need for more financial support by international organizations and foundations.

It is recommended that in most regions of the world, increased financial commitment by the public sector be a requisite for long-term tree improvement research. However cooperation between public and private agencies and foundations will be needed for the development of long-range research programs in tree improvement with reasonably secure funding sources.

FORESTRY SCIENCES

43. R.T. Prinsley (ed.): *The Role of Trees in Sustainable Agriculture.* 1993 ISBN 0-7923-2030-1
44. S.M. Jain, P.K. Gupta and R.J. Newton (eds.): *Somatic Embryogenesis in Woody Plants,* Volume 3: Gymnosperms. 1995 ISBN 0-7923-2938-4
45. S.M. Jain, P.K. Gupta and R.J. Newton (eds.): *Somatic Embryogenesis in Woody Plants,* Volume 1: History, Molecular and Biochemical Aspects, and Applications. 1995 ISBN 0-7923-3035-8
46. S.M. Jain, P.K Gupta and R.J. Newton (eds.): *Somatic Embryogenesis in Woody Plants,* Volume 2: Angiosperms. 1995 ISBN 0-7923-3070-6
Set ISBN (Volumes 44–46) 0-7923-2939-2
47. F.L. Sinclair (ed.): *Agroforestry: Science, Policy and Practice.* Selected Papers from the Agroforestry Sessions of the IUFRO 20th World Congress (Tampere, Finland, 6–12 August 1995). 1995 ISBN 0-7923-3696-8
48. J.H. Goldammer and V.V. Furyaev (eds.): *Fire in Ecosystems of Boreal Eurasia.* 1996 ISBN 0-7923-4137-6
49. M.R. Ahuja, W. Boerjan and D.B. Neale (eds.): *Somatic Cell Genetics and Molecular Genetics of Trees.* 1996 ISBN 0-7923-4179-1
50. H.L. Gholz, K. Nakane and H. Shimoda (eds.): *The Use of Remote Sensing in the Modeling of Forest Productivity.* 1996 ISBN 0-7923-4278-X
51. P. Bachmann, M. Köhl and R. Päivinen (eds.): *Assessment of Biodiversity for Improved Forest Planning.* Proceedings of the Conference on Assessment of Biodiversity for Improved Planning (Monte Verità, Switzerland, 1996). 1998 ISBN 0-7923-4872-9
52. G.M.J. Mohren, K. Kramer and S. Sabaté (eds.): *Impacts of Global Change on Tree Physiology and Forest Ecosystems.* Proceedings of the International Conference on Impacts of Global Change on Tree Physiology and Forest Ecosystems (Wageningen, The Netherlands, 26–29 November 1996). 1997 ISBN 0-7923-4921-0
53. P.K.R. Nair and C.R. Latt (eds.): *Directions in Tropical Agroforestry Research.* Selected Papers from the Symposium on Tropical Agroforestry organized in connection with the annual meeting of USA (Indianapolis, U.S.A., 5 November 1996). 1998 ISBN 0-7923-5035-9
54. K. Sassa (ed.): *Environmental Forest Science.* Proceedings of the IUFRO Division 8 Conference Environmental Forest Science. (Kyoto, Japan, 19–23 October 1998). 1998 ISBN 0-7923-5280-7
55. S.M. Jain, P.K Gupta and R.J. Newton (eds.): *Somatic Embryogenesis in Woody Plants,* Volume 4. 1999 ISBN 0-7923-5340-4
56. J.R. Boyle, J.K. Vinjum, K. Kavanagh, E.J. Jensen (eds.): *Planted Forests: Contribution to the Quest for Sustainable Societies.* 1999 ISBN 0-7923-5468-0
57. K. von Gadow and G. Hui: *Modelling Forest Development.* 1999 ISBN 0-7923-5488-5
58. J. Abildtrup, F. Helles, P. Holten-Andersen, J.F. Larsen and B.J. Thorsen (eds.): *Modern Time Series Analysis in Forest Products Markets.* 1999 ISBN 0-7923-5524-5
59. S.M. Jain, P.K. Gupta and R.J. Newton (eds.): *Somatic Embrygenesis in Woody Plants,* Volume 5. 1999 ISBN 0-7923-5553-9
60. D. Auclair and C. Dupraz (eds.): *Agroforestry for Sustainable Land-Use.* Fundamental Research and Modelling with Emphasis on Temperate and Mediterranean Applications. 1999 ISBN 0-7923-5799-X

FORESTRY SCIENCES

61. F. Helles, P. Holten-Andersen and L. Wichmann (eds.): *Multiple Use of Forests and Other Natural Resources*. Aspects of Theory and Application. 1999 ISBN 0-7923-5859-7
62. A. Yoshimoto and K. Yukutake (eds.): *Global Concerns for Forest Resource Utilization*. Sustainable Use and Management. 1999 ISBN 0-7923-5968-2
63. C. Mátyás (ed.): *Forest Genetics and Sustainability*. 1999 ISBN 0-7923-6011-7
64. S.M. Jain and S.C. Minocha (eds.): *Molecular Biology of Woody Plants*, Volume 1. 1999 ISBN 0-7923-6012-5

KLUWER ACADEMIC PUBLISHERS – DORDRECHT / BOSTON / LONDON